Lecture Notes in Physics

Springer
Berlin
Heidelberg
New York
Barcelona
Hong Kong
London
Milan
Paris
Tokyo

Physics and Astronomy ONLINE LIBRARY
http://www.springer.de/phys/

Editorial Policy

The series *Lecture Notes in Physics* (LNP), founded in 1969, reports new developments in physics research and teaching -- quickly, informally but with a high quality. Manuscripts to be considered for publication are topical volumes consisting of a limited number of contributions, carefully edited and closely related to each other. Each contribution should contain at least partly original and previously unpublished material, be written in a clear, pedagogical style and aimed at a broader readership, especially graduate students and nonspecialist researchers wishing to familiarize themselves with the topic concerned. For this reason, traditional proceedings cannot be considered for this series though volumes to appear in this series are often based on material presented at conferences, workshops and schools (in exceptional cases the original papers and/or those not included in the printed book may be added on an accompanying CD ROM, together with the abstracts of posters and other material suitable for publication, e.g. large tables, colour pictures, program codes, etc.).

Acceptance

A project can only be accepted tentatively for publication, by both the editorial board and the publisher, following thorough examination of the material submitted. The book proposal sent to the publisher should consist at least of a preliminary table of contents outlining the structure of the book together with abstracts of all contributions to be included.
Final acceptance is issued by the series editor in charge, in consultation with the publisher, only after receiving the complete manuscript. Final acceptance, possibly requiring minor corrections, usually follows the tentative acceptance unless the final manuscript differs significantly from expectations (project outline). In particular, the series editors are entitled to reject individual contributions if they do not meet the high quality standards of this series. The final manuscript must be camera-ready, and should include both an informative introduction and a sufficiently detailed subject index.

Contractual Aspects

Publication in LNP is free of charge. There is no formal contract, no royalties are paid, and no bulk orders are required, although special discounts are offered in this case. The volume editors receive jointly 30 free copies for their personal use and are entitled, as are the contributing authors, to purchase Springer books at a reduced rate. The publisher secures the copyright for each volume. As a rule, no reprints of individual contributions can be supplied.

Manuscript Submission

The manuscript in its final and approved version must be submitted in camera-ready form. The corresponding electronic source files are also required for the production process, in particular the online version. Technical assistance in compiling the final manuscript can be provided by the publisher's production editor(s), especially with regard to the publisher's own Latex macro package which has been specially designed for this series.

Online Version/ LNP Homepage

LNP homepage (list of available titles, aims and scope, editorial contacts etc.):
http://www.springer.de/phys/books/lnpp/

LNP online (abstracts, full-texts, subscriptions etc.):
http://link.springer.de/series/lnpp/

Y. Aurégan A. Maurel V. Pagneux
J.-F. Pinton (Eds.)

Sound–Flow Interactions

Springer

Editors

Yves Aurégan
Vincent Pagneux
Université du Maine, UMR CNRS 6613
Laboratoire d'Acoustique de l'Université du Maine
Av. O. Messiaen
72017 Les Mans, Cedex 9, France

Jean-Francois Pinton
Ecole normale supérieure de Lyon, UMR CNRS 1325
Laboratoire de Physique
46 allée d'Italie
69007 Lyon, France

Agnes Maurel
ESPCI, UMR CNRS 7587
Lab. Ondes et Acoustique
10 rue Vauquelin
75005 Paris, France

Cover Picture: (see figure 1 page 141, contribution of F. Lund in this volume)

Library of Congress Cataloging-in-Publication Data

Sound-flow interactions / Y. Auregan ... [et al.].
p. cm. -- (Lecture notes in physics, ISSN 0075-8450 ; 586)
Includes bibliographycal references.
ISBN 3540433325 (alk. paper)
1. Acoustic streaming. I. Auregan, Y. (Yves), 1962-II. Series.
QC243.3.A25 S68 2002
534--dc2l

2002021074

ISSN 0075-8450
ISBN 3-540-43332-5 Springer-Verlag Berlin Heidelberg New York

Springer-Verlag Berlin Heidelberg New York
a member of BertelsmannSpringer Science+Business Media GmbH

http://www.springer.de

Typesetting: Camera-ready by the authors/editor
Camera-data conversion by Steingraeber Satztechnik GmbH Heidelberg
Cover design: *design & production*, Heidelberg

Printed on acid-free paper
54/3111 - 5 4 3 2 1

Preface

The Cargèse Summer School "Sound–flow interactions" was held in the Institut d'Etudes Scientifiques de Cargèse in Corsica, France from 19th June to 1st July, 2000.

The understanding of sound and flow interactions has made some remarkable progress since the pioneering works of the Russian and British schools, in the 1950s. In addition, the growing availability during the past 10 years of sophisticated computer/electronics/materials techniques allows for the development of a growing number of applications as well as the possibility of addressing new fundamental problems. The coupling between acoustic waves and flow motion is basically nonlinear, so that the sound propagation and generation is modified by the flow and the flow can also be modified by the sound. As a result, this problem is investigated in many different scientific communities, such as applied mathematics, acoustics and fluid mechanics, among others. In our opinion, the time had come to try to gather the researchers in the different communities together in a tutorial environemnt. So, this school brought together worldwide specialists in order to present various aspects of sound–flow interactions, and share expertise and methodologies so as to promote cross-fertilisation.

The basic knowledge in the area is introduced by A. Hirschberg and C. Schram. He presents the aeroacoustics of internal flow in a very lively way with a lot of illustration devices. He introduces aeroacoustic analogies and applications like musical instruments, the Rijke tube, speech production *etc.* M.S. Howe introduces the theory of vortex sound in a very didactic way. From Lighthill's acoustic analogy, he shows how vorticity and entropy fluctuations can be seen as sources of sound. Then, using the compact Green's functions, he shows how to compute the vortex sound. As an example of the method presented, he applies this theory to pressure transients generated by high-speed trains. F. Lund gives the basic equations of sound–flow interactions. Then he introduces very clearly the scattering of sound because of vorticity and gives the most recent results on ultrasound propagation through a disordered flow. V. Ostashev presents geometrical acoustics in moving media and the important practical problem of sound propagation in turbulence (atmosphere, ocean). A. Fabrikant examines the plasma–hydrodynamics analogies including the resonant wave-flow interaction in shear flows, waves of negative energy and over-reflection and acoustic oscillators in fluid flows. P. J. Morrison describes the dynamics of the continuous spectrum which occurs in shear flow. The results are interpreted in the context of infinite dimensional Hamiltonian systems theory. G. Chagelishvili presents new linear mechanisms of acoustic wave generation in smooth shear flows using a non-modal study. N. Peake presents fluid–structure interactions in the presence of mean flows, including the problems of instability and causality. Finally, W. Lauterborn presents nonlinear acoustics with applications to sonoluminescence and to acoustic chaos.

In this Cargèse Summer School, 54 students from 12 nations, and 11 lecturers from 7 nations participated.

Aknowledgements. The Summer School and this publication would not have been possible without:

- financial support from the European Union, the Centre National de la Recherche Scientifique, the Ministère des Affaires Etrangères, the Ministère de l'Education Nationale, de la Recherche et de la Technologie and the Groupement de Recherche "Turbulence";
- the guidance of Elisabeth Dubois–Violette, director of the Institut d'Etudes Scientifiques de Cargèse;
- the help of Chantal Ariano, Nathalie Bedjai, Brigitte Cassegrain, Pierre-Eric Grossi and the whole team in preparing and hosting of this school.

Finally, we wish to thank the lecturers for giving so much time in preparing the lectures and writing them up, as well as making themselves available for discussions during the school.

Le Mans, Paris, Lyon
September 2001

Yves Aurégan[1],
Agnès Maurel[2],
Vincent Pagneux[1],
Jean-François Pinton[3].

[1] Laboratoire d'Acoustique de l'Université du Maine, UMR CNRS 6613, Av. O Messiaen, 72085 Le Mans Cedex 9, France
[2] Laboratoire Ondes et Acoustique, UMR CNRS 7587, ESPCI, 10 rue Vauquelin, 75005 Paris, France
[3] Laboratoire de Physique, UMR CNRS 1325, Ecole Normale Supérieure de Lyon, 46 allée d'Italie, 69007 Lyon, France

Some of the lecturers of the Cargèse School, from left to right: M.S. Howe, A. Hirschberg, P. Morrison, W. Lauterborn, V. Ostashev, A. Fabrikant, N. Peake, T. Colonius *(Photo C. Schram)*

Some of the participants of the Cargèse School *(Photo C. Schram)*

Table of Contents

A Primitive Approach to Aeroacoustics
Avraham Hirschberg, Christophe Schram 1
1 Introduction 1
2 Fluid Dynamics 2
3 Lighthill's Analogy 4
4 Jet Noise 7
5 Thermo-Acoustics 9
6 Acoustical Energy 10
7 Rijke-Tube 11
8 Vortex-Sound Theory 14
9 Choice of the Green's Function 17
10 Howe's Energy Corollary 20
11 The Open Pipe Termination of an Unflanged Pipe 21
12 Whistler-Nozzle and Human Whistling 25
13 Conclusion 27
References 28

Lectures on the Theory of Vortex–Sound
Michael S. Howe 31
1 Aerodynamic Sound 31
1.1 Lighthill's Acoustic Analogy (1952) 31
1.2 Aerodynamic Sound from Low-Mach-Number Turbulence of Uniform Mean Density 34
1.3 Aerodynamic Sound from Low-Mach-Number Turbulence of Variable Mean Density 35
2 Vorticity and Entropy Fluctuations as Sources of Sound 37
2.1 The Rôle of Vorticity in Lighthill's Theory 37
2.2 Acoustic Analogy in Terms of the Total Enthalpy 39
2.3 Vorticity and Entropy Sources 40
3 Fundamental Solutions of the Wave Equation 43
3.1 The Helmholtz Equation 43
3.2 The Wave Equation 46
4 General Solution of the Inhomogeneous Wave Equation 47
4.1 General Solution in the Frequency-Domain 47

4.2 General Solution in the Time-Domain 49
5 Compact Green's Functions 49
5.1 Time-Harmonic Problems 50
5.2 Time-Domain Problems 52
6 Influence of Solid Bodies on the Generation of Aerodynamic Sound 58
6.1 Integral Representation of Aerodynamic Sound 58
6.2 Curle's Theory Applied to Compact Bodies 60
6.3 Vortex Sound 61
6.4 Influence of Vortex Shedding 62
7 Vortex-Airfoil Interaction Noise 66
7.1 Blade-Vortex Interactions in Two-Dimensions 67
7.2 Three-Dimensional Interactions 70
7.3 Blade-Vortex Interactions in Three Dimensions 71
8 Vorticity Production by Sound 74
8.1 Interaction of Sound with a Solid Surface 75
8.2 Interactions at a Trailing Edge 78
8.3 Leading Edge Interactions 80
8.4 Application to Resonant Oscillations 83
9 Perforated Screens 85
9.1 Rayleigh Conductivity 85
9.2 The Bias Flow Aperture 87
9.3 Reflection and Transmission of Sound by a Perforated Screen. 91
10 Compression Wave Generated when a Train Enters a Tunnel 94
10.1 The Governing Equations 96
10.2 Linear Theory 98
10.3 Contribution from the Surface Dipoles 102
10.4 Contribution from the Exit Flow Vorticity 103
11 The Flared Portal 104
11.1 Suppression of the Micro-pressure Wave 104
11.2 The Function φ^* 105
11.3 Optimal Flaring 106
11.4 Comparison with Measurements 107
References 110

Sound–Vortex Interaction in Infinite Media
Fernando Lund 112
1 Introduction 112
2 The Basics – Vortex Sound 113
2.1 Incompressible Flow 113
2.2 Slightly Compressible Flow 115
2.3 Incompressible Vortex Dynamics 117
2.4 Vortex Sound 119
2.5 Bibliographical Notes 121
3 Scattering of Sound by Vorticity 122

3.1 Kelvin's Theorem ... 122
3.2 Scattering to Order M ... 125
3.3 Bibliographical Notes ... 129
4 Two Dimensions ... 131
4.1 Vanishing Circulation ... 131
4.2 Surface Waves ... 132
4.3 Shallow Water ... 135
4.4 Deeper Water ... 139
4.5 Experimental Results ... 140
4.6 Bibliographical Notes ... 141
5 Multiple Scattering ... 143
5.1 Equations to Order M^2 ... 144
5.2 Appendix ... 148
5.3 Example: a "Gas" of Vortices ... 151
5.4 Two Dimensions ... 152
5.5 Three Dimensions ... 154
5.6 Bibliographical Notes ... 154
6 Loose Ends ... 156
6.1 Optical Theorem ... 156
6.2 Validity of the Born Approximation ... 158

Geometrical Acoustics in Moving Media
Vladimir E. Ostashev ... 160
1 Introduction ... 160
2 Linearized Equations of Fluid Dynamics ... 161
3 Solution of Linearized Equations of Fluid Dynamics ... 162
3.1 Debye Series ... 162
3.2 Eikonal Equation ... 163
3.3 Dispersion Equation ... 163
3.4 Transport Equation ... 163
3.5 Conservation of Acoustic Energy ... 163
4 Geometrical Acoustics in a Three-Dimensional Inhomogeneous Medium ... 164
4.1 Group and Phase Velocities ... 164
4.2 Ray Path ... 165
4.3 Eikonal and Time of Sound Propagation ... 165
4.4 Amplitude of the Sound Pressure ... 166
5 Geometrical Acoustics in a Stratified Medium ... 166
5.1 Refraction Law for the Unit Vector **n** ... 166
5.2 Refraction Law for the Unit Vector **s** ... 167
5.3 The Ray Path ... 168
5.4 Eikonal ... 168
References ... 168

Sound Propagation and Scattering in Random Moving Media
Vladimir E. Ostashev 169
1 Introduction 169
2 Equations for a Sound Wave in a Random Moving Medium 170
2.1 Helmholtz-Type Equation 170
2.2 Parabolic Equation 171
2.3 Parabolic Equation in a Refractive Medium 172
3 Statistical Description of Random Inhomogeneities 172
3.1 Correlation Functions and Spectral Densities 172
3.2 Energy, Inertial, and Dissipation Subranges 173
3.3 Kolmogorov Spectrum 173
3.4 Gaussian Spectrum 174
3.5 von Kármán Spectrum 174
4 Scattering of Sound 177
4.1 Sound Scattering Cross-Section 177
4.2 von Kármán Spectrum 178
5 Line-of-Sight Sound Propagation in a Random Moving Medium 178
5.1 Statistical Moments of a Sound Field 179
5.2 Effective Correlation Function and Spectral Density 180
5.3 Markov Approximation 180
5.4 Rytov Method 180
5.5 Formulas for Statistical Moments of a Sound Field 181
5.6 Coherence Function 183
6 Interference of the Direct and Ground-Reflected Waves 184
6.1 Mean Squared Sound Pressure 185
6.2 Comparison with Experimental Data 186
7 Sound Propagation Near Impedance Ground in a Refractive, Turbulent Atmosphere 187
7.1 Mean Sound Field 188
7.2 Coherence Function 188
References 190

Plasma–Hydrodynamic Analogy for Waves and Vortices in Shear Flows
Anatoly Fabrikant 192
1 Research Philosophy 192
2 Resonant Wave-Flow Interactions 194
3 Negative Wave Energy and Over-Reflection 199
4 Acoustic Analogues of Microwave Devices: Monotron and Klystron 201
5 Wave-Vortex Interactions 206
6 Analogies Table 208

New Linear Mechanisms of Acoustic Wave Generation in Smooth Shear Flows (Nonmodal Study)
George D. Chagelishvili . . . 210
1 Introduction . . . 210
1.1 Basic Problems in Shear Flows . . . 210
1.2 Essence of the Difficulties of the Modal Approach . . . 211
1.3 Onto Nonmodal Approach . . . 211
2 Mathematics of the Study . . . 213
2.1 Types of Disturbances . . . 216
2.2 Onto the Value of Shear Rate . . . 217
3 The Dynamics of SFH of Acoustic Waves . . . 217
3.1 Velocity Field of SFH of Acoustic Waves . . . 218
3.2 The Dynamics of SFH of Acoustic Waves: Low Shear Rates . . 220
3.3 The Dynamics of SFH of Acoustic Waves: Moderate Shear Rates . . . 222
4 The Dynamics of the SFH of Vortices . . . 223
4.1 Conversion of Vortices to Waves . . . 224
4.2 The Nature of the Conversion Phenomenon . . . 226
4.3 Onto the Amplitude of the Generated Wave SFH at Conversion . . . 227
4.4 Demonstration of the Conversion Phenomenon by Direct Numerical Calculation . . . 227
4.5 Onto the Vortex-Wave Interaction . . . 230
5 Mutual Transformation of Waves . . . 230
5.1 Mechanical Analogy . . . 231
5.2 Qualitative Description of the Resonance Transformation of Waves . . . 231
6 Conclusion . . . 234
References . . . 235

Singular Eigenfunctions and an Integral Transform for Shear Flow
Philip J. Morrison . . . 238
1 Introduction . . . 238
2 Shear Flow . . . 238
2.1 Equilibrium and Linearization . . . 239
2.2 Singular Eigenmodes and the Continuous Spectrum . . . 240
3 The Integral Transform–Coordinate Change . . . 243
3.1 Integral Transform Pair . . . 243
3.2 Solution . . . 245
4 Hamiltonian Interpretation . . . 245
References . . . 246

Structural Acoustics with Mean Flow
Nigel Peake 248
1 Basic Problem 248
2 Solution Technique 250
3 Solution for Fluid-Loaded Plate with Mean Flow 253
3.1 Wave Energy 254
4 Modifications to the Basic Model 257
4.1 Plate Curvature 257
4.2 Nonlinear Motion 259
4.3 Effects of a Boundary Layer 262
5 Conclusion 263
References 264

Nonlinear Acoustics and Acoustic Chaos
Werner Lauterborn 265
1 Introduction 265
2 Origin of Nonlinearity 266
3 Equation of State 266
4 Simple Nonlinear Waves 268
5 Shock Waves 270
6 Sonoluminescence 272
7 Chaos Physics 276
8 Acoustic Chaos 280
References 283

Subject Index 285

List of Contributors

George D. Chagelishvili
Centre for Plasma Astrophysics, Abastumani Astrophysical Observatory, Ave. A. Kazbegi 2a, 380060 Tbilisi, Georgia
and
Space Research Institute, Str. Profsoyuznaya 84/32, 117810 Moscow, Russia
georgech123@yahoo.com

Anatoly Fabrikant
KLA-Tencor, One Technology Dr., Milpitas, CA 95035, USA
fabr3@home.com

Avraham Hirschberg
Eindhoven University of Technology, Dept. of App. Physics, J.M. Burgerscentrum, Postbus 513, 5600 MB Eindhoven, The Netherlands
A.Hirschberg@tue.nl

Michael S. Howe
Boston University, College of Engineering, 110 Cummington Street, Boston, MA 02215, USA
mshowe@bu.edu

Werner Lauterborn
Georg-August-Universität Göttingen, Drittes Physikalisches Institut, Buergerstr. 42-44, 37073 Göttingen, Germany
W.Lauterborn@dpi.physik.uni-goettingen.de

Fernando Lund
Centro para la Investigación Interdisciplinaria Avanzada en Ciencia de los Materiales
and
Departamento de Física, Facultad de Ciencias Físicas y Matemáticas, Universidad de Chile, Casilla 487-3, Santiago, Chile
flund@dirac.dfi.uchile.cl

Philip J. Morrison
Physics Department C1600, University of Texas at Austin, Austin, TX 78712, USA

Vladimir E. Ostashev
NOAA, Environmental Technology Laboratory, 325 Broadway, Boulder, CO 80305
and
Department of Physics, New Mexico State University, Las Cruces, NM, USA
Vladimir.Ostashev@noaa.gov

Nigel Peake
DAMTP, University of Cambridge, Silver St., Cambridge CB3 9EW, UK
N.Peake@damtp.cam.ac.uk

Christophe Schram
von Karman Institute for Fluid Dynamics, Environmental and Applied Fluid Dynamics Dept.,
72 chaussée de Waterloo,
1640 Rhode-Saint-Genèse,
Belgium
schram@vki.ac.be

A Primitive Approach to Aeroacoustics

Avraham Hirschberg[1] and Christophe Schram[2]

[1] Eindhoven University of Technology, Dept. of App. Physics, J.M. Burgerscentrum, Postbus 513, 5600 MB Eindhoven, The Netherlands. Tel: 00(31)402473110, Fax: 00(31)402464151, Email: A.Hirschberg@tue.nl

[2] von Karman Institute for Fluid Dynamics, Environmental and Applied Fluid Dynamics Dept., 72 chaussée de Waterloo, 1640 Rhode-Saint-Genèse, Belgium. Tel: 00(32)23599611, Fax: 00(32)23599600, Email: schram@vki.ac.be

Abstract. A primitive approach to aeroacoustics is presented using the aeroacoustic analogies. This aproach is illustrated by a few examples like musical instruments, the Rijke tube, speech production *etc.*

1 Introduction

Due to the non-linearity of the governing equations it is very difficult to predict the sound production by fluid flows. This occurs typically at high speed flows, for which non-linear inertial terms in the equation of motion are much larger than the viscous terms (high Reynolds numbers). Direct simulation of such flows is very difficult. As sound production represents only a very minute fraction of the energy in the flow, the accuracy of numerical simulations is usually not sufficient to predict the sound generation. This is particularly dramatic in free space and at low subsonic speeds. The fact that the sound field is in some sense a small perturbation of the flow, can however be used to obtain approximate solutions.

Aeroacoustics provides such approximations and at the same time a definition of the acoustical field as an extrapolation of an ideal reference flow which we will specify later. The difference between the actual flow and the reference flow is identified as a source of sound. This idea was introduced by Lighthill ([21], [22]) which calls this an *analogy.* A second key idea of Lighthill [22] is the use of an integral equation as a formal solution. When approximations are used in such an integral equation, random errors are smoothed out.

Considering confined flows, the acoustical energy can accumulate into resonant modes. The acoustical particle displacement velocity can become of the same order of magnitude as the main flow velocity. Yet the ideas of the analogy will appear to remain useful.

The recent publication of Fedorchenko [8] contains severe criticisms on the analogy and the way the analogy has been used. Tam [46] also provides an example of how the analogy can be used to draw erroneous conclusions.

We will also discuss some examples of such problems. The key issue is that the analogy is only useful as an approximation. Such an approximation is based on some intuition and usually empirical observations. We could say that Lighthill has proposed an intelligent way to carry out dimensional analysis. It is a joy for engineers to discover semi-empirical relationships between relevant parameters under circumstances where exact solutions are not available. This kind of results are usually obtained by using rather crude approximations. In fact, ignoring subtleties does help. We would like to introduce the reader to this primitive approach to aeroacoustics by considering a few examples:

- the sound production by a subsonic isothermal turbulent free jet (Sect. 4),
- the thermo-acoustical oscillation of the Rijke-tube (Sect. 7),
- the aero-acoustical response of an open pipe termination (Sect. 11), and
- the human whistling (Sect. 12).

Before each example we will discuss some basic theory:

- fluid dynamics (Sect. 2),
- Lighthill's analogy (Sect. 3),
- thermo-acoustics (Sect. 5),
- acoustical energy (Sect. 6),
- vortex-sound theory (Sect. 8).

2 Fluid Dynamics

We consider the motion of fluids in the continuum approximation. This means that quantities such as the velocity $\boldsymbol{v}$ and the density ϱ can be described by fields, smooth functions of space and time coordinates $(\boldsymbol{x}, t)$. In a non-relativistic approach the mass is conserved. Hence the rate of change of the density observed while moving with the fluid velocity is equal to minus the dilatation rate ([1], [47], [20]):

$$\frac{1}{\varrho}\frac{\mathrm{D}\varrho}{\mathrm{D}t} = -\nabla . \boldsymbol{v} \tag{1}$$

where the Lagrangian time derivative $\mathrm{D}\varrho/\mathrm{D}t$ is related to the Eulerian time derivative $\partial\varrho/\partial t$ by:

$$\frac{\mathrm{D}\varrho}{\mathrm{D}t} = \frac{\partial \varrho}{\partial t} + (\boldsymbol{v} \, . \nabla)\varrho \, . \tag{2}$$

Substitution of definition (2) into (1) yields the mass conservation law applied to a fixed infinitesimal volume element:

$$\frac{\partial \varrho}{\partial t} + \nabla . \, (\varrho \boldsymbol{v}) = 0 \, . \tag{3}$$

Using Einstein summation convention we can write this for cartesian coordinates $\boldsymbol{x} = (x_1, x_2, x_3)$ as:

$$\frac{\partial \varrho}{\partial t} + \frac{\partial \varrho v_i}{\partial x_i} = 0 \, . \tag{4}$$

The repetition of an index indicates a summation:

$$\frac{\partial \varrho v_i}{\partial x_i} = \frac{\partial \varrho v_1}{\partial x_1} + \frac{\partial \varrho v_2}{\partial x_2} + \frac{\partial \varrho v_3}{\partial x_3}\,. \tag{5}$$

Please note that we do not have in a non relativistic approximation any mass source term. In agreement with this approximation we apply the second law of Newton to a fluid particle:

$$\varrho \frac{\mathrm{D}\boldsymbol{v}}{\mathrm{D}t} = -\nabla . \boldsymbol{P} + \boldsymbol{f} \tag{6}$$

where $\boldsymbol{f}$ is the density of the force field acting on the bulk of the fluid, while $-\nabla \boldsymbol{P}$ corresponds to the forces acting on the fluid surface expressed in terms of a stress tensor $\boldsymbol{P}$. Using the mass-conservation law (4) we obtain the momentum equation in conservation form:

$$\frac{\partial \varrho v_i}{\partial t} + \frac{\partial \varrho v_i v_j}{\partial x_j} = -\frac{\partial P_{ij}}{\partial x_j} + f_i\,. \tag{7}$$

Please note that it is rather odd that Fedorchenko [8] does introduce a mass source term in a non-relativistic approach and worries about its consequences for the momentum conservation law.

The isotropic part of this tensor corresponds to the effect of the hydrostatic pressure p:

$$P_{ij} = p\,\delta_{ij} - \sigma_{ij}\,. \tag{8}$$

The deviation σ_{ij} from the hydrostatic behavior corresponds usually in a simple fluid to the effect of viscosity. For a simple fluid σ_{ij} is symmetrical. Hence (3) and (7) form a set of four equations with eleven unknowns $(\varrho, p, v_i, \sigma_{ij})$. Using the energy equation would add one equation and at least one more unknown. In order to obtain a number of equations equal to the number of unknowns we have to use empirical information. This is commonly done by using so-called constitutive equations. A widely used approximation is to assume a fluid at a state of local thermodynamic equilibrium. This implies that there exists a functional relationship between the pressure p, the density ϱ and the specific entropy s (entropy per unit of mass). This is called an equation of state:

$$p = p(\varrho, s)\,. \tag{9}$$

In differential form we have:

$$\frac{\mathrm{D}p}{\mathrm{D}t} = c^2 \frac{\mathrm{D}\varrho}{Dt} + \left(\frac{\partial p}{\partial s}\right)_\varrho \frac{\mathrm{D}s}{\mathrm{D}t} \tag{10}$$

where c is identified as the isentropic equilibrium speed of sound:

$$c = \sqrt{\left(\frac{\partial p}{\partial \varrho}\right)_s}\,. \tag{11}$$

The entropy equation can be written as [47]:

$$\varrho T \frac{Ds}{Dt} = -\nabla . \boldsymbol{q} + \boldsymbol{\sigma} : (\nabla \boldsymbol{v}) + Q_w \tag{12}$$

where T is the absolute temperature, $\boldsymbol{q}$ the heat flux and Q_w a density of rate of energy production in the bulk of the fluid. In the audio frequency and amplitude range it appears that sound propagation in ambient air is approximately isentropic ([35], [7]).

3 Lighthill's Analogy

We use as starting point the exact mass and momentum equations (3) and (7). We take the time derivative of the mass equation (3) and subtract from this the divergence of the momentum equation (7) to obtain:

$$\frac{\partial^2 \varrho}{\partial t^2} = \frac{\partial^2(\varrho v_i v_j - \sigma_{ij})}{\partial x_i \partial x_j} + \frac{\partial^2 p}{\partial x_i^2} - \frac{\partial f_i}{\partial x_i} . \tag{13}$$

We can subtract on both sides of this equation a term $c_0^2(\partial^2 \varrho / \partial x_i^2)$ were c_0 is a velocity. This provides the right hand side of the equation the form of the wave equation of d'Alembert:

$$\frac{\partial^2 \varrho}{\partial t^2} - c_0^2 \frac{\partial^2 \varrho}{\partial x_i^2} = \frac{\partial^2(\varrho v_i v_j - \sigma_{ij})}{\partial x_i \partial x_j} + \frac{\partial^2 (p - c_0^2 \varrho)}{\partial x_i^2} - \frac{\partial f_i}{\partial x_i} . \tag{14}$$

This is in general a useless equation. The velocity c_0 could be the speed of light or one centimeter per century... The equation would still be exact. Furthermore we had four exact equations with eleven unknowns, and we are left with a single equation and as many unknowns. Clearly this seems a dead end.

The key idea is to compare this equation with the equation which we obtain under normal audio frequency and amplitude conditions for the perturbation of a uniform and stagnant fluid in the state (ϱ_0, p_0). Under such circumstances we can neglect friction losses and heat conduction for distances of propagation comparable to the wavelength. We introduce the perturbations:

$$\varrho' = \varrho - \varrho_0 \tag{15}$$

$$p' = p - p_0 \tag{16}$$

$$v_i' = v_i - v_{i,0} \tag{17}$$

where in the specific case discussed $v_{i,0} = 0$. The linearized equations of mass and momentum become:

$$\frac{\partial \varrho'}{\partial t} + \varrho_0 \frac{\partial v_i'}{\partial x_i} = Q_m \tag{18}$$

$$\varrho_0 \frac{\partial v_i'}{\partial t} + \frac{\partial p'}{\partial x_i} = f_i \,. \tag{19}$$

We have introduced here a mass source term Q_m which we specify further by assuming that it does not involve any momentum injection and that the injected fluid has the same properties as the reference fluid (ϱ_0, p_0). Under such circumstances the reference fluid remains uniform and stagnant so that the equation of state (10) can be written as:

$$p' = c_0^2 \,\varrho' \,. \tag{20}$$

Due to linearization in combination with the assumption of a uniform fluid, the convective effects drop. Please note that this will not occur if ϱ_0 or p_0 are not uniform ([35]). Eliminating the velocity v_i' by subtracting the divergence of (19) from the time derivative of (18) yields:

$$\frac{\partial^2 \varrho'}{\partial t^2} - c_0^2 \frac{\partial^2 \varrho'}{\partial x_i^2} = \frac{\partial Q_m}{\partial t} - \frac{\partial f_i}{\partial x_i} \,. \tag{21}$$

When we are listening to the sound of a flute in a concert hall, the sound field around our ears is accurately described by the homogeneous wave equation in which we have dropped the source terms. In the case of the equation of Lighthill (14) linearization and neglecting visco-thermal dissipation would also yield this equation if we identify c_0 as *the speed of sound in the stagnant uniform fluid surrounding the listener.* We can further define the perturbations $\varrho' = \varrho - \varrho_0$ and $p' = p - p_0$ as the differences between the local values of ϱ and p and the values of these quantities in the reference fluid surrounding the listener (Fig. 1). This is a generalization of (15) and (16) in which the perturbations do not need to be small. As ϱ_0 and p_0 are like c_0 constants we can write Lighthill's analogy as ([10], [4]):

$$\frac{\partial^2 \varrho'}{\partial t^2} - c_0^2 \frac{\partial^2 \varrho'}{\partial x_i^2} = \frac{\partial^2 (\varrho v_i v_j - \sigma_{ij})}{\partial x_i \partial x_j} + \frac{\partial^2 (p' - c_0^2 \varrho')}{\partial x_i^2} - \frac{\partial f_i}{\partial x_i} \tag{22}$$

and identify the right hand side of this equation as a source of sound.

This procedure can be considered as a generalization of Rayleigh's approach to scattering theory [37]. We consider the case of the scattering of a wave on non-homogeneities in fluid density which have a small spatial extension compared to the wavelength of the incident wave p'_{in}. When the unperturbed fluid is stagnant and the density difference $\varrho - \varrho_0$ is small one can assume a uniform mean pressure p_0. The entropy will however not be uniform so that the speed of sound c will vary. The main difference between this situation and the earlier discussion of the acoustics of a stagnant uniform fluid is that the convective effects in the equation of state (10) cannot be neglected. In linearized form and neglecting the pressure gradient term ∇p_0 we find:

$$\frac{\partial p'}{\partial t} = c^2 \left(\frac{\partial \varrho'}{\partial t} + \boldsymbol{v}' . \nabla \varrho \right) \tag{23}$$

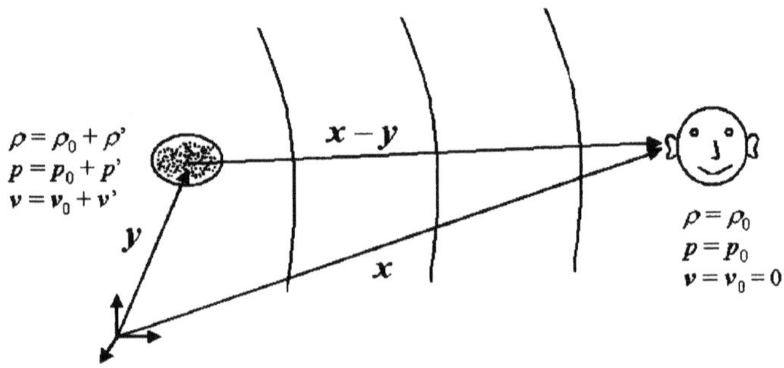

Fig. 1. Source and listener in the analogy of Lighthill. While in the source region the density fluctuations are not necessarily small, around the listener we assume an acoustical behavior with $\varrho' \ll \varrho_0$.

and the corresponding mass and momentum conservation laws in the absence of source terms ($Q_m = 0$ and $f_i = 0$):

$$\frac{1}{\varrho c^2}\frac{\partial p'}{\partial t} + \nabla . \boldsymbol{v}' = 0 \tag{24}$$

and:

$$\frac{\partial \boldsymbol{v}'}{\partial t} + \frac{1}{\varrho}\nabla p' = 0 \tag{25}$$

where the quantities ϱ and c designate the time average of the density and speed of sound. They are therefore time independent but vary in space. Elimination of $\boldsymbol{v}$ by subtracting the divergence of (25) from the time derivative of (24) yields:

$$\frac{\partial}{\partial t}\frac{1}{\varrho c^2}\frac{\partial p'}{\partial t} - \nabla . \left(\frac{1}{\varrho}\nabla p'\right) = 0\,. \tag{26}$$

By addition on both sides of the same factor

$$\frac{1}{\varrho_0}\left(\frac{1}{c_0^2}\frac{\partial^2 p'}{\partial t^2} - \nabla^2 p'\right) \tag{27}$$

we obtain:

$$\frac{1}{c_0^2}\frac{\partial^2 p'}{\partial t^2} - \nabla^2 p' = \frac{1}{c_0^2}\frac{\partial}{\partial t}\left(\frac{\varrho c^2 - \varrho_0 c_0^2}{\varrho c^2}\frac{\partial p'}{\partial t}\right) - \nabla . \left(\frac{\varrho - \varrho_0}{\varrho}\nabla p'\right)\,. \tag{28}$$

The first source term is identified as a monopole source term due to the difference in compressibility of the reference fluid $1/\varrho_0 c_0^2$ surrounding the listener and the local fluid compressibility $1/\varrho c^2$ in the source region. The second term is a dipole sound source due to the Archimedes force on the non-uniform fluid particle as a result of its acceleration $\partial \boldsymbol{v}'/\partial t = -(\nabla p')/\varrho$. A further step is to split the acoustic field into an incident wave p'_{in} which satisfies the homogeneous wave equation and a scattered wave p'_{sc} which we assume to be small compared to the incident wave. An approximate solution can then be calculated by neglecting the effect of p'_{sc} on the source term:

$$\frac{1}{c_0^2}\frac{\partial^2 p'_{sc}}{\partial t^2} - \nabla^2 p'_{sc} \simeq \frac{1}{c_0^2}\frac{\partial}{\partial t}\left(\frac{\varrho c^2 - \varrho_0 c_0^2}{\varrho c^2}\frac{\partial p'_{in}}{\partial t}\right) - \nabla . \left(\frac{\varrho - \varrho_0}{\varrho}\nabla p'_{in}\right) . \quad (29)$$

Another limit case is obtained by neglecting all the source terms in Lighthill's analogy except for the force term $-\nabla . \boldsymbol{f}$. This is a very useful approximation when considering the sound production by moving objects. The force $\boldsymbol{f}$ represents then the reaction of the object to the force applied by the fluid flow. This corresponds to the principle of Gutin which is discussed by Goldstein [10].

4 Jet Noise

In the previous section we introduced Lighthill's analogy and related it to Rayleigh's scattering and Gutin's principle. Lighthill's ambition was to apply this approach to the much more delicate problem of the sound production by a free turbulent flow [22]. He argued that when one considers an isothermal subsonic free jet of air in air all the sources terms in equation (22) are negligible except for $\partial^2 \varrho v_i v_j / \partial x_i \partial x_j$. It is rather obvious that in the absence of walls and of external force field $f_i = 0$. The fact that the other terms vanish is not as trivial as suggested by Lighthill [22]. A more careful discussion of the validity of this approximation was provided much later by Morfey ([26],[27]) and Obermeier ([32],[33], [19]). Lighthill's intuition appears not to be bad.

At this point we note that a source term involving a second spatial derivative such as $\partial^2 \varrho v_i v_j / \partial x_i \partial x_j$ is likely to behave as a quadrupole. This simply is due to the fact that since there is no external force to sustain a dipole and that we have neglected monopole terms we are left with a quadrupole as leading order term. Of course one can always find a pathological case in which the leading order term in the radiated sound field is an octupole [8], but this has little practical implications. If we now ignore the quadrupole character of the sound field and use a characteristic length scale D in the flow field to predict the sound strength we would find $\partial / \partial x_i \sim 1/D$. If we assume a compact sound source, we would find a radiated sound pressure scaling as the stagnation pressure of the flow $p' \sim \varrho_0 U_0^2$ and as a consequence a sound source power scaling as U_0^4. This is contradicted by the experiments. If we ignore the retarded time effects and simply integrate the source taking into account the presence of spatial derivatives the sound production vanishes due to the limited size of the sound region ([4]).

The most convenient way to obtain a better approximation for the effect of retarded time differences across the source region is to start from the integral formulation:

$$\varrho'(\boldsymbol{x},t) = \int_{-\infty}^{t}\int_{V} G_0(t,\boldsymbol{x}|\tau,\boldsymbol{y}) \frac{\partial^2 \varrho v_i v_j}{\partial y_i \partial y_j}\, d\boldsymbol{y} d\tau \tag{30}$$

with the free space Green's function given by:

$$G_0(t,\boldsymbol{x}|\tau,\boldsymbol{y}) = \frac{\delta(t-\tau-|\boldsymbol{x}-\boldsymbol{y}|/c_0)}{4\pi c_0^2|\boldsymbol{x}-\boldsymbol{y}|} \tag{31}$$

which has the nice symmetry property:

$$\frac{\partial G_0}{\partial x_i} = -\frac{\partial G_0}{\partial y_i} \tag{32}$$

for differentiation with respect to the source coordinate $\boldsymbol{y}$ and the observer coordinate $\boldsymbol{x}$ (see Fig. 2). By partial integration we can move the space deriva-

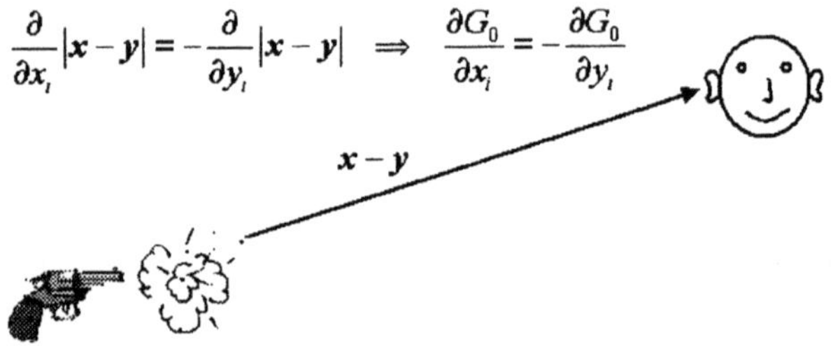

Fig. 2. Symmetry of differentiation of the free space Green's function with respect to observer's and source coordinates. In free space, moving the source towards the observer is equivalent to moving the observer to the source.

tive from the source term $\partial^2 \varrho v_i v_j/\partial y_i \partial y_j$ (which we do not know accurately) towards the well known Green's function. Using the symmetry property (32) we can replace the derivatives of G_0 with respect to the source coordinate $\boldsymbol{y}$ by derivatives with respect to the observer coordinate $\boldsymbol{x}$. As the integration is performed on the source coordinate, we can move this spatial derivatives out of the integral, and obtain the integral formulation

$$\varrho'(\boldsymbol{x},t) = \frac{\partial^2}{\partial x_i \partial x_j}\int_{-\infty}^{t}\int_{V} \frac{\delta(t-\tau-|\boldsymbol{x}-\boldsymbol{y}|/c_0)}{4\pi c_0^2|\boldsymbol{x}-\boldsymbol{y}|}\, \varrho v_i v_j\, d\boldsymbol{y} d\tau\,. \tag{33}$$

When the observer is in the far field, the relevant length scale there is the acoustical wavelength. Assuming that the sound production is dominated by the time scale D/U_0 in the flow, we can replace each derivative by a multiplication by a factor $U_0/(c_0 D)$. Assuming $U_0/c_0 << 1$ appears then to be equivalent to the assumption of a compact flow region.

Neglecting other, more subtle, retarded time effects ([10], [23]) we find that the sound source pressure will scale as U_0^4 and the power as U_0^8, which is the famous law of Lighthill.

As stressed Powell in [37] this is one of the few examples in which theory predicted something before it was observed experimentally. We insist again on the fact that we found a quadrupole because we used the fact that we know that there are no external forces, which is used as argument to exclude any dipole term. A conceptually similar discussion will follow in Sect. 8 where an alternative formulation of Lighthill's analogy, known as vortex sound theory, will be developed.

Yet, (33) is not a formal solution of our problem, since the density on its right-hand side contains the unknown ($\varrho = \varrho_0 + \varrho'$). To obtain an integral solution, Lighthill further argues that for isothermal free jets, the relative fluctuations of density in the source region scale with the square of the Mach number of the flow. The density in the right-hand side of (33) can then be approximated by the reference density ϱ_0, and in the low Mach number limit, we end up with the integral solution:

$$\varrho'(\boldsymbol{x},t) = \frac{\partial^2}{\partial x_i \partial x_j} \int_{-\infty}^{t} \int_V \frac{\delta(t-\tau-|\boldsymbol{x}-\boldsymbol{y}|/c_0)}{4\pi c_0^2 |\boldsymbol{x}-\boldsymbol{y}|} \varrho_0 v_i v_j \, d\boldsymbol{y} d\tau \,. \tag{34}$$

5 Thermo-Acoustics

A success of Lighthill's law was to stress the essential difference between the sound radiated by isothermal and non-isothermal free jets. This triggered fundamental research. Initially it was expected that heat-transfer would induce a monopole sound source. This was not found. Morfey [26] and Obermeier [32] understood that this was a result of the lack of difference in compressibility between the cold and the hot gas in a flow where the mean pressure is uniform (see (28)). The relevant sound source appears to be the force of Archimedes due to the acceleration of fluid particles with a density deviating from the surrounding. This results into a dipole sound source which in Lighthill's analogy (22) can be identified as the term $\partial^2(p' - c_0^2\varrho')/\partial x_i^2$. Superficial inspection of (22) would suggest that a monopole sound source cannot be induced by a flow. This is contradicted by Rayleigh's scattering theory (28). Furthermore we expect unsteady combustion or unsteady heat transfer from a wall to act as a monopole sound source. The key issue is that the analogy is not unique. In particular we have presented an analogy in which the acoustic field was expressed in terms of density fluctuations ϱ'. It is possible to use the pressure fluctuations p' as aero-acoustical variable. We simply add to both sides of (13) the term $(1/c_0^2)(\partial^2 p'/\partial t^2)$ and obtain:

$$\frac{1}{c_0^2}\frac{\partial^2 p'}{\partial t^2} - \frac{\partial^2 p'}{\partial x_i^2} = \frac{\partial^2(\varrho v_i v_j - \sigma_{ij})}{\partial x_i \partial x_j} - \frac{\partial f_i}{\partial x_i} + \frac{\partial^2}{\partial t^2}\left(\frac{p'}{c_0^2} - \varrho'\right) \tag{35}$$

which of course is as exact and as useless as Lighthill's formulation (22), as long as we do not introduce any approximation. If we neglect convective effects we can write in linear approximation:

$$\frac{p'}{c_0^2} - \varrho' = \frac{1}{c_0^2}\left(\frac{\partial p}{\partial s}\right)_\varrho s' . \tag{36}$$

The entropy equation (12) provides then a link between this sound source term and heat transfer $\boldsymbol{q}$ or heat release by combustion Q_w. It appears therefore that using p' as aero-acoustical variable is most convenient when we study thermo-acoustical phenomena ([23], [4]).

Seeking for an optimal choice of the aero-acoustical variable for a particular application is one of the concerns in aeroacoustics ([6], [30]). This problem is particularly acute for supersonic flows in which the flow is not compact, so that convective effects on wave propagation are important. Making a choice of a aeroacoustic analogy providing separation of "source" from "convective effect" is very difficult ([10], [23]).

6 Acoustical Energy

When considering linear perturbations of a uniform stagnant reference fluid we can easily derive an energy equation for the perturbations. We add the product of the mass equation (18) by p'/ϱ_0 to the scalar product of the momentum equation (19) with the velocity v_i'. We find:

$$\frac{\partial E}{\partial t} + \frac{\partial I_i}{\partial x_i} = p'\frac{Q_m}{\varrho_0} + f_i v_i' \tag{37}$$

where the acoustical energy E is defined by:

$$E = \frac{(p')^2}{2\varrho_0 c_0^2} + \frac{\varrho_0 (v_i')^2}{2} \tag{38}$$

and the intensity I_i is defined by:

$$I_i = p' v_i' . \tag{39}$$

In integral form and averaged over a period of oscillation for periodic perturbations we find for the acoustical power $< P_{ac} >$ produced by the flow:

$$< P_{ac} > = \int_S < I_i n_i > dS = \int_V < p'\frac{Q_m}{\varrho_0} > dV + \int_V < f_i v_i' > dV \tag{40}$$

where the brackets indicate the time averaging and n_i is the outer-unit normal vector to the surface S delimiting the integration volume V. The first volume integral corresponds to the work performed by volume injection Q_m/ϱ_0 while the second integral corresponds to the work performed by the bulk forces on the acoustical field. Please note here again that the injected fluid has the mean density ϱ_0 (otherwise the entropy would not remain uniform). While this equation seems obvious, its derivation is quite hazardous. We have use linearized equations to obtain an equation for quantities E and I_i which are of second order in the perturbations [20]. This approach provides a correct result in this particular case. It should not be used in general ([35], [10], [31]). We will, however, show that this equation provides some insight into the behavior of a strange device called the Rijke-tube.

7 Rijke-Tube

The Rijke-tube ([41], [38], [25], [11]) is a glass pipe of internal diameter $D = 2.5$ cm and with a length $L = 30$ cm. A metal gauze is placed in across the pipe section at a distance $L/4$ from one of the open ends. The gauze is formed by wires with a diameter $d = 0.3$ mm and has about 1 mm mesh size. The tube is held vertically with the metal gauze at the lower part (Fig. 3). The gauze is heated electrically. When a critical temperature is reached the pipe starts producing a loud sound at its fundamental resonance frequency. This corresponds roughly to $f = c_0/(2L)$. When a flame is used to heat up the gauze, the candle or the lighter should be removed in order to allow the pipe to sound. The first theoretical analysis of this phenomenon was provided by Rayleigh [38]. We follow to a large extent his discussion.

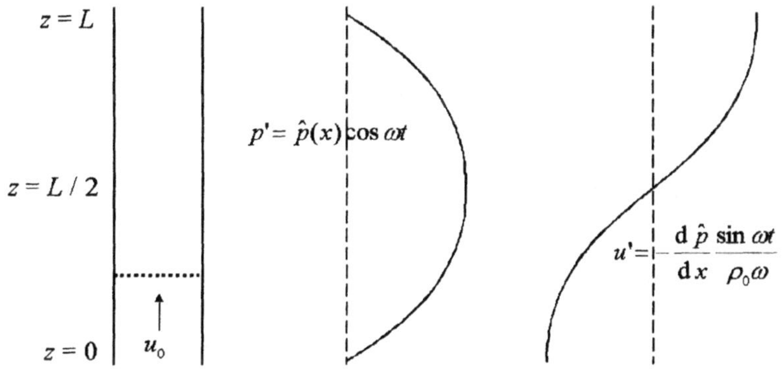

Fig. 3. Rijke-tube and the acoustic standing wave pattern.

The oscillation is an example of self sustained oscillation due to modulation of the heat transfer from the gauze to the air by the acoustic field in the pipe. From our previous discussion we expect that this sound source has a monopole character (see (35)). The heating of the gas causes an expansion which acts as a volume injection Q_m/ϱ_0. The power generated by the source is therefore given by:

$$< P_{ac} >= \int_S < I_i n_i > dS = \int_V < p' \frac{Q_m}{\varrho} > dV\,. \tag{41}$$

Before proceeding we should propose a model for the heat transfer from the gauze to the ambient air. We use Newton's approximation for the heat flux $q_n = \boldsymbol{q}.\boldsymbol{n}$, where $\boldsymbol{n}$ is the unit normal to the surface of the gauze:

$$q_n = \alpha(T_w - T) \tag{42}$$

where α is the heat transfer coefficient, T_w is the wall temperature and T the ambient air temperature at some distance from the wall. We assume that T_w is constant due to the thermal inertia of the gauze. The ambient temperature T is assumed to be modulated by the adiabatic compression corresponding to the acoustic pressure fluctuations p'. The heat transfer coefficient is determined by the flow through the gauze. The steady component of this flow with a velocity U_0 is due to the upwards draft of the hot air in the upper part of the Rijke tube, which acts as a chimney. In addition to this we have an acoustical modulation of the velocity which is in first order approximation related to the acoustical pressure fluctuations by the linearized momentum equation (19).

We follow Rayleigh [38] by considering first the behavior of the system when the gauze is placed at the middle of the pipe where there is an acoustical velocity node. The modulation of heat transfer is in that case due to the adiabatic temperature change T' associated with the acoustic pressure fluctuations p'. This tends to reduce the heat transfer rate q_n when the pressure p' increases. If we neglect memory effects the corresponding volume source is opposite in phase to the pressure fluctuations. Following (41) we expect that the gauze will only absorb acoustical energy. This seems to be confirmed by experiment. The pipe does not sound when the gauze is placed in the middle.

It seems logical to assume that the modulation in heat transfer α is due to a modulation of the heat transfer coefficient by the acoustical velocity fluctuations u'. The largest fluctuations are found near the open pipe terminations. As those positions correspond to pressure nodes $p' = 0$, the source cannot transfer any energy to the acoustical resonant mode of the pipe (see (41)). Hence the position halfway between the middle of the pipe and an open end is a compromise. At this position we have both acoustical pressure and velocity fluctuations.

Until now we have followed Rayleigh's discussion [38]. We want now to understand why the gauze does sustain the pipe oscillation when placed at the lower part of the pipe but does not do it when it is placed at the upper part. At this point the discussion of Rayleigh is not very clear and we propose an alternative. The first observation which we should do is that if the heat transfer coefficient would react instantaneously to velocity fluctuations the heat transfer would be $\pi/2$ out of phase with p', which is a direct consequence of the momentum equation (19). If p' has a $\cos(2\pi ft)$ time dependence, u' will have a $\sin(2\pi ft)$ time dependence. Hence a quasi steady theory for the response of the heat transfer cannot predict any acoustical energy transfer to the acoustical field. The average over an acoustical oscillation period vanishes ($\int_0^{1/f}\cos(2\pi ft)\sin(2\pi ft)dt = 0$).

A plausible explanation is that the heat transfer has some memory time τ. If we assume that the heat transfer is determined by thin boundary layers between the gauze and the surrounding, an increase in velocity will reduce the boundary layer thickness. However the remaining hot fluid from the "old" boundary layer has to be evacuated by convection which will last for a time τ of the order of magnitude of ([40]):

$$\tau \simeq \frac{2d}{U_0}. \tag{43}$$

As shown in Fig. 4 the time delay between pressure fluctuations and velocity fluctuations for a gauze placed at $z = L/4$ is a quarter of an oscillation period $1/(4f)$ so that optimal sounding conditions are found for:

$$\frac{fd}{U_0} \simeq \frac{1}{8}. \tag{44}$$

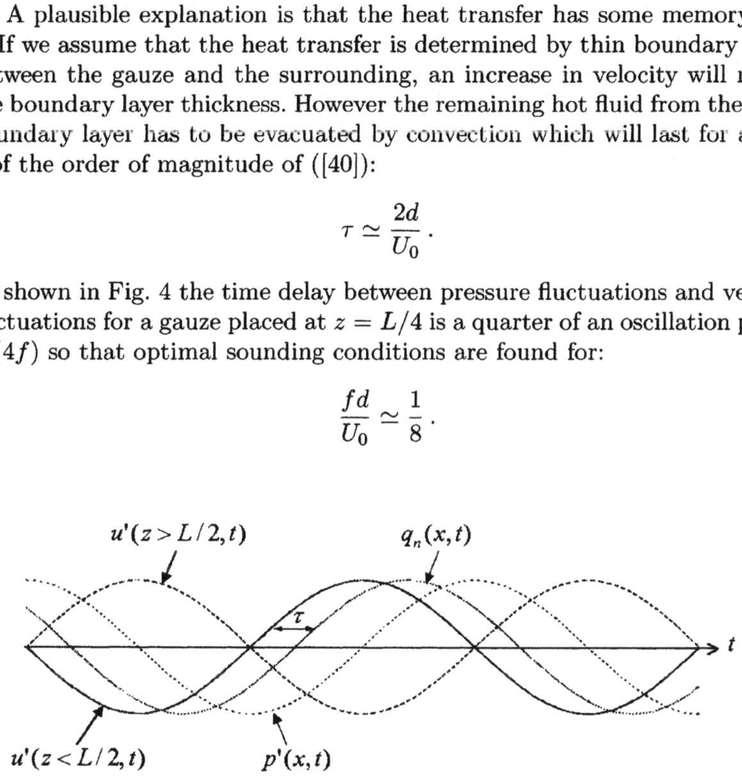

Fig. 4. Time dependence of the pressure, acoustical velocity and of the heat transfer at a gauze placed in the lower half of a Rijke-tube.

A gauze placed in the upper part of the pipe will experience a velocity fluctuation opposite in phase to that found at the lower part of the pipe (Fig. 3). The pipe aspires and expires from both open ends. This can also be verified mathematically by calculating the velocity field corresponding to a pressure distribution $p' = \hat{p}\sin(2\pi fz/c_0)\cos(2\pi ft)$. This implies a time lag of $3/4f$. Hence for a gauze placed at the upper side of the pipe ($z = 3L/4$) the flow velocity needed to sound the tube should be three times lower than for the gauze placed at the lower part. In such a case the thermal boundary layer around the gauze are thicker, which reduces the heat transfer and the sound source. While it is not impossible to make a Rijke tube sound with a gauze placed in its upper part, this appears not to be possible for the particular conditions considered in our experiment.

Please note that in our discussion we used a linear theory. This implies that the acoustical power delivered by the source will be proportional to the square of the acoustical amplitude because Q_m is proportional to u'. As the losses are also proportional to the square of the acoustical amplitude, such a linear theory cannot predict a finite oscillation amplitude. A finite amplitude is established as a result of non-linear effects. In the case of the Rijke tube, it is believed that when acoustical particle displacement become of the order of magnitude of the gauze wire diameter d the hot air from the wake of the gauze flows back to the gauze during part of the oscillation period [11]. This blocks the heat transfer. The amplitude of the particle displacement is given by $u'/(2\pi f)$. Because at the grid position $z = L/4$ we have

$$u' \simeq \frac{p'}{\varrho_0 c_0} \tag{45}$$

we find:

$$\frac{p'}{\varrho_0 c_0} \simeq 2\pi f d\,. \tag{46}$$

To our great surprise this is verified by experiments.

8 Vortex-Sound Theory

Vortex-sound theory was developed by Powell [36] for subsonic flows in free-field conditions. Howe [17] realized that such an elegant formalism was probably much more general, and proposed a generalization to internal flows and arbitrary Mach numbers. In analogy to Prandtl's: "Vortices are the muscles of the flow" we have the Müller and Obermeier [29] statement: "Vortices are the voice of the flow". The power of the "vortex-sound" theory comes from two facts:

- the first one is that we have an intuition for vortex dynamics because of the conservation of the circulation of vortices in frictionless flows. The study of the movement of vortices corresponds to a kinematic problem (application of the Biot-Savart induction law [44]), which is much easier than the solution of the dynamical problem. This is quite similar to the role of vorticity when explaining the lift of an airfoil. By using a Kutta condition for tangential flow separation at the trailing edge of the airfoil, the problem of calculating the flow becomes purely kinematic. The lift is then deduced by application of the Bernoulli equation to calculate the pressure distribution on a control surface. In vortex sound theory, we also assume some existing knowledge of the vortical distribution in the flow and deduce the aero-acoustical sound production from this knowledge ;

- the second fact is that in common flows the region over which the vorticity is non vanishing has usually a spatial extent much smaller than does the region where Lighthill's source is non vanishing. It is therefore much easier to bound the flow regions that are significant to sound production using vortex sound theory than with the original Lighthill's formulation.

The derivation of vortex sound theory, as initiated by Powell ([36]), follows the same philosophy as the one that Lighthill applied in establishing his analogy. It consists again in rearranging the equations of fluid mechanics in a single inhomogeneous wave equation, the right-hand side of which being considered as the source of sound. The difference with respect to Lighthill's approach stands in the use of the vectorial identity

$$\nabla\left(\frac{1}{2}\,|\boldsymbol{v}|^2\right) = (\boldsymbol{v}.\nabla)\,\boldsymbol{v} - (\nabla\times\boldsymbol{v})\times\boldsymbol{v} \tag{47}$$

to introduce the vorticity $\boldsymbol{\omega} = \nabla\times\boldsymbol{v}$. Using (47), in absence of external forces and neglecting the effect of viscosity, the momentum equation (6) can be rewritten:

$$\varrho\,\frac{\partial\boldsymbol{v}}{\partial t} + \varrho\nabla\left(\frac{1}{2}\,|\boldsymbol{v}|^2\right) + \varrho\,(\boldsymbol{\omega}\times\boldsymbol{v}) + \nabla p = 0\,. \tag{48}$$

Using the definition (15) and (16) of the perturbation density and pressure, and taking into account the fact that ϱ_0 and p_0 are constant, we can write:

$$\frac{\partial^2\varrho}{\partial t^2} - \frac{\partial^2 p}{\partial x_i^2} = \frac{\partial^2\varrho'}{\partial t^2} - \frac{\partial^2 p'}{\partial x_i^2}\,. \tag{49}$$

For an isentropic flow with uniform entropy (homentropic), the speed of sound is defined by (20) and (49) can be written:

$$\frac{\partial^2\varrho}{\partial t^2} - \frac{\partial^2 p}{\partial x_i^2} = \frac{\partial^2\varrho'}{\partial t^2} - c_0^2\,\frac{\partial^2\varrho'}{\partial x_i^2}\,. \tag{50}$$

We inject the continuity equation (3) and the momentum equation (48) in the left-hand side of (50), to end up with a formulation equivalent to Lighthill's analogy (22):

$$\begin{aligned}\frac{\partial^2\varrho'}{\partial t^2} - c_0^2\,\frac{\partial^2\varrho'}{\partial x_i^2} = \frac{\partial}{\partial x_i}\,(\varrho\,(\boldsymbol{\omega}\times\boldsymbol{v})_i) + \frac{\partial^2}{\partial x_i^2}\left(\frac{1}{2}\,\varrho|\boldsymbol{v}|^2\right)\\ - \frac{\partial}{\partial x_i}\left(v_i\,\frac{\partial\varrho}{\partial t}\right) - \frac{\partial}{\partial x_i}\left(\frac{1}{2}\,|\boldsymbol{v}|^2\,\frac{\partial\varrho}{\partial x_i}\right)\,.\end{aligned} \tag{51}$$

For homentropic flows, the relative variations of the density are purely kinetic, so that the four terms on the right-hand side of (51) scale as $1 : 1 : Sr\,M^2 : M^2$ ([36]), where Sr is the Strouhal number and M is the Mach number. The last two terms can therefore be dropped for compact flows ($M \ll 1$) with a Strouhal number $Sr = O(1)$, and we finally have

$$\frac{\partial^2 \varrho'}{\partial t^2} - c_0^2 \frac{\partial^2 \varrho'}{\partial x_i^2} = \frac{\partial}{\partial x_i} \left(\varrho \left(\boldsymbol{\omega} \times \boldsymbol{v} \right)_i \right) + \frac{\partial^2}{\partial x_i^2} \left(\frac{1}{2} \varrho |\boldsymbol{v}|^2 \right) \tag{52}$$

The contribution of the second term is negligible at low Mach number for a compact source region, and because of conservation of the kinetic energy in a non-viscous flow ([4]). Another way to get rid of this second term is to include it in the aero-acoustical variable, which becomes then the stagnation enthalpy ([17]), as will be shown in Sect. 10.

Retaining the first source term of (52), we find a dipolar nature of the source $\partial(\varrho(\boldsymbol{\omega} \times \boldsymbol{v}))/\partial x_i$ compared to the quadrupole of Lighthill $\partial^2(\varrho v_i v_j) / \partial x_i \partial x_j$. We have seen that a dipole in Lighthill's analogy (22) is induced by external forces. If there is no external force then the acoustical field produced by the first source term of (52) should have at most a quadrupole as leading order term. We will now illustrate how, using the free field Green's function and taking the variation of the retarded time over the compact source region into account, we can recover the expected quadrupolar acoustical field.

We obtain an integral formulation of (52) using the same free field Green's function G_0 (31) as in Sect. 4, and using its properties of symmetry (32):

$$\varrho'(\boldsymbol{x}, t) = \frac{\partial}{\partial x_i} \int_{-\infty}^{t} \int_{V} \frac{\delta(t - \tau - |\boldsymbol{x} - \boldsymbol{y}|/c_0)}{4\pi c_0^2 |\boldsymbol{x} - \boldsymbol{y}|} \varrho \left(\boldsymbol{\omega} \times \boldsymbol{v} \right)_i d\boldsymbol{y} d\tau \tag{53}$$

and performing the integration in time gives

$$\varrho'(\boldsymbol{x}, t) = \frac{\partial}{\partial x_i} \int_{V} \left[\frac{\varrho \left(\boldsymbol{\omega} \times \boldsymbol{v} \right)_i}{4\pi c_0^2 |\boldsymbol{x} - \boldsymbol{y}|} \right]_{\tau = t^*} d\boldsymbol{y} \tag{54}$$

where $t^* = t - |\boldsymbol{x} - \boldsymbol{y}|/c_0$ is the retarded time. Assuming a compact source and that in the far field we can make the simplifications

$$|\boldsymbol{x} - \boldsymbol{y}| \simeq |\boldsymbol{x}|$$

and

$$t^* \simeq t - \frac{|\boldsymbol{x}|}{c_0} .$$

The integral in (54) reduces to

$$\frac{1}{4\pi c_0^2 |\boldsymbol{x}|} \int_{V} \left[\varrho (\boldsymbol{\omega} \times \boldsymbol{v})_i \right]_{\tau = t^*} d\boldsymbol{y} \tag{55}$$

where the integral vanishes, owing to the conservation of the impulse in absence of non-conservative external forces ([44]). Clearly, if our aim was to predict the radiated sound field, something has been missed. A more subtle approach consists in keeping the first two terms of the Taylor expansion of the retarded time over the source region. In the far field, we make the approximation

$$|\boldsymbol{x}-\boldsymbol{y}| = |\boldsymbol{x}| - \frac{\boldsymbol{x}.\boldsymbol{y}}{|\boldsymbol{x}|} + \ldots$$

and a function g of the retarded time $t - |\boldsymbol{x}-\boldsymbol{y}|/c_0$ can be approximated at first order by:

$$g\left(t - \frac{|\boldsymbol{x}-\boldsymbol{y}|}{c_0}\right) = g\left(t - \frac{|\boldsymbol{x}|}{c_0}\right) + \frac{\boldsymbol{x}.\boldsymbol{y}}{c_0|\boldsymbol{x}|}\frac{\partial}{\partial t} g\left(t - \frac{|\boldsymbol{x}|}{c_0}\right) + \ldots .$$

We have also the plane wave approximation for the far field derivative:

$$\frac{\partial}{\partial x_i} \simeq -\frac{x_i}{c_0|\boldsymbol{x}|}\frac{\partial}{\partial t}$$

and (54) becomes

$$\varrho'(\boldsymbol{x},t) = -\frac{x_i}{4\pi c_0^3|\boldsymbol{x}|^2}\frac{\partial}{\partial t}\int\limits_V \left[\varrho(\boldsymbol{\omega}\times\boldsymbol{v})_i\right]_{\tau=t^*} d\boldsymbol{y}$$
$$-\frac{x_i x_j}{4\pi c_0^4|\boldsymbol{x}|^3}\frac{\partial^2}{\partial t^2}\int\limits_V \left[y_j\,\varrho(\boldsymbol{\omega}\times\boldsymbol{v})_i\right]_{\tau=t^*} d\boldsymbol{y} \tag{56}$$

where, as before, the first integral vanishes in virtue of the conservation of the impulse ([44]).

Following the same arguments that led to (34), we make the expression (56) explicit by approximating $\varrho \simeq \varrho_0$ and find Powell's integral solution

$$\varrho'(\boldsymbol{x},t) = -\frac{\varrho_0\, x_i x_j}{4\pi c_0^4|\boldsymbol{x}|^3}\frac{\partial^2}{\partial t^2}\int\limits_V \left[y_j\,(\boldsymbol{\omega}\times\boldsymbol{v})_i\right]_{\tau=t^*} d\boldsymbol{y} \tag{57}$$

which corresponds to a quadrupolar radiation field.

The discussion above stresses that in spite of the dipolar character of the first source term in the wave equation (52) its sound radiation is quadrupolar because we explicitly disregarded the dipole contribution and used a sufficiently accurate series expansion of the retarded time to retain the quadrupole term.

9 Choice of the Green's Function

In the previous paragraph we limited our discussion of vortex sound to the free field conditions. The theory can be applied to confined flows. In such a case we should reconsider the choice of the Green's function. The Green's function is defined as a solution of the wave equation with an impulsive source term:

$$\frac{\partial^2 G}{\partial t^2} - c_0^2 \frac{\partial^2 G}{\partial x_i^2} = \delta(t-\tau)\,\delta(\boldsymbol{x}-\boldsymbol{y})\,. \tag{58}$$

This definition should be completed by initial and boundary conditions. The initial conditions follow from the causality condition ([28]). The boundary conditions can actually be chosen arbitrarily. One can for example use the free space Green's function G_0 (31) in the presence of walls. In that case surface terms appear in the integral equation which take reflections of the waves on the surface into account. For compact objects in free space (small compared to the wavelength), the physical interpretation of those surface integrals is quite straightforward. This corresponds to the so-called Curle's integral formulation ([5]). This allows a simple derivation of Gutin's principle for compact rigid bodies, relating the force F_j exerted by the flow on the body to the radiated field in far field approximation ([4]):

$$\varrho'(\boldsymbol{x},t) = -\frac{x_j}{4\pi|\boldsymbol{x}|^2 c_0^3}\frac{\partial}{\partial t}F_j\left(t-\frac{|\boldsymbol{x}|}{c_0}\right) \tag{59}$$

where we neglected the sound produced by the Lighthill quadrupole in the bulk of the flow.

An alternative procedure would be to consider a Green's function which satisfies exactly the same boundary conditions as the "acoustical field" on the surface. This leads to a tailored Green's function. For a rigid body the normal component of the flow at the surface should then match the wall velocity. In that case surface terms only appear as a result of visco-thermal effects. Friction and heat conduction are often negligible, so that when the object is compact the flow can be approximated by a locally incompressible potential flow around the object. Using the reciprocity of the Green's function, Howe ([17]) shows that this Green's function can be calculated in far field approximation from the incompressible potential flow induced by a uniform flow in the direction of the observer.

In a confined flow such as a clarinet we can have a similar freedom in the choice of the Green's function. It is often said that a clarinet behaves as closed-open pipe. This means that the internal acoustical field is dominated by the resonant modes of such a closed-open pipe model of the clarinet. Therefore it may be convenient to develop the Green's function in terms of these pipe modes in which we neglect friction and radiation losses. When plane wave propagation prevails in the pipe this is simply a spatial Fourier analysis of the distribution of the amplitude of the acoustical pressure along the pipe. The sources of sound in such a model are a monopole representing a time dependent air flow passing through the reed opening into the pipe, the visco-thermal losses at the walls and a radiation monopole at the open end.

In Fig. 5 we give a schematic representation of the clarinet in which we only indicate the monopole Q_m at the reed end of the instrument which in the analogy is represented as a closed end. Of course if the reed end would actually be closed we could not blow into the clarinet. The monopole Q_m can include any model of the reed, player's mouth and respiratory system. Often one uses a very simple model in which the player's mouth pressure is assumed to be constant. This simplistic model can lead to quite an insight but should not be expected to explain the clarinet behavior in all playing conditions. When this model fails, because the mouth acoustics is significant one could consider the use of a more extended "tailored" Green's function. Such a Green's function would include both the pipe of the instrument and the player's respiratory system. In such an approach the source of sound is no longer a monopole because there is no significant change in fluid density when the flow passes through the reed. The sound source can be identified as being the dipole due to the vortex shedding in the free jet formed by flow separation at the reed. The one-dimensional representation of this fluctuating dipole would be a fluctuating pressure difference between the mouth and the pipe in an acoustical model (ignoring further any fluid flow or wall movement!).

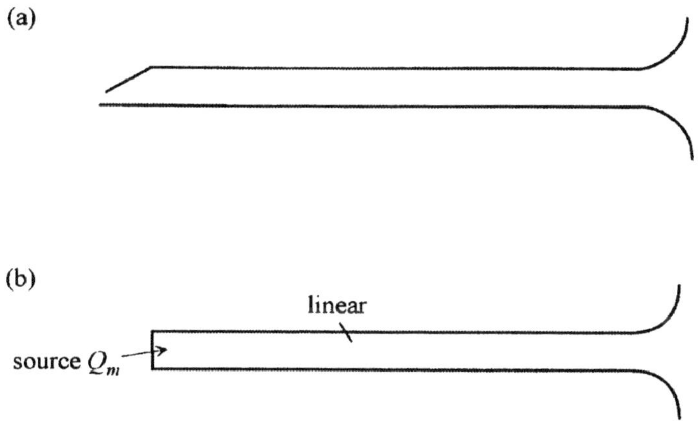

Fig. 5. Aeroacoustic analogy in which the clarinet (a) is represented as a closed pipe (b). The monopole Q_m can include the modelling of the reed, player's mouth and respiratory system.

In the choice of the optimal Green's function we can therefore use some physical intuition. In some cases the most efficient choice of the Green's function is however dictated by numerical considerations. The development in modes of a closed-open pipe without friction is an excellent approach in the frequency domain when for example we consider steady oscillation of a clarinet (a sustained note). In musical acoustics the attack transient is essential for the perception of the sound ([9]). In dynamical models used for real-time sound synthesis it appears that the optimal choice of the Green's function is that of a semi-infinite tube (terminated by the clarinet pipe). This choice avoids the multiple reflections in the Green's function which would give a very slow numerical convergence for time domain calculations ([24]).

10 Howe's Energy Corollary

We will explicit an energy corollary which has a limited range of validity, but provides an almost intuitive understanding of the interaction between vortices and an acoustic field. This energy corollary was proposed by Howe ([15], [16]). The key of Howe's approach is to *define* the acoustic field by considering a Helmholtz decomposition of the velocity field, $\boldsymbol{v}$, into the gradient of a scalar potential, φ, and the curl of a stream function, $\boldsymbol{\psi}$:

$$\boldsymbol{v} = \nabla\varphi + \nabla \times \boldsymbol{\psi} \,. \tag{60}$$

The acoustic velocity $\boldsymbol{u}$ is the gradient of the unsteady component φ' of the scalar potential:

$$\boldsymbol{u} = \nabla\varphi' \,. \tag{61}$$

As a next step we consider Crocco's formulation of the equation of Euler for a homentropic (constant and uniform entropy) flow:

$$\frac{\partial \boldsymbol{v}}{\partial t} + \nabla B = -\boldsymbol{\omega} \times \boldsymbol{v} + \frac{\boldsymbol{f}}{\varrho} \tag{62}$$

where B is the total enthalpy :

$$B = \int \frac{dp}{\varrho} + \frac{|\boldsymbol{v}|^2}{2} \tag{63}$$

ϱ is the density, p is the pressure, $\boldsymbol{\omega} = \nabla \times \boldsymbol{v}$ is the vorticity and $\boldsymbol{f}$ is the density of an external force. When we have a potential flow ($\boldsymbol{\omega} = 0$) without external force ($\boldsymbol{f} = 0$) we are left with an acoustic flow superposed on a steady potential flow. This indicates that the Coriolis force:

$$\boldsymbol{f}_c = -\varrho(\boldsymbol{\omega} \times \boldsymbol{v}) \tag{64}$$

acts as an external force on the acoustical flow field $\boldsymbol{u}$. We also see that in a homentropic flow the total enthalpy, B, is a natural acoustic variable. If we neglect wave-propagation effects in the source region (where $\boldsymbol{\omega} \neq 0$), and mean-flow convection in the wave region (where $\boldsymbol{\omega} = 0$) we can use Kirchhoff's definition of the acoustical energy. In such a case, the time average $< \mathcal{P} >$ of the power transfer from the vorticity field to the acoustical field is simply the time average of the integral of the work performed by the Coriolis force on the acoustical field:

$$< \mathcal{P} > = < \int_V \boldsymbol{f}_c \,.\, \boldsymbol{u} \, dV > \tag{65}$$

where V is the source volume (where $\boldsymbol{\omega} \neq 0$) and the brackets $<>$ indicate a time average. In the case of a 1-D pipe flow (outside the source region) we can represent a compact vortical source region as a time-dependent discontinuity ΔB_s in the total enthalpy. This source is calculated by subtracting from the actual enthalpy difference, ΔB, the difference ΔB_{pot} which is induced by a potential flow across the source region. In the calculation of ΔB and ΔB_{pot} we should use the same mass flow Φ_m through the pipe termination. We have therefore:

$$\Delta B_s = \Delta B - \Delta B_{pot} \, . \tag{66}$$

The corresponding energy corollary is:

$$< \mathcal{P} > = < \Phi_m \Delta B_s > \, . \tag{67}$$

Both the original equation of Howe and the 1-D formulation stress that vortices are associated with a force or dipole source of sound, which can only transfer energy to an oscillating acoustical flow when there is an acoustical velocity at the source ($\boldsymbol{u} \neq 0$). This corresponds to the fact that a violin string is more easily bowed in the middle of the string where there are large string displacements, than very close to the bridge where the displacements are much smaller. This is the complement to the situation encountered in the Rijke-tube. In that case a volume source could not perform any work when placed at a pressure node, close to the open end of the tube.

11 The Open Pipe Termination of an Unflanged Pipe

Let us consider a thin-walled pipe with a mean-flow velocity U_0 directed towards an unflanged open pipe termination. The pipe cross section has a radius a. The speed of sound, c_0, is uniform. We will discuss the reflection at the pipe termination of harmonic waves with a low frequency, f, travelling down the flow in the pipe. These acoustical plane waves induce (just) upstream of the pipe termination a uniform acoustical velocity $\boldsymbol{u} = (u_a, 0, 0)$ along the pipe axis. As we consider low Helmholtz numbers, $He = (af/c_0) \ll 1$, the acoustical flow at the pipe exit is a locally incompressible potential flow as sketched in Fig. 6. Near the sharp edges of the open pipe termination the potential flow is singular, which implies very large ratios of the local acoustical velocity $|\boldsymbol{u}|$ and the acoustical velocity u_a at some distance upstream of the pipe termination. At a distance of the order of the pipe radius a (downstream of the pipe outlet) the acoustical flow is already close to a spherical symmetrical flow with a velocity decreasing quadratically with the distance from the origin (locally incompressible flow). In the absence of mean flow, the inertia of this acoustical flow just outside the pipe induces a difference in the scalar potential, which is often represented as an end correction ([35]).

In the presence of a mean flow U_0, leaving the pipe, a free jet is formed. The shear layer separating the jet flow from the environment is a region of

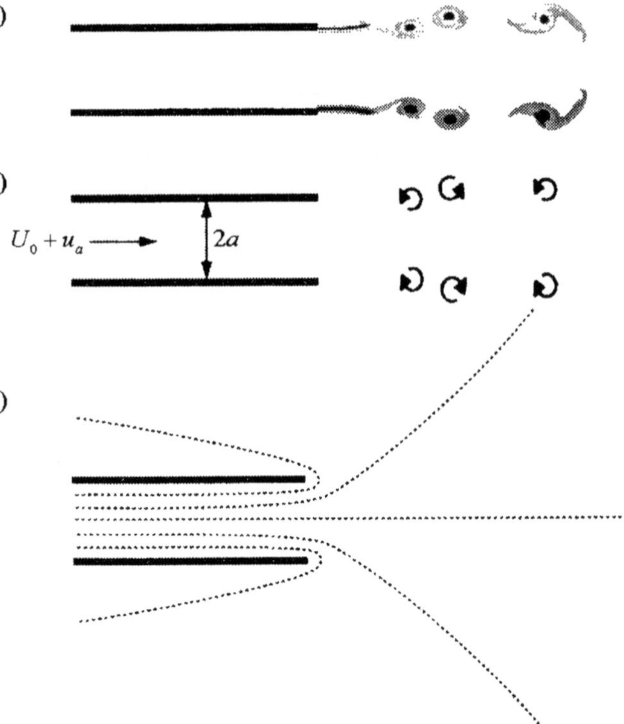

Fig. 6. Acoustically forced flow at an unflanged open pipe exit a) actual vorticity distribution in the flow (PIV measurements) b) modelled flow c) potential flow.

high vorticity, ω. At low Strouhal numbers, $Sr_a = (fa/U_0) \ll 1$, we can assume a quasi-stationary response of the free jet to the acoustic flow. All the kinetic-energy flux of the flow, $\Phi_m \frac{1}{2}(U_0+u_a)^2$, is dissipated by turbulence in the jet while the pressure in the jet is equal to the stationary pressure of the surrounding (as for any stationary subsonic free jet). Comparison of this flow with a potential flow implies that $\Delta B_s = -\frac{1}{2}(U_0 + u_a)^2$. In a linear approximation we therefore find $< \mathcal{P} >= -\varrho_0 U_0 \pi a^2 < u_a^2 >$. Using this result to calculate the reflection coefficient $R_B = |B^-/B^+|$ for an incident wave of amplitude B^+ yields excellent results (see Fig. 7). This provides a vortex-sound interpretation of a well-known result from the literature: $R_B = (c_0 - U_0)/(c_0 + U_0)$.

At higher frequencies, when $Sr_a = fa/U_0 = O(1)$, it is not obvious that acoustical energy will be absorbed upon reflection at an open pipe termination. The key to understanding sound production is the fact that the acoustical velocity $\boldsymbol{u}$ changes sign from pipe outwards to pipe inwards each period of oscillation. We now focus on the vorticity perturbation generated by the

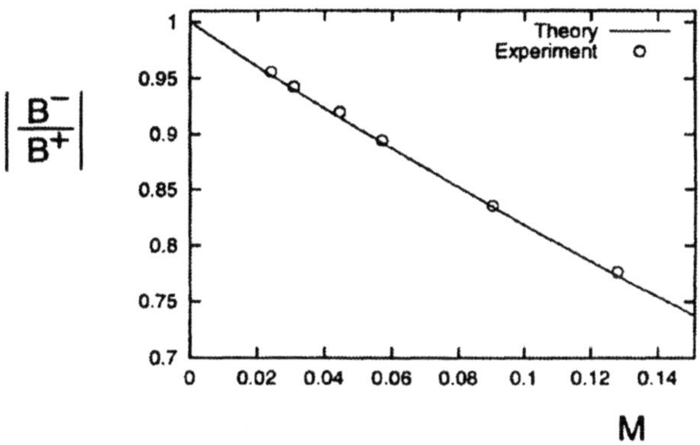

Fig. 7. Reflection coefficient $R_B = |B^-/B^+|$ at an unflanged open pipe termination as a function of the mean flow Mach number $M = U_0/c_0$.

acoustic flow at a certain moment of the period in the shear layer of the free jet. This vorticity perturbation $\boldsymbol{\omega}'$ is generated at the edge of the pipe exit where flow separation occurs. The vorticity perturbation is convected away with the fluid in the shear layer at about half the mean flow velocity U_0. As illustrated in Fig. 8, the sign of the vorticity is such that it induces a flow at the edge that opposes the driving acoustic flow around the edge. The amount of vorticity shed is such that tangential flow separation is maintained (Kutta condition). Inspection of the sign of $\boldsymbol{\omega}'$, $\boldsymbol{u}$ and $\boldsymbol{v}$ indicates that at the moment the vorticity is generated $(\boldsymbol{f}'_c \cdot \boldsymbol{u}) = -\varrho(\boldsymbol{\omega}' \times \boldsymbol{v}) \cdot \boldsymbol{u} < 0$ which implies following Howe's formula (65) that there is a transfer of energy from the acoustical flow to the vortical flow. We accept this result easily. The acoustical flow generates a vortical perturbation, and should pay a price for it. We also see that the interaction is strong because the convective velocity $\boldsymbol{v}$ is normal to the local acoustical velocity $\boldsymbol{u}$. Furthermore, due to the singularity of the acoustical (potential) flow we have locally relatively large acoustical velocities at the sharp edge.

The big surprise comes from the fact that both $\boldsymbol{\omega}'$ and $\boldsymbol{v}$ do not change sign while $\boldsymbol{u}$ does change sign. Hence, after half a period $(T = 1/f)$ of oscillation the perturbation of vorticity which we have focused on produces acoustical energy because $(\boldsymbol{f}'_c \cdot \boldsymbol{u}) > 0$. As we travel away from the edge, $\boldsymbol{v}$ remains approximately constant while $\boldsymbol{u}$ and the angle between $\boldsymbol{u}$ and $\boldsymbol{v}$ both decrease with the distance from the pipe outlet. This implies that if $\boldsymbol{\omega}'$ remains constant, the production of acoustical energy will not be sufficient to obtain a net production of acoustical energy after one period of oscillation. The mean power, $< \mathcal{P} >$, would thus be negative. In practice, however, the initial vorticity perturbation initiates an instability of the shear layer. This results in an increase of the magnitude of the time-dependent vorticity perturbation. At low amplitudes, when linear theory can be applied, the

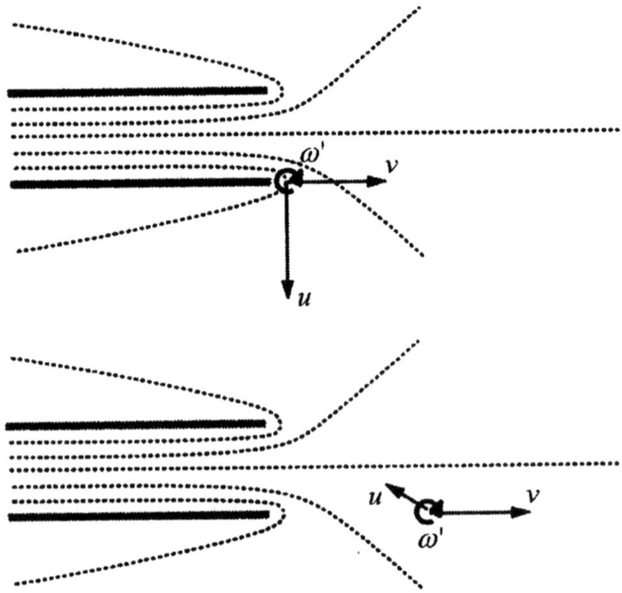

Fig. 8. Vorticity perturbation generated at the edge at the moment it is generated and half a period of oscillation later.

perturbation increases exponentially with the distance from the edge ([39], [16]).

In many practical conditions the amplitude, $u/U_0 \geq 10^{-2}$, is much too large to use a linear theory. A reasonable alternative is to describe the shear layer in terms of discrete vortex shedding. It appears from experiments that a new vortex is shed at each instant for which the acoustic flow turns from pipe inwards to pipe outwards. The vortex structure accumulates all the vorticity shed at the sharp edge while it travels parallel to the pipe axis with a velocity of the order of $U_0/2$. The magnitude of the circulation $|\Gamma|$ of the vortex increases about linearly in time $d|\Gamma|/dt \simeq U_0^2/2$. This is much weaker than the exponential growth predicted by linear theory. This limited growth of the vorticity perturbation implies a net sound absorption for an unflanged pipe termination [34].

We should note here that a free vortex cannot have a time dependent circulation ([17], [44]). Assuming such a time dependence introduces a spurious force on the flow. By using Howe's analogy we simply ignore the acoustical effect of this spurious force ([45]). Furthermore, Howe's energy equation is also an integral formulation which will tend to smooth out errors.

12 Whistler-Nozzle and Human Whistling

The analysis of the acoustical response of an unflanged open pipe termination has indicated that in the presence of a mean outflow, U_0, the net sound absorption is the result of a combination of an initially large sound absorption upon vortex shedding, followed by sound production half a period later. If we could reduce the initial absorption and increase the later production, we can expect that a net sound production is possible (in order to reduce self-sustained oscillations we can use spoilers to increase the initial sound absorption, [3]). The whistler-nozzle configuration shown in Fig. 9 satisfies both these requirements, and does produce sound at specific Strouhal numbers [12]. The sharp edge of the unflanged pipe termination of Fig. 6 is replaced by a

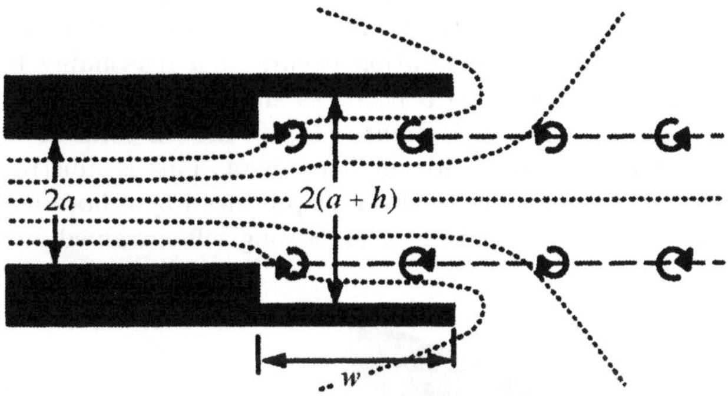

Fig. 9. Whistler-nozzle configuration.

stepwise increase in pipe radius from a to $a + h$ at a distance w from the end of the pipe. We call this pipe segment (of length w) the "collar" of the nozzle. Vortex shedding is now initiated at the edge of the step, where the acoustical flow presents a 90° sharp bend rather than the original 180° sharp bend at the edge of the open pipe termination. This already considerably reduces the singularity of the acoustical flow at the separation point. Furthermore, the collar bends the acoustical streamlines back in the direction of the pipe axis which corresponds to the direction of convection of the vortices. This implies that within the collar, the interaction between the acoustical field and the vortices is weak ($\boldsymbol{f}_c \cdot \boldsymbol{u} \simeq 0$). The absorption is reduced to a small interaction region near the separation point. A strong interaction between the vortices and the acoustical flow occurs again at the end of the collar because the diverging acoustical streamlines cut through the path of the vortex. If the travel time of vortices along the collar is of the order of one half oscillation period, we do expect a net production. This corresponds to a critical value $Sr_c = wf/U_0$ of the Strouhal number. Increasing the step size h reduces the interaction of the vortices with their images in the wall of the collar. As these images slow down the vortex convection, we observe that the critical Strouhal number Sr_c increases with h ([18],[13]).

It appears that also higher-order hydrodynamic modes are possible, corresponding to larger travel times of the vortices in the collar. This implies the presence of more than one vortex at the time in the collar. Up to four vortices ($n = 4$) have been observed. Five vortices ($n = 5$) seems to be an upper limit for the number of vortices observed in many self-sustained oscillations ([42], [43], [2]).

In the original papers on the whistler-nozzle it was argued that the sound production was due to the impingement of the oscillating shear layers of the free jet on the edges of the pipe termination at the end of the collar ([18],[43]). Vortex-sound theory indicates that such an impingement is not necessary for sound production. This is confirmed by the fact that rounding the edges at the end of the collar increases the sound production (20%) rather than reducing it [13]. This leads us to the idea that we may not need any sharp edge at all, as long as flow separation occurs at a reasonably fixed point. Removing all edges leads us to a horn, as shown in Fig. 10. Experiments by Powell [2], Peters [34] and Hofmans [14] do indeed confirm that a horn can produce sound at critical Strouhal numbers. This is illustrated by the experimental data on the acoustical power produced by a horn as a function of the Strouhal number shown in Fig. 11. Strong self-sustained oscillations of

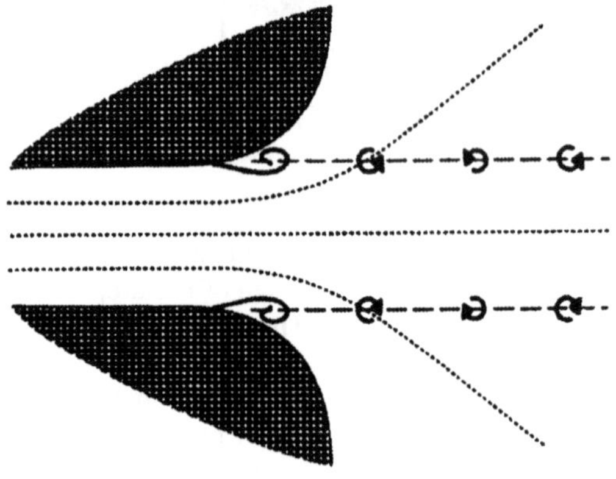

Fig. 10. Horn at the end of a pipe.

the flow in a whistler-nozzle appear when the acoustical energy accumulates in standing waves of a resonator attached to the nozzle. In such cases the acoustical resonator imposes the oscillation frequency, f, and the pulsations show a maximum when the flow is adjusted to meet the critical Strouhal number condition (see the previous section). Such self-sustained oscillations have been observed when a pipe segment (length L) terminated by a whistler-nozzle is placed downstream of a settling chamber ([12], [18], [13]). When the first pipe mode is excited ($f = c/2L$) one observes acoustical velocities at the nozzle of the order of magnitude $u/U_0 = O(0.1)$. When a resonant closed side branch is placed upstream of the pipe segment pulsation amplitudes of the order of magnitude $u/U_0 = O(0.4)$ are observed [13]. In this case, the acoustical energy is produced both at the Whistler nozzle and at the junction between the closed side branch and the main pipe. The most common example

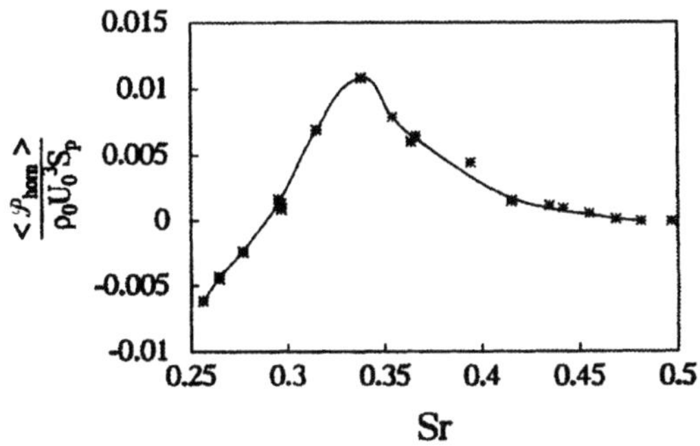

Fig. 11. Acoustical power produced by a horn as a function of the Strouhal number.

of such self-sustained oscillations is human whistling. Following Wilson [48] human whistling is due to a coupling between vortex shedding at our lips and the Helmholtz resonance frequency of our mouth. This is illustrated in Fig. 12. The Strouhal condition for optimal whistling is easily observed by first fixing the geometrical configuration (lip and tongue positions) and then gradually increasing the volume flow. A certain critical flow velocity has to be reached before whistling appears. Above a certain flow velocity, whistling disappears again. Young children often blow too hard when they try to whistle... This disappearance is due to the fact that at high flow rates the vortices are convected far outside the lips (horn) before they can produce sound. When performing this experiment we also observe that the system does not oscillate exactly at a constant frequency. The system tries instead to keep the Strouhal number as close as possible to the optimal Strouhal number by increasing gradually the oscillation frequency as we increase the flow velocity. Large variations in oscillation frequency are only possible when the acoustical mode has a low quality factor, Q ($Q/2\pi$ is the number of oscillation periods after which, in the absence of a sound source, the amplitude of a mode has decayed by a factor $\exp(0.5)$).

13 Conclusion

We hope you enjoyed and that you will use the aeroacoustical analogies.

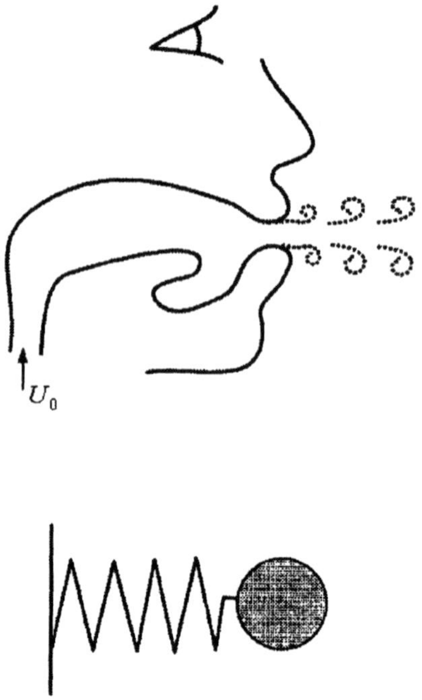

Fig. 12. Human whistling is induced by the coupling of vortex shedding at our lips, along with acoustical oscillations of our mouth. The lowest resonance frequency corresponds to the Helmholtz resonance, at which the mouth volume acts as an acoustical spring, and the narrowing at the lips acts as an acoustical mass.

References

1. C.K. Batchelor: *An Introduction to Fluid Dynamics* (Cambridge University Press, Cambridge 1967)
2. W.K. Blake, A. Powell: 'The development of contemporary views of flow-tone generation'. In: *Recent Advances in Aeroacoustics*, ed. by A. Krothapalli and C.A. Smith (Springer-Verlag, NY 1986) pp. 248–325
3. J.C. Bruggeman, A. Hirschberg, M.E.H. van Dongen, A.P.J. Wijnands: J. Sound Vib. **150**(3), 371–393 (1991)
4. D.G. Crighton, A.P. Dowling, J.E. Ffowcs Williams, M. Heckl, F.G. Leppington: *Modern Methods in Analytical Acoustics* (Springer-Verlag, London 1992)
5. N. Curle: Proc. Roy. Soc. (London) Ser. A **264**, 321-342 (1969)
6. P.E. Doak: Acoust Phys **44**, 677-685 (1995)
7. A.P. Dowling and J.E. Ffowcs Williams: *Sound and Sources of Sound* (Ellis Horwood, Chichester 1983)
8. A.T. Fedorchenko: J. Sound Vib. **232**(4), 719–782 (2000)

9. N.H. Fletcher and T.D. Rossing: *The Physics of Musical Instruments* 2nd edn. (Springer, New York 1998)
10. M.E. Goldstein: *Aeroacoustics* (McGraw Hill, New York 1976)
11. M.A. Heckl: J. Sound Vib. **124**, 117–133 (1988)
12. W.G. Hill Jr., P.R. Greene: Trans. ASME J. Fluids Eng. **99**(3), 520–525 (1977)
13. A. Hirschberg, J.C. Bruggeman, A.P.J. Wijnands, N. Smits: Acustica **68**, 157–160 (1989)
14. G.C.J. Hofmans: Vortex Sound in Confined Flows. PhD Thesis, Technische Universiteit Eindhoven, Eindhoven (1998)
15. M.S. Howe: J. Sound Vib. **70**, 407–411 (1980)
16. M.S. Howe: I.M.A. J. Appl. Math. **32**, 187–203 (1984)
17. M.S. Howe: *Acoustics of Fluid-Structure Interactions* (Cambridge University Press, Cambridge 1998)
18. A.K.M.F. Hussain and M.A. Hasan: J. Fluid Mech. **134**, 431 (1983)
19. A. Iafrati, G. Riccardi: J. Sound Vib. **196** 129-146 (1996)
20. L.D. Landau and E.M. Lifchitz: *Course of Theoretical Physics Volume 6: Fluid Mechanics*, 2nd edn. (Pergamon, Oxford 1987)
21. M.J. Lighthill: Proc. Roy. Soc. **A**211, 564–587 (1952)
22. M.J. Lighthill: Proc. Roy. Soc. **A**222, 1–32 (1954)
23. G.M. Lilley: 'Jet Noise Classical Theory and Experiments'. In: *Aeroacoustics of Flight Vehicles, Theory and Practice, Volume 1: Noise Sources* ed. by H.H. Hubbard (Acoustical Society of America, New York 1995) pp. 211-289
24. M.E. McIntyre, R.T. Schumacher, J. Woodhouse: J. Acoust. Soc. Am. **74**, 1325-1345 (1983)
25. H.J. Merk: Applied Scientific Research **A6**, 402-420 (1956-1957)
26. C.L. Morfey: J. Sound Vib. **31**, 391-397 (1973)
27. C.L. Morfey: J. Sound Vib. **48**, 95-111 (1984)
28. P.M. Morse, H. Feshbach: *Methods of Theoretical Physics* (MacGraw Hill, New York 1953)
29. E.A. Müller, F. Obermeier: Fluid Dyn. Res. **3**, 43–51 (1988)
30. R.E. Musafir: 'A Discussion on the Structure of Aeroacoustic Wave Equations'. In *Proceedings of the 4th Congress on Acoustics, Marseille, France April, 1997* (Teknea, Toulouse 1997) pp. 923-926
31. M.K. Myers: J. Sound Vib. **109**, 277-284 (1986)
32. F. Obermeier: J. Sound Vib. **41**, 463-472 (1975)
33. F. Obermeier: J. Sound Vib. **99**, 111-120 (1985)
34. M.C.A.M. Peters, A. Hirschberg, A.J. Reijnen, A.P.J. Wijnands: J. Fluid Mech. **256**, 499–534 (1993)
35. A. Pierce: *Acoustics, An Introduction to its Physical Principles and Applications* (Acoustical Society of America, New York 1989)
36. A. Powell: J. Acoust. Soc. Am. **36**(1), 177–195 (1964)
37. A. Powell: Trans. ASME, J. Vib. Ac. **112**, 145–159 (1990)
38. J.W.S. Rayleigh: *The Theory of Sound* (Dover, New York 1945)
39. S.W. Rienstra: J. Fluid Mech. **108**, 443–460 (1981)
40. S.W. Rienstra: *An Introduction to Acoustics* (Report IWDE 01-03, Technische Universiteit Eindhoven, 2001)
41. P.L. Rijke: Annalen der Physik **107**, 339-343 (1859)
42. D. Rockwell, E. Naudasher: Trans. ASME, J. Fluids Eng. **100**, 152–165 (1978)
43. D. Rockwell: AIAA Journal **21**(5), 645–664 (1983)

44. P.J. Saffman: *Vortex Dynamics* (Cambridge Monographs on Mechanics and Applied Mathematics, Cambridge 1992)
45. C. Schram, A. Hirschberg: *to be submitted*
46. C.K.W. Tam: 'On the failure of the acoustic analogy theory to identify the correct noise sources'. In: *Proc. 7th AIAA/CEAS Aeroac. Conf., Maastricht, 28–30 May 2001*
47. P.A. Thompson: *Compressible Fluid Dynamics* (MacGraw-Hill, New York 1972)
48. T.A. Wilson, G.S. Beavers, M.A. de Coster, D.K. Holger, D.Regenfuss: J. Acoust. Soc. Am. **50**, 366–372 (1971)

Lectures on the Theory of Vortex–Sound

Michael S. Howe

Boston University, College of Engineering 110 Cummington Street, Boston MA 02215, USA

Abstract. The theory of vortex sound is introduced. From Lighthill's acoustic analogy, it is shown how vorticity and entropy fluctuations can be seen as sources of sound. The use of the compact Green's functions is introduced to compute the vortex sound. As an example of the method presented, this theory is applied to pressure transients generated by high-speed trains.

1 Aerodynamic Sound

The sound generated by vorticity in an unbounded fluid is called *aerodynamic sound.* Most unsteady flows of technological interest are of high Reynolds number and turbulent, and the acoustic radiation is a very small by-product of the motion. The turbulence is usually produced by fluid motion relative to solid boundaries or by the instability of free shear layers separating a high speed flow (such as a jet) from a stationary environment. In this lecture the influence of boundaries on the production of sound as opposed to the production of vorticity will be ignored. The aerodynamic sound problem then reduces to the study of mechanisms that convert kinetic energy of rotational motions into acoustic waves involving longitudinal vibrations of fluid particles. There are two principal source types in free vortical flows: a quadrupole, whose strength is determined by the unsteady Reynolds stress, and a dipole, which is important when mean mass density variations occur within the source region.

1.1 Lighthill's Acoustic Analogy (1952)

The theory of aerodynamic sound was developed by Lighthill, who reformulated the Navier-Stokes equation into an exact, inhomogeneous wave equation whose source terms are important only within the turbulent (vortical) region. Sound is expected to be such a very small component of the whole motion that, once generated, its back-reaction on the main flow is usually negligible. In a first approximation the motion in the source region may then be determined by neglecting the production and propagation of the sound. This would be inappropriate if the Mach number M is large enough for compressibility to be important in the source flow, when the source flow is coupled to a resonator, such as an organ pipe, or when bubbles are present in the case of liquids.

Consider the sound generated by a finite region of rotational flow in an unbounded fluid at rest at infinity. Let us compare the equations for the *density* fluctuations in the real flow with those for an ideal, linear acoustic medium which coincides with the real fluid at large distances from the sources. The difference between these equations will be shown to be equivalent to a distribution of sources in the ideal acoustic medium, whose radiation field is the same as that in the real flow and may therefore be calculated by the methods of linear acoustics. To do this, body forces are neglected, and the momentum equation is written in the form

$$\partial(\rho v_i)/\partial t = -\partial \pi_{ij}/\partial x_j, \tag{1}$$

where π_{ij} is the **momentum flux** tensor

$$\pi_{ij} = \rho v_i v_j + (p - p_o)\delta_{ij} - \sigma_{ij}, \tag{2}$$

and ρ, $\mathbf{v}$, p, p_o, σ_{ij} are respectively the density, velocity, pressure, uniform pressure at infinity, and the viscous stress tensor.

By integrating over a fixed region V, it can be seen that (1) equates the rate of change of momentum in V to the action of the pressure and viscous stresses on its boundary and to the convection of momentum across the boundary at a rate determined by the Reynolds stress $\rho v_i v_j$. In an ideal, linear acoustic medium, momentum transfer is produced solely by the pressure, i.e., $\pi_{ij} \to \pi^o_{ij} = (p - p_o)\delta_{ij} \equiv c_o^2(\rho - \rho_o)\delta_{ij}$, where ρ_o and c_o are the mean density and sound speed. By eliminating ρv_i between (1) (with π_{ij} replaced by π^o_{ij}) and the continuity equation

$$\frac{\partial(\rho - \rho_o)}{\partial t} + \frac{\partial(\rho v_i)}{\partial x_i} = 0.$$

we obtain the equation of linear acoustics for the perturbation density $\rho - \rho_o$,

$$\left(\partial^2/c_o^2\partial t^2 - \nabla^2\right)\left[c_o^2(\rho - \rho_o)\right] = 0. \tag{3}$$

In the absence of externally applied forces or moving boundaries, this equation has only the trivial solution $\rho - \rho_o = 0$, because the radiation condition ensures that sound waves cannot enter from infinity.

The sound generated in the real fluid may now be seen to be exactly equivalent to that produced in an ideal, stationary acoustic medium that is forced by the stress distribution $T_{ij} = \pi_{ij} - \pi^o_{ij}$. T_{ij} is the **Lighthill stress tensor**

$$T_{ij} = \rho v_i v_j + \left((p - p_o) - c_o^2(\rho - \rho_o)\right)\delta_{ij} - \sigma_{ij}. \tag{4}$$

The Reynolds stress $\rho v_i v_j$ is nonlinear, and is significant only within the rotational source region. The second term is the *excess* of momentum transfer by the pressure over that in an ideal ("linear") fluid of density ρ_o and sound speed c_o. This is caused by wave amplitude nonlinearity, and by mean density variations in the source flow. The viscous stress tensor σ_{ij} is linear in the perturbation quantities, and properly accounts for the attenuation of the sound; in most applications the Reynolds number in the source region is sufficiently large that σ_{ij} can be neglected, and attenuation in the radiation zone is usually ignored in a first approximation.

Lighthill's **acoustic analogy** equation for the production of aerodynamic sound is obtained by first re-writing (1) as the momentum equation for an ideal, stationary fluid of density ρ_o and sound speed c_o subject to the externally applied stress T_{ij}:

$$\partial(\rho v_i)/\partial t + \partial\left(c_o^2(\rho - \rho_o)\right)/\partial x_i = -\partial T_{ij}/\partial x_j.$$

Elimination of the momentum density ρv_i between this and the continuity equation yields **Lighthill's equation**, which is the exact, nonlinear counterpart of (3):

$$\left(\frac{1}{c_o^2}\frac{\partial^2}{\partial t^2} - \nabla^2\right)[c_o^2(\rho - \rho_o)] = \frac{\partial^2 T_{ij}}{\partial x_i \partial x_j}. \qquad (5)$$

The problem of calculating the aerodynamic sound is therefore formally equivalent to solving this equation for the radiation into a stationary, ideal fluid produced by a distribution of *quadrupole* sources whose strength per unit volume is the Lighthill stress tensor T_{ij}.

The solution of Lighthill's equation should strictly be regarded as an alternative (integral equation) formulation of the Navier-Stokes equation, that provides a useful representation of the sound when T_{ij} is known. T_{ij} accounts not only for the generation of sound, but also for self-modulation due to acoustic nonlinearity, convection by the flow, refraction due to sound speed variations, and attenuation due to thermal and viscous actions. Nonlinear effects on propagation and dissipation are usually sufficiently weak to be neglected within the source region, although they may affect propagation to a distant observer. Convection and refraction of sound within and near the source flow can be important, for example when the sources are contained in a turbulent shear layer, or are adjacent to a large, quiescent region of fluid whose mean thermodynamic properties differ from those in the radiation zone. Effects of this kind are accounted for by contributions to T_{ij} that are *linear* in the perturbation quantities relative to a mean background flow.

Thus, with the exception of flows amenable to special treatment, the practical utility of Lighthill's equation rests on the hypothesis that all of these effects, which actually depend on the *compressibility* of the source flow, can be ignored, and that adequate predictions of the aerodynamic sound are obtained by taking for T_{ij} an estimate based on the equations of motion of an incompressible fluid. This approximation is likely to be acceptable when $M^2 \ll 1$, and when the wavelength of the sound is much larger than the dimension of the source region. The remainder of this section is devoted to a consideration of such cases.

1.2 Aerodynamic Sound from Low-Mach-Number Turbulence of Uniform Mean Density

When the mean density and sound speed are uniform, the variations in ρ produced by low Mach number, high Reynolds number velocity fluctuations are of order $\rho_o M^2$, and $\rho v_i v_j \approx \rho_o v_i v_j$. Similarly, $p - p_o - c_o^2(\rho - \rho_o) \approx (p - p_o)(1 - c_o^2/c^2) \sim O(\rho_o v^2 M^2) \ll \rho_o v^2$. Thus, $T_{ij} \approx \rho_o v_i v_j$, when viscous stresses are neglected, and the solution of Lighthill's equation becomes

$$p(\mathbf{x},t) \approx \frac{\partial^2}{\partial x_i \partial x_j} \int \frac{\rho_o v_i v_j(\mathbf{y}, t - |\mathbf{x}-\mathbf{y}|/c_o)}{4\pi|\mathbf{x}-\mathbf{y}|} d^3\mathbf{y}$$

$$\approx \frac{x_i x_j}{4\pi c_o^2 |\mathbf{x}|^3} \frac{\partial^2}{\partial t^2} \int \rho_o v_i v_j(\mathbf{y}, t - |\mathbf{x}-\mathbf{y}|/c_o)\, d^3\mathbf{y}, \quad |\mathbf{x}| \to \infty, \qquad (6)$$

where $p(\mathbf{x},t) = c_o^2(\rho - \rho_o)$ is the perturbation pressure in the far field. Quantitative predictions can be made from this equation provided the behavior of the Reynolds stress is known.

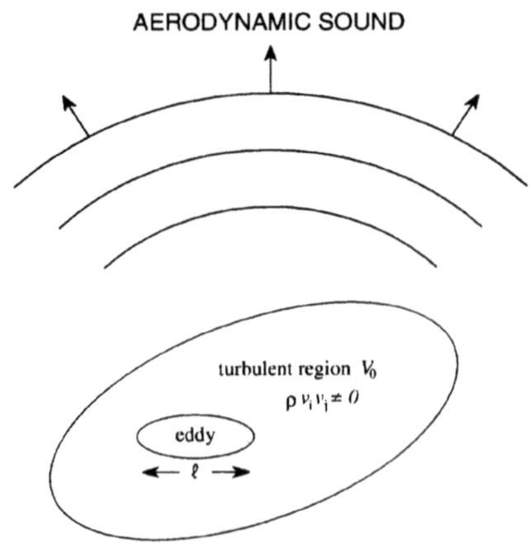

Fig. 1. Aerodynamic sound in an unbounded fluid.

The order of magnitude of p can be estimated in terms of the characteristic velocity v and length scale ℓ (of the energy containing eddies) in the source region (Fig. 1). ℓ is determined by the scale of the mechanism responsible for turbulence production, such as the width of a jet mixing layer. Fluctuations in $v_i v_j$ occurring in different regions of the turbulent flow separated by distances $> O(\ell)$ will tend to be statistically independent, and the sound may therefore be considered to be generated by a collection of

V_o/ℓ^3 independent eddies, where V_o is the volume occupied by the turbulence. The dominant frequency of the motion $\sim v/\ell$, so that the wavelength of the radiated sound $\sim \ell/M \gg \ell$ $(M \sim v/c_o)$, and each eddy is therefore acoustically *compact.* Hence, the acoustic pressure generated by a single eddy is of order $p \sim (\ell/|\mathbf{x}|)\rho_o v^2 M^2$, and the acoustic power it radiates $\sim 4\pi|\mathbf{x}|^2 p^2/\rho_o c_o \approx \ell^2 \rho_o v^8/c_o^5 = \ell^2 \rho_o v^3 M^5$. This is Lighthill's "eighth power" law. The total acoustic power is $\Pi_a \approx (V_o/\ell^3)(\ell^2 \rho_o v^3 M^5) = \rho_o v^3 M^5 V_o/\ell$. Dimensional arguments and experiment indicate that the rate of decay of the turbulence kinetic energy $\sim O(V_o \rho_o v^3/\ell)$. In a statistically steady state this must equal the rate Π_o, say, at which energy is supplied to the flow by the action of external forces. The *efficiency* Π_a/Π_o with which this energy is converted into sound is therefore proportional to M^5, confirming Lighthill's hypothesis that the flow generated sound is an infinitesimal by-product of the motion.

1.3 Aerodynamic Sound from Low-Mach-Number Turbulence of Variable Mean Density

When the mean density in the source region is not constant, the Reynolds stress quadrupoles are augmented by a distribution of *dipoles*, the dipole strength being equal to the hydrodynamic force experienced by a fluid particle of density ρ relative to the force the same particle would experience had its density been equal to ρ_o. This dipole is 'hidden' within the Lighthill quadrupole T_{ij} and might therefore be expected to have the same radiation efficiency as the Reynolds stresses. However, the component of T_{ij} responsible for the dipole is $c_o^2(\rho - \rho_o)$, which is larger than the Reynolds stress $\rho v_i v_j$ by a factor of order $O(1/M^2)$ provided the mean value of $\rho - \rho_o \neq 0$ in the source region.

To obtain a precise estimate of the dipole radiation, the solution of (5) is written

$$p(\mathbf{x},t) \approx \frac{1}{4\pi c_o^2}\frac{\partial^2}{\partial t^2}\int \left[\rho v_i v_j\right]\frac{(x_i - y_i)(x_j - y_j)d^3\mathbf{y}}{|\mathbf{x}-\mathbf{y}|^3} + \frac{1}{4\pi}\int \frac{\partial^2}{\partial t^2}\left[\frac{p - p_o}{c_o^2} - (\rho - \rho_o)\right]\frac{d^3\mathbf{y}}{|\mathbf{x}-\mathbf{y}|}, \quad |\mathbf{x}-\mathbf{y}| \to \infty \qquad (7)$$

where the square brackets [ ] indicate evaluation at the retarded time $t - |\mathbf{x}-\mathbf{y}|/c_o$.

The density of a fluid particle changes by $(p - p_o)/c^2$ when isentropically compressed by departures of p from the ambient pressure p_o, where c is the local speed of sound. Let us assume that the motion of each fluid particle is isentropic, but that the mean density, which will be denoted by $\bar{\rho}(\mathbf{x},t)$, varies from point to point in the source region, where $(\bar{\rho} - \rho_o)/\rho_o$ need not necessarily be small. This means that $(p - p_o)/c^2 = \rho - \bar{\rho}$ and $D\bar{\rho}/Dt = 0$.

The Reynolds stress on the right of (7) must be replaced by $\bar{\rho} v_i v_j$, and the integrand of the second term in (7) becomes

$$\frac{\partial^2}{\partial t^2}\left\{\frac{p-p_o}{c_o^2}-(\rho-\rho_o)\right\}=\frac{\partial^2}{\partial t^2}\left\{(p-p_o)\left(\frac{1}{c_o^2}-\frac{1}{c^2}\right)\right\}-\frac{\partial^2}{\partial t^2}(\bar{\rho}-\rho_o).$$

The first term on the right is already in a form suitable for estimating the acoustic radiation. However, the second term would yield zero when retarded time variations are neglected over a coherent source region of constant mass. To deal with this the relation $D(\bar{\rho}-\rho_o)/Dt=0$ is used to show that

$$\frac{\partial^2}{\partial t^2}(\bar{\rho}-\rho_o)=-\frac{\partial}{\partial x_j}\left((\bar{\rho}-\rho_o)\frac{Dv_j}{Dt}\right)+\frac{\partial^2}{\partial x_i \partial x_j}((\bar{\rho}-\rho_o)v_i v_j)$$
$$+\frac{\partial}{\partial t}((\bar{\rho}-\rho_o)\operatorname{div}\mathbf{v})-\frac{\partial}{\partial x_j}(v_j(\bar{\rho}-\rho_o)\operatorname{div}\mathbf{v}). \qquad (8)$$

This is further simplified by substitution from the momentum equation

$$Dv_j/Dt=(-1/\rho)\partial p/\partial x_j,$$

and from the continuity equation

$$\operatorname{div}\mathbf{v}=(-1/\rho c^2)Dp/Dt\approx -D[(p-p_o)/\rho c^2]/Dt.$$

The amplitude of sound produced by the final term on the right of (8) is $O(M)$ relative to the preceding one and is neglected.

Inserting these results into (7), and discarding the overbar on $\bar{\rho}$, we obtain

$$p(\mathbf{x},t)\approx\frac{x_i x_j}{4\pi c_o^2|\mathbf{x}|^3}\frac{\partial^2}{\partial t^2}\int\left[\rho_o v_i v_j\right]d^3\mathbf{y}+\frac{\rho_o x_j}{4\pi c_o|\mathbf{x}|^2}\frac{\partial}{\partial t}\int\left[\left(\frac{1}{\rho_o}-\frac{1}{\rho}\right)\frac{\partial p}{\partial y_j}\right]d^3\mathbf{y}$$
$$+\frac{\rho_o}{4\pi|\mathbf{x}|}\frac{\partial^2}{\partial t^2}\int\left[\left(\frac{1}{\rho_o c_o^2}-\frac{1}{\rho c^2}\right)(p-p_o)\right]d^3\mathbf{y},\quad |\mathbf{x}|\to\infty, \qquad (9)$$

where quantities in square braces are evaluated at time $t-|\mathbf{x}-\mathbf{y}|/c_o$. The new terms in this formula, in addition to the Reynolds stress radiation of (6), are respectively non-zero when the mean density in the source region $\rho\neq\rho_o$, and when $\rho_o c_o^2\neq\rho c^2$, and are the sound fields of dipole and monopole sources.

All of these source types occur in the turbulent mixing region of a hot gas jet exhausting into cold air: 'hot spots' or 'entropy inhomogeneities' behave as scattering centers at which dynamic pressure fluctuations are converted directly into sound. The dipole source strength is proportional to $(1/\rho-1/\rho_o)\nabla p$, i.e., to the difference between the acceleration of fluid of density ρ in the jet and that which fluid of ambient mean density ρ_o would experience in the same pressure gradient. For an ideal gas with $p-p_o\sim O(\rho_o v^2)$ in the jet, the order of magnitude of the dipole sound from an eddy of scale ℓ is

$$p\sim(\ell/|\mathbf{x}|)(1-\rho_o/\rho)\rho_o v^2 M\approx(\ell/|\mathbf{x}|)(\Delta T/T)\rho_o v^2 M,$$

where T denotes temperature, and $\Delta T/T \approx (\rho - \rho_o)/\rho$ is the fractional temperature difference between the hot spot and its environment, which can be large. The sound power $\sim \ell^2(\Delta T/T)^2\rho_o v^3 M^3$ exceeds that from a Reynolds stress quadrupole by a factor of order $(\Delta T/T)^2/M^2$. Thus, 'entropy noise' may be an important component of the noise of a hot gas jet at very low Mach numbers.

The final term on the right of (9) is a monopole whose strength is determined by the difference between the adiabatic compressibilities ($K_s = 1/\rho c^2$) in the source region and in the ambient medium. In an ideal gas $K_s = 1/\gamma p$, so that this source is generally small, although it may be significant during the turbulent mixing of gases with different values of the specific heat ratios γ. It can be very important in multiphase flows where, for example, the presence of small air bubbles in water often leads to an immense increase in the turbulence generated noise.

2 Vorticity and Entropy Fluctuations as Sources of Sound

2.1 The Rôle of Vorticity in Lighthill's Theory

At low Mach numbers the velocity defining the Reynolds stress quadrupole in Lighthill's equation can be determined by regarding the source flow as incompressible, and using the Biot-Savart induction law of hydrodynamics to express $\mathbf{v}$ in terms of the vorticity $\boldsymbol{\omega}$.

Consider an acoustically compact, homentropic vorticity distribution $\boldsymbol{\omega}$ of scale ℓ centered on the coordinate origin in a medium of density ρ_o and sound speed c_o. The velocity $\mathbf{v} = \mathbf{u} + \nabla\varphi$, where $\mathbf{u}$ is the incompressible velocity field defined in terms of $\boldsymbol{\omega}$ by the Biot-Savart induction formula, which implies that $u \sim O(1/|\mathbf{x}|^3)$ as $|\mathbf{x}| \to \infty$. Since $\operatorname{div}\mathbf{u} = 0$, the value of φ is determined by the compressibility of the fluid. In the source region $p - p_o \sim \rho_o u^2$, and the characteristic frequency of the source flow $\sim u/\ell$. Thus, $Dp/Dt \sim \rho_o u^3/\ell$, and the continuity equation $\nabla^2\varphi + D\rho/\rho Dt \approx \nabla^2\varphi + (1/\rho_o c_o^2)Dp/Dt = 0$ implies that

$$\nabla\varphi = O(uM^2), \quad M = u/c_o \text{ for } |\mathbf{x}| \sim \ell. \tag{10}$$

Let us now write

$$\partial^2(u_i u_j)/\partial x_i \partial x_j = \operatorname{div}(\boldsymbol{\omega} \wedge \mathbf{u}) + \nabla^2\left(u^2/2\right) \tag{11}$$

and express the solution (6) of Lighthill's equation in the form

$$\begin{aligned} p(\mathbf{x},t) &= p_1(\mathbf{x},t) + p_2(\mathbf{x},t), \\ p_1(\mathbf{x},t) &= \frac{-\rho_o x_i}{4\pi c_o|\mathbf{x}|^2}\frac{\partial}{\partial t}\int \Big[(\boldsymbol{\omega}\wedge\mathbf{u})_i\Big] d^3\mathbf{y}, \\ p_2(\mathbf{x},t) &= \frac{\rho_o}{4\pi c_o^2|\mathbf{x}|}\frac{\partial^2}{\partial t^2}\int \Big[\frac{1}{2}u^2\Big] d^3\mathbf{y}, \quad |\mathbf{x}|\to\infty, \end{aligned} \tag{12}$$

where $[\;]$ denotes evaluation at time $t - |\mathbf{x}-\mathbf{y}|/c_o$. When retarded time variations across the source region are neglected the identity (11) and the divergence theorem imply that $\int[\boldsymbol{\omega}\wedge\mathbf{u}]d^3\mathbf{y} \equiv \mathbf{0}$, since $u \sim O(1/|\mathbf{y}|^3)$ as $|\mathbf{y}| \to \infty$. To estimate the value of the first integral in (12) it is therefore necessary to expand $(\boldsymbol{\omega}\wedge\mathbf{u})(t - |\mathbf{x}-\mathbf{y}|/c_o)$ in powers of the retarded time element $\mathbf{x}.\mathbf{y}/c_o|\mathbf{x}|$. The first term in the expansion yields,

$$p_1(\mathbf{x},t) \approx \frac{-\rho_o x_i x_j}{4\pi c_o^2|\mathbf{x}|^3}\frac{\partial^2}{\partial t^2}\int y_i(\boldsymbol{\omega}\wedge\mathbf{u})_j(\mathbf{y}, t-|\mathbf{x}|/c_o)\, d^3\mathbf{y} \quad \sim O(\ell/|\mathbf{x}|)\rho_o u^2 M^2 \tag{13}$$

where $\partial/\partial t \sim u/\ell$.

The order of magnitude of $p_2(\mathbf{x},t)$ is estimated by first writing the momentum equation in the form,

$$\partial\mathbf{u}/\partial t + \nabla\left(\int dp/\rho + v^2/2 + \partial\varphi/\partial t\right) = -\boldsymbol{\omega}\wedge\mathbf{u} - \boldsymbol{\omega}\wedge\nabla\varphi.$$

Take the scalar product with $\mathbf{u}$

$$\partial(\frac{1}{2}u^2)/\partial t + \operatorname{div}\left(\mathbf{u}(\int dp/\rho + \frac{1}{2}v^2 + \partial\varphi/\partial t)\right) = -\mathbf{u}.\boldsymbol{\omega}\wedge\nabla\varphi,$$

and integrate over the whole of space. The contribution from the divergence vanishes because $\mathbf{u}(\int dp/\rho + \frac{1}{2}v^2 + \partial\varphi/\partial t)$ tends to zero at least as fast as $1/|\mathbf{y}|^3$ as $|\mathbf{y}| \to \infty$. Hence, using (10)

$$\frac{\partial}{\partial t}\int \frac{1}{2}u^2(\mathbf{y},t)d^3\mathbf{y} = -\int(\mathbf{u}.\boldsymbol{\omega}\wedge\nabla\varphi)(\mathbf{y},t)d^3\mathbf{y} \sim \ell^2 u^3 M^2, \tag{14}$$

where the final estimate is really a crude upper bound, that takes no account of the details of the interactions between the vorticity and irrotational velocity.

Thus,

$$p_2(\mathbf{x},t) \sim O\left((\ell/|\mathbf{x}|)\rho_o u^2 M^4\right),$$

and by comparison with (13) we see that $p_2 \sim O(M^2)p_1 \ll p_1$ when $M \ll 1$. The component $\operatorname{div}(\rho_o\boldsymbol{\omega}\wedge\mathbf{v})$ of $\partial^2(\rho_o v_i v_j)/\partial x_i\partial x_j$ is therefore the principal source of sound at low Mach numbers.

2.2 Acoustic Analogy in Terms of the Total Enthalpy

The analysis in §2.1 and the discussion of §1.3 show that the dominant acoustic sources at low Mach numbers are vorticity and entropy fluctuations. When Lighthill's equation is cast in a form that emphasizes vorticity and entropy as the sound sources, it will is necessary to select an independent acoustic variable that reduces to $\partial\varphi/\partial t$ in irrotational regions, where $\varphi(\mathbf{x},t)$ is the velocity potential governing the motion in the irrotational regions of the fluid. Bernoulli's equation suggests that the total enthalpy $B = w + \frac{1}{2}v^2$ is the appropriate choice, where w is the specific enthalpy ($= \int dp/\rho(p)$ in homentropic flow). B is constant in steady irrotational flow, and at large distances from the acoustic sources perturbations $B = -\partial\varphi/\partial t$ represents acoustic waves.

To reformulate Lighthill's equation in terms of B for a *homogeneous* fluid, whose chemical composition is the same everywhere, we start from Crocco's equation

$$\partial \mathbf{v} \partial t + \nabla B = -\boldsymbol{\omega} \wedge \mathbf{v} + T\nabla s + \boldsymbol{\sigma} \tag{15}$$

$$\sigma_i = (1/\rho)\partial\sigma_{ij}/\partial x_j, \tag{16}$$

where s is the specific entropy, and re-arrange the continuity equation to read

$$\frac{1}{\rho c^2}\frac{Dp}{Dt} + \operatorname{div}\mathbf{v} = \frac{\beta T}{c_p}\frac{Ds}{Dt}, \tag{17}$$

where $D\rho/Dt$ has been eliminated by means of the relations

$$d\rho = \frac{dp}{c^2} + \left(\frac{\partial\rho}{\partial s}\right)_p ds \quad \text{and} \quad \left(\frac{\partial\rho}{\partial s}\right)_p = \left(\frac{\partial\rho}{\partial T}\right)_p \left(\frac{\partial T}{\partial s}\right)_p \equiv \frac{-\beta\rho T}{c_p},$$

β and c_p being respectively the coefficient of expansion and the specific heat at constant pressure. For an ideal gas $\beta = 1/T$.

Subtract the divergence of (15) from the time derivative of (17):

$$\frac{\partial}{\partial t}\left(\frac{1}{\rho c^2}\frac{Dp}{Dt}\right) - \nabla^2 B = \operatorname{div}(\boldsymbol{\omega}\wedge\mathbf{v} - T\nabla s - \boldsymbol{\sigma}) + \frac{\partial}{\partial t}\left(\frac{\beta T}{c_p}\frac{Ds}{Dt}\right). \tag{18}$$

This equation is already in a form that strongly suggests that the terms on the right hand side constitute the most important sources of flow generated sound, because they are unchanged when the compressibility $\mathrm{K}_s = 1/\rho c^2 \to 0$, and accordingly dominate the hydrodynamic far field of B which must ultimately match the outgoing sound waves.

The first term on the left of (18) is expanded as follows:

$$\frac{\partial}{\partial t}\left(\frac{1}{\rho c^2}\frac{Dp}{Dt}\right) = \frac{D}{Dt}\left(\frac{1}{\rho c^2}\frac{\partial p}{\partial t}\right) + \frac{1}{\rho c^2}\frac{\partial v_j}{\partial t}\frac{\partial p}{\partial x_j}$$
$$+ v_j\left\{\frac{\partial}{\partial t}\left(\frac{1}{\rho c^2}\right)\frac{\partial p}{\partial x_j} - \frac{\partial}{\partial x_j}\left(\frac{1}{\rho c^2}\right)\frac{\partial p}{\partial t}\right\}. \tag{19}$$

The last term on the right will be discarded. It vanishes when the adiabatic compressibility $K_s = 1/\rho c^2$ is a function of the pressure alone, i.e., when the motion is homentropic, or in the particular case of an ideal gas where $K_s = 1/\gamma p$ and $\gamma = c_p/c_v$ = constant. Variations of γ in a real gas are significant only in the presence of large temperature gradients; problems of this kind are best treated separately. In liquids the compressibility may be regarded as constant; acoustic sources involving variations of K_s are important only when gas bubbles are present. The other terms on the right of (19) are transformed as follows. First:

$$\frac{1}{\rho}\frac{\partial p}{\partial t} = \frac{1}{\rho}\frac{Dp}{Dt} - \frac{1}{\rho}\mathbf{v}.\nabla p = \frac{DB}{Dt} - T\frac{Ds}{Dt} - \mathbf{v}.\boldsymbol{\sigma}, \tag{20}$$

and second, from (15),

$$\frac{\partial v_j}{\partial t}\frac{\partial p}{\partial x_j} = -\nabla p.\left(\nabla B + \boldsymbol{\omega} \wedge \mathbf{v} - T\nabla s - \boldsymbol{\sigma}\right). \tag{21}$$

Substituting from (19) into (18), we arrive at the acoustic analogy equation for the total enthalpy

$$\begin{aligned}\left(\frac{D}{Dt}\left(\frac{1}{c^2}\frac{D}{Dt}\right) - \frac{\nabla p.\nabla}{\rho c^2} - \nabla^2\right) B = \\ \left(\frac{\partial}{\partial \mathbf{x}} + \frac{\nabla p}{\rho c^2}\right).(\boldsymbol{\omega} \wedge \mathbf{v} - T\nabla s - \boldsymbol{\sigma}) \\ + \frac{\partial}{\partial t}\left(\frac{\beta T}{c_p}\frac{Ds}{Dt}\right) + \frac{D}{Dt}\left\{\frac{1}{c^2}\left(T\frac{Ds}{Dt} + \mathbf{v}.\boldsymbol{\sigma}\right)\right\}.\end{aligned} \tag{22}$$

2.3 Vorticity and Entropy Sources

The nonlinear operator on the left of this equation is identical with that governing the propagation of sound in irrotational, homentropic flow. All terms on the right hand side vanish in irrotational regions, and in the absence of such terms and of boundary motions, B = constant. The radiation condition ensures that the terms on the right may be identified as acoustic sources. They are confined to the region in which $\boldsymbol{\omega} \neq \mathbf{0}$ and where $\nabla s \neq \mathbf{0}$. The wave operator on the left describes propagation of the sound through the non-uniform flow although, as for Lighthill's equation, it will not usually be permissible to neglect the interaction of the aerodynamically generated sound with the vorticity and entropy gradients when the source flow is very extensive. The following special cases should be noted:

(i) High Reynolds number, homentropic flow. When dissipation is neglected and s = constant (22) becomes

$$\left(\frac{D}{Dt}\left(\frac{1}{c^2}\frac{D}{Dt}\right) - \frac{1}{\rho}\nabla\cdot(\rho\nabla)\right) B = \frac{1}{\rho}\mathrm{div}(\rho\boldsymbol{\omega} \wedge \mathbf{v}). \tag{23}$$

At low Mach numbers when the flow is at rest at infinity (where $\rho = \rho_o$ and $c = c_o$) further simplification is possible by (a) neglecting nonlinear effects of propagation and the scattering of sound by the vorticity, and (b) taking $c = c_o$, and $\rho = \rho_o$. Then

$$\left(\frac{1}{c_o^2}\frac{\partial^2}{\partial t^2} - \nabla^2\right) B = \operatorname{div}(\boldsymbol{\omega} \wedge \mathbf{v}), \tag{24}$$

and in the far field the acoustic pressure is given by the linearized approximation

$$p(\mathbf{x}, t) \approx \rho_o B(\mathbf{x}, t). \tag{25}$$

(ii) Non-homentropic source flow. When the source region is not homentropic, it is important to distinguish between terms on both sides of (22) that account principally for *scattering* of the sound, and those (on the right) that can unequivocally be recognized as sources. For compact sources we can argue that the correct source terms are those remaining when the fluid is temporarily taken to be *incompressible.* These are the sources on the right of (18). When dissipation is ignored, (22) then reduces to

$$\left\{\left(\frac{\partial}{\partial t} + \mathbf{U}\cdot\nabla\right)\left[\frac{1}{c^2}\left(\frac{\partial}{\partial t} + \mathbf{U}\cdot\nabla\right)\right] - \frac{1}{\rho}\nabla\cdot(\rho\nabla)\right\} B =$$
$$\operatorname{div}(\boldsymbol{\omega} \wedge \mathbf{v} - T\nabla s) + \frac{\partial}{\partial t}\left(\frac{\beta T}{c_p}\frac{Ds}{Dt}\right). \tag{26}$$

This approximation is useful when the background mean flow may be regarded as irrotational (steady flow past a streamlined body, for example) at mean velocity $\mathbf{U}(\mathbf{x})$ and density and sound speed $\rho(\mathbf{x})$ and $c(\mathbf{x})$. At very small Mach numbers it simplifies further to

$$\left(\frac{1}{c_o^2}\frac{\partial^2}{\partial t^2} - \nabla^2\right) B = \operatorname{div}(\boldsymbol{\omega} \wedge \mathbf{v} - T\nabla s) + \frac{\partial}{\partial t}\left(\frac{\beta T}{c_p}\frac{Ds}{Dt}\right). \tag{27}$$

The final entropy source on the right of (26) and (27) represents the production of sound by unsteady *heating* of the fluid. According to the equation of continuity (27) this source is equivalent to a volume monopole of strength,

$$q(\mathbf{x}, t) = \frac{\beta T}{c_p}\frac{Ds}{Dt}. \tag{28}$$

Spinning vortex pair (Fig.2): Two parallel line vortices each of circulation Γ and distance 2ℓ apart rotate about the x_3-axis midway between them at angular velocity $\Omega = \Gamma/4\pi\ell^2$ in fluid of mean density ρ_o. At time t their positions in the x_1x_2-plane are $\mathbf{x} = \pm\mathbf{s}(t) \equiv \pm\ell(\cos\Omega t, \sin\Omega t)$. Then

$$\operatorname{div}(\boldsymbol{\omega} \wedge \mathbf{v}) = \frac{\Gamma^2}{2\pi\ell^2}\frac{\partial^2}{\partial x_i \partial x_j}(s_i(t)s_j(t)\delta(\mathbf{x})) + \ldots, \quad \mathbf{x} = (x_1, x_2)$$

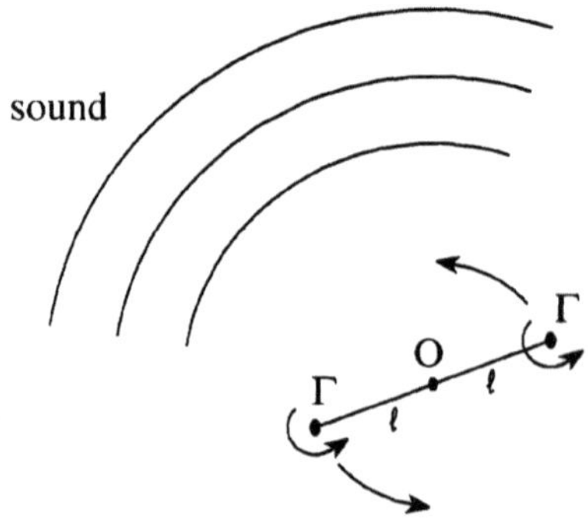

Fig. 2. Spinning vortices.

where the terms omitted are higher order multipoles. If $\Gamma/c_o\ell \ll 1$ (compact source condition), the acoustic pressure is given by

$$p \approx -4\sqrt{\frac{\pi\ell}{r}}\rho_o U^2 M^{3/2} \cos\left(2\theta - 2\Omega(t - r/c_o) + \frac{\pi}{4}\right), \quad \Omega r/c_o \to \infty,$$

where $U = \Gamma/4\pi\ell$, $M = U/c_o$ and (r,θ) are polar coordinates of (x_1, x_2). The power radiated per unit length of vortex $\sim \ell\rho_o U^3 M^4$, which is the same as for 'two-dimensional' turbulence.

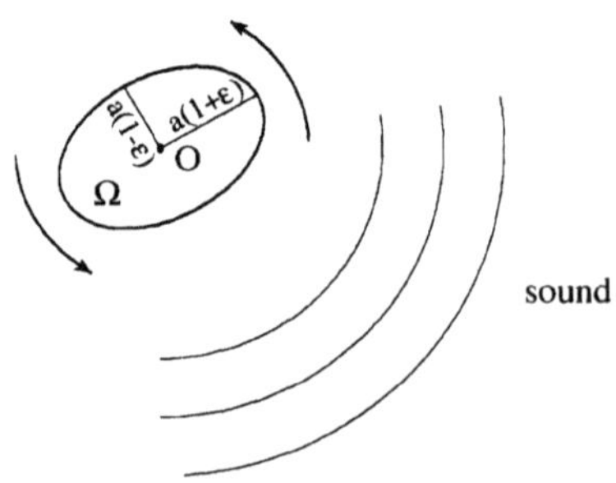

Fig. 3. Kirchhoff's vortex.

Kirchhoff's spinning vortex (Fig.3): Consider the sound produced by a spinning, columnar vortex of elliptic cross-section defined by the polar equation $r = a\,(1 + \epsilon\cos(2\theta - \Omega t/2))$, where $\epsilon \ll 1$ and Ω is the uniform vorticity in the core. The ellipse rotates at angular velocity $\frac{1}{4}\Omega$, and the velocity distribution within the core is given by,

$$\mathbf{v} = (v_1, v_2) = -\frac{1}{2}\Omega r\Big(\sin\theta + \epsilon\sin(\theta - \Omega t/2), -\cos\theta + \epsilon\cos(\theta - \Omega t/2)\Big).$$

The vortex is equivalent to the two-dimensional quadrupole $\mathrm{div}(\boldsymbol{\omega} \wedge \mathbf{v}) \approx \partial^2 \left(\mathrm{t}_{ij}\delta(x_1)\delta(x_2)\right)/\partial x_i \partial x_j$, where

$$\mathrm{t}_{ij} = \frac{\epsilon\pi\Omega^2 a^4}{8}\begin{pmatrix}\cos(\Omega t/2) & \sin(\Omega t/2)\\ \sin(\Omega t/2) & -\cos(\Omega t/2)\end{pmatrix}$$

and that the acoustic pressure is,

$$p \approx -\frac{\epsilon}{8}\sqrt{\frac{2\pi a}{r}}\rho_o U^2 M^{3/2}\cos\left(2\theta - \frac{\Omega}{2}(t - r/c_o) + \frac{\pi}{4}\right), \quad \Omega r/c_o \to \infty,$$

where $U = \frac{1}{2}a\Omega$ is the maximum mean flow speed (at the edge of the core) and $M = U/c_o$. The acoustic frequency is twice the rotation frequency of the ellipse, and the directivity has a characteristic four-lobe shape of a 'lateral' quadrupole.

3 Fundamental Solutions of the Wave Equation

To investigate the production of sound it is convenient to consider first sources whose strengths vary periodically with time. This is because the wave equation is linear, and solutions corresponding to the different frequency components of the source may be superposed, permitting the full solution to be expressed as a Fourier series or integral.

3.1 The Helmholtz Equation

Consider the sound radiated into an unbounded, stationary fluid from a time-harmonic volume source $q(\mathbf{x},t) = q(\mathbf{x},\omega)\mathrm{e}^{-i\omega t}$ of radian frequency ω. The velocity potential $\varphi(\mathbf{x},t) = \varphi(\mathbf{x},\omega)\mathrm{e}^{-i\omega t}$ evidently oscillates at the same frequency, and $\varphi(\mathbf{x},\omega)$ satisfies the inhomogeneous Helmholtz equation

$$(\nabla^2 + \kappa_o^2)\varphi = q, \tag{29}$$

where

$$\kappa_o = \omega/c_o \tag{30}$$

is the **acoustic wavenumber**. We adopt the shorthand notation that (except where confusion arises) φ and q represent the **frequency-domain** quantities $\varphi(\mathbf{x},\omega)$ and $q(\mathbf{x},\omega)$. In an unbounded fluid the solution of (29) must satisfy the **radiation condition** that energy delivered to the fluid by the source radiates away from the source, i.e. the solution must exhibit *outgoing wave behavior.*

The solution for a unit point source $q(\mathbf{x},\omega) = \delta(\mathbf{x}-\mathbf{y}) \equiv \delta(x_1-y_1)\delta(x_2-y_2)\delta(x_3-y_3)$ is called the frequency-domain **Green's function** $G(\mathbf{x},\mathbf{y};\omega)$, which satisfies

$$(\nabla^2 + \kappa_o^2)G = \delta(\mathbf{x}-\mathbf{y}). \tag{31}$$

The identity $q(\mathbf{x},\omega) = \int_V q(\mathbf{y},\omega)\delta(\mathbf{x}-\mathbf{y})d^3\mathbf{y}$, where the integration is taken over a volume V containing the sources, and the principle of superposition permits the solution of the general problem (29) to be written

$$\varphi(\mathbf{x},\omega) = \int_V G(\mathbf{x},\mathbf{y};\omega)q(\mathbf{y},\omega)d^3\mathbf{y}. \tag{32}$$

Let us solve (31) by the method of Fourier transforms. The n-dimensional Fourier space-transform $f(\mathbf{k})$ of a function $f(\mathbf{x})$ of $\mathbf{x} = (x_1, x_2, \ldots, x_n)$ satisfies the reciprocal relations

$$f(\mathbf{k}) = \frac{1}{(2\pi)^n}\int_{-\infty}^{\infty} f(\mathbf{x})\mathrm{e}^{-i\mathbf{k}.\mathbf{x}}d^n\mathbf{x}, \quad f(\mathbf{x}) = \int_{-\infty}^{\infty} f(\mathbf{k})\mathrm{e}^{i\mathbf{k}.\mathbf{x}}d^n\mathbf{k}, \tag{33}$$

where $\mathbf{k}$ is the n-dimensional *wavenumber vector* $\mathbf{k} = (k_1, k_2, \ldots, k_n)$.

Take the Fourier transform of (31) by applying the integral operator $(1/2\pi)^3\int_{-\infty}^{\infty}(\bullet)\mathrm{e}^{-i\mathbf{k}.\mathbf{x}}d^3\mathbf{x}$. By integration by parts (which is permissible for any generalized function such as $G(\mathbf{x},\mathbf{y};\omega)$) we write

$$\begin{aligned}\int_{-\infty}^{\infty} \nabla^2 G(\mathbf{x},\mathbf{y};\omega)\mathrm{e}^{-i\mathbf{k}.\mathbf{x}}d^3\mathbf{x} &= -k^2\int_{-\infty}^{\infty} G(\mathbf{x},\mathbf{y};\omega)\mathrm{e}^{-i\mathbf{k}.\mathbf{x}}d^3\mathbf{x} \\ &\equiv -(2\pi)^3 k^2 G(\mathbf{k},\mathbf{y};\omega),\end{aligned}$$

where $k = |\mathbf{k}|$, to obtain

$$G(\mathbf{k},\mathbf{y};\omega) = \frac{-\mathrm{e}^{-i\mathbf{k}.\mathbf{y}}}{(2\pi)^3(k^2-\kappa_o^2)}. \tag{34}$$

The inverse transform (the second of equations (33)) yields

$$G(\mathbf{x},\mathbf{y};\omega) = \frac{-1}{(2\pi)^3}\int_{-\infty}^{\infty}\frac{\mathrm{e}^{i\mathbf{k}.(\mathbf{x}-\mathbf{y})}}{(k^2-\kappa_o^2)}d^3\mathbf{k}. \tag{35}$$

To evaluate the integral introduce spherical polar coordinates (k,θ,ϕ) for $\mathbf{k}$, where the latitude θ is measured from the source-observer direction of $\mathbf{x}-\mathbf{y}$, and $d^3\mathbf{k} = k^2\sin\theta d\theta d\phi dk$. Performing the integrations with respect to θ and ϕ, we find

$$G(\mathbf{x},\mathbf{y};\omega) = \frac{i}{(2\pi)^2|\mathbf{x}-\mathbf{y}|}\int_0^{\infty}\frac{k}{(k^2-\kappa_o^2)}\left(\mathrm{e}^{ik|\mathbf{x}-\mathbf{y}|}-\mathrm{e}^{-ik|\mathbf{x}-\mathbf{y}|}\right)dk. \tag{36}$$

The remaining integral is undefined for *real* values of κ_o because of the pole on the real axis at $k = |\kappa_o|$. The pole can be avoided by indenting the path into the complex plane to pass either above or below the singularity. Two alternative solutions of equation (31) are obtained in this way, but only one, where path runs below or above the pole according as $\kappa_o \gtrless 0$, exhibits outgoing wave behaviour.

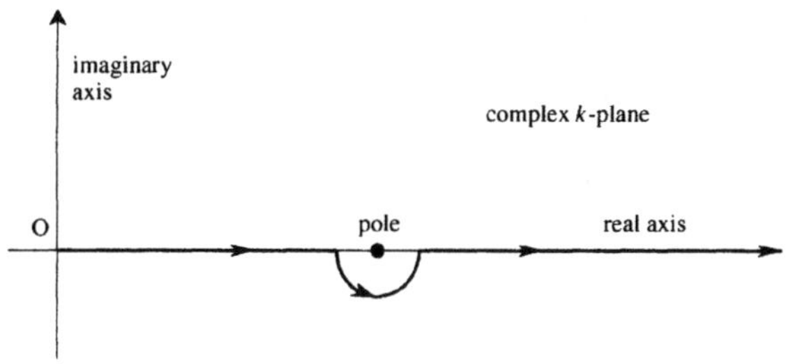

Fig. 4. Path of integration in the wavenumber plane when $\kappa_o > 0$.

To prove this suppose that $\kappa_o > 0$. Indent the integration path to pass below the pole at $k = \kappa_o$ (Fig. 4), and consider the separate contributions to the integral in (36) from the two exponential terms in the integrand. For the first, Cauchy's theorem permits the path of integration to be rotated through 90° onto the positive imaginary axis, on which the integrand decays exponentially as $k \to +i\infty$. The pole at $k = \kappa_o$ is crossed during this rotation, giving a residue contribution $i\pi \mathrm{e}^{i\kappa_o|\mathbf{x}-\mathbf{y}|}$ to the value of the integral. The path of integration for the second exponential may be similarly rotated onto the *negative* imaginary axis, this time without encountering any singularities. The two integrals on the positive and negative imaginary axes are equal and opposite, and therefore

$$G(\mathbf{x},\mathbf{y};\omega) = \frac{-\mathrm{e}^{i\kappa_o|\mathbf{x}-\mathbf{y}|}}{4\pi|\mathbf{x}-\mathbf{y}|}. \tag{37}$$

When the exponential time factor is restored

$$G(\mathbf{x},\mathbf{y};\omega)\,\mathrm{e}^{-i\omega t} = -\mathrm{e}^{-i\omega(t-|\mathbf{x}-\mathbf{y}|/c_o)}/4\pi|\mathbf{x}-\mathbf{y}|,$$

which represents a continuous wave propagating radially outwards at speed c_o (with increasing time) from the source at $\mathbf{y}$. The amplitude decreases inversely with distance $|\mathbf{x}-\mathbf{y}|$, in accordance with the inverse square law of energy conservation. When $|\mathbf{x}-\mathbf{y}| \to \infty$ the wave "crests" (surfaces of constant *phase*) become locally plane with wavelength $2\pi/\kappa_o$.

Conversely, when the integration path in (36) passes *above* the pole at $k = \kappa_o > 0$, the solution is of the form (37), but with the sign of κ_o reversed. This represents a spherical wave converging towards the source in violation of the radiation condition.

The outgoing wave solution is also obtained from the condition that dissipation within the fluid (which gradually transforms acoustic energy into heat) causes the wave to decay faster than $1/|\mathbf{x}-\mathbf{y}|$ at large distances from the source. This will be the case if, *ab initio*, κ_o is imagined to be assigned a small positive imaginary part which (for $\omega > 0$) shifts the pole off the real axis into the upper half-plane.

This is effectively identical to a formal method proposed independently by Landau and Lighthill for obtaining the *causal* solution of time-harmonic wave problems. The frequency ω is temporarily assigned a small positive imaginary part $i\epsilon$, which is equivalent to considering the radiation from a source that is 'switched-on' in the infinite past, whose amplitude increases slowly in proportion to $e^{\epsilon t}$. The exponential growth of the solution as $t \to +\infty$ is of no significance because causality ensures that it cannot influence the behavior at finite times. When $\kappa_o > 0$ the pole is displaced to $k = \kappa_o + i\epsilon/c_o$ in the upper half plane, and the integration is performed as before, leading to (37). The method is applicable generally to time harmonic wave problems governed by linear equations. However, misleading predictions can occur for systems that are linearly unstable. For example, the linearized treatment of the motion produced by a source in a steady flow over an elastic surface exhibits linear instabilities that grow by extracting energy from the mean flow. These instabilities are not captured by the Landau-Lighthill procedure, because it assumes that all of the perturbation energy is supplied by the source. In these circumstances a causal solution can be derived by considering an initial value problem.

Example: In *two*-dimensions Green's function $G(\mathbf{x}, \mathbf{y}; \omega)$ for the Helmholtz equation

$$(\nabla_2^2 + \kappa_o^2)G = \delta(\mathbf{x} - \mathbf{y}), \quad \mathbf{x} = (x_1, x_2), \quad \mathbf{y} = (y_1, y_2),$$

is given by

$$G(\mathbf{x}, \mathbf{y}; \omega) = \frac{-i}{4}\mathrm{H}_0^{(1)}(\kappa_o|\mathbf{x} - \mathbf{y}|), \tag{38}$$

where $\mathrm{H}_0^{(1)}$ is a Hankel function. The result may be derived by integrating the three-dimensional Green's function (37) over $-\infty < y_3 < \infty$. It represents a cylindrical disturbance whose behavior at large distances from the source is given by,

$$G(\mathbf{x}, \mathbf{y}; \omega) \approx \frac{-e^{i(\kappa_o|\mathbf{x}-\mathbf{y}| + \frac{\pi}{4})}}{\sqrt{8\pi\kappa_o|\mathbf{x} - \mathbf{y}|}}, \quad |\mathbf{x} - \mathbf{y}| \to \infty.$$

Example: In *one* dimension

$$G(x, y; \omega) = \frac{-ie^{i\kappa_o|x-y|}}{2\kappa_o} \tag{39}$$

3.2 The Wave Equation

Green's function for the wave equation is the solution $G(\mathbf{x}, \mathbf{y}, t - \tau)$ with outgoing wave behavior of

$$(\partial^2/c_o^2\partial t^2 - \nabla^2)G = \delta(\mathbf{x} - \mathbf{y})\delta(t - \tau). \tag{40}$$

The right hand side is an impulsive point source which vanishes except at $t = \tau$.

The equation is solved by superposition, by observing that $\delta(t-\tau) = (1/2\pi)\int_{-\infty}^{\infty} e^{-i\omega(t-\tau)}d\omega$, and therefore that (31) is transformed into (40) by application of the integral operator $(-1/2\pi)\int_{-\infty}^{\infty}(\bullet)e^{-i\omega(t-\tau)}d\omega$. Hence,

$$G(\mathbf{x},\mathbf{y},t-\tau) = \frac{-1}{2\pi}\int_{-\infty}^{\infty} G(\mathbf{x},\mathbf{y};\omega)e^{-i\omega(t-\tau)}d\omega, \tag{41}$$

so that in **three dimensions** the solution (37) of (31) supplies

$$\begin{aligned} G(\mathbf{x},\mathbf{y},t-\tau) &= \frac{1}{8\pi^2|\mathbf{x}-\mathbf{y}|}\int_{-\infty}^{\infty} e^{-i\omega(t-\tau-|\mathbf{x}-\mathbf{y}|/c_o)}d\omega \\ &= \frac{1}{4\pi|\mathbf{x}-\mathbf{y}|}\delta(t-\tau-|\mathbf{x}-\mathbf{y}|/c_o). \end{aligned} \tag{42}$$

This vanishes for $t < \tau$, in accordance with the causality principle, and represents an impulsive, spherically symmetric wave expanding from the source at $\mathbf{y}$ at the speed of sound.

The causal solution in unbounded fluid of the inhomogeneous wave equation

$$(\partial^2/c_o^2\partial t^2 - \nabla^2)\varphi = f(\mathbf{x},t) \tag{43}$$

is obtained by superposition, by observing that

$$f(\mathbf{x},t) = \iint_{-\infty}^{\infty} f(\mathbf{y},\tau)\delta(\mathbf{x}-\mathbf{y})\delta(t-\tau)d^3\mathbf{y}d\tau,$$

and therefore that

$$\varphi(\mathbf{x},t) = \iint_{-\infty}^{\infty} f(\mathbf{y},\tau)G(\mathbf{x},\mathbf{y},t-\tau)d^3\mathbf{y}d\tau \tag{44}$$

$$= \frac{1}{4\pi}\int_{-\infty}^{\infty}\frac{f(\mathbf{y},t-|\mathbf{x}-\mathbf{y}|/c_o)}{|\mathbf{x}-\mathbf{y}|}d^3\mathbf{y}. \tag{45}$$

The integral formula (45) is called a **retarded potential**, and represents φ at position $\mathbf{x}$ and time t as a linear superposition of contributions from sources at positions $\mathbf{y}$ which radiated at the earlier times $t-|\mathbf{x}-\mathbf{y}|/c_o$, where $|\mathbf{x}-\mathbf{y}|/c_o$ is the time of travel of sound waves from $\mathbf{y}$ to $\mathbf{x}$.

4 General Solution of the Inhomogeneous Wave Equation

4.1 General Solution in the Frequency-Domain

The solution of the wave equation in the presence of arbitrary source distributions and arbitrary surfaces S involves the Green's function for the *reciprocal*

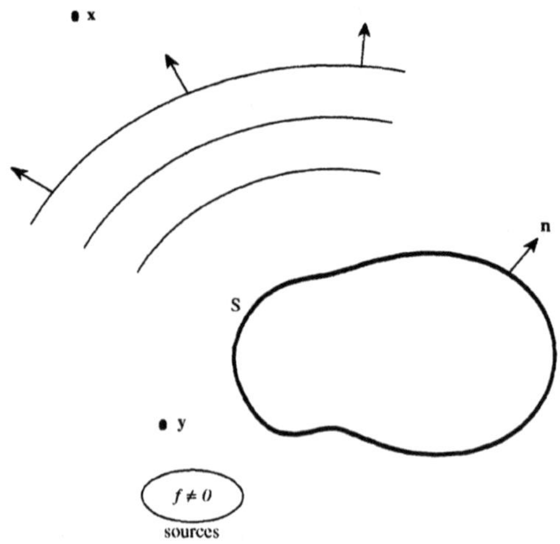

Fig. 5. Solution of the wave equation in the presence of a boundary S.

problem (Fig. 5). Consider first the frequency domain form of (43), i.e. the inhomogeneous Helmholtz equation,

$$(\nabla^2 + \kappa_o^2)\varphi = -f(\mathbf{x}, \omega), \tag{46}$$

where $f(\mathbf{x}, \omega) = (1/2\pi)\int_{-\infty}^{\infty} f(\mathbf{x}, t)\mathrm{e}^{i\omega t}dt$ is the Fourier time transform of $f(\mathbf{x}, t)$.

Green's function $G(\mathbf{y}, \mathbf{x}; \omega) \equiv G(\mathbf{x}, \mathbf{y}; \omega)$ is *any* solution with outgoing wave behavior of the reciprocal problem in which the source is placed at the point $\mathbf{x}$ where the solution $\varphi(\mathbf{x}, \omega)$ is to be found:

$$(\nabla_{\mathbf{y}}^2 + \kappa_o^2)G(\mathbf{x}, \mathbf{y}; \omega) = \delta(\mathbf{x} - \mathbf{y}). \tag{47}$$

The notation implies that $G(\mathbf{x}, \mathbf{y}; \omega)$ is sought as a function of $\mathbf{y}$ for fixed $\mathbf{x}$.

Equations (46) and (47) are now combined in the following way: (i) replace $\mathbf{x}$ by $\mathbf{y}$ in (46) and multiply by $G(\mathbf{x}, \mathbf{y}; \omega)$, (ii) subtract the product of (47) and $\varphi(\mathbf{y}, \omega)$. Integrate with respect to $\mathbf{y}$ over the whole of the fluid and apply the divergence theorem to obtain

$$\varphi(\mathbf{x}, \omega) = \oint_{\mathrm{S}} \left\{ G(\mathbf{x}, \mathbf{y}; \omega)\frac{\partial \varphi}{\partial y_j}(\mathbf{y}, \omega) - \varphi(\mathbf{y}, \omega)\frac{\partial G}{\partial y_j}(\mathbf{x}, \mathbf{y}; \omega) \right\} n_j d\mathrm{S}(\mathbf{y})$$
$$- \int G(\mathbf{x}, \mathbf{y}; \omega) f(\mathbf{y}, \omega) d^3\mathbf{y}, \tag{48}$$

where the unit normal $\mathbf{n}$ on S is directed into the fluid. The radiation condition ensures that there are no additional contributions from a surface integral at infinity.

Equation (48) is applicable for any outgoing solution G of (47). The surface integral can be evaluated if φ and $\partial\varphi/\partial y_n$ are known on S, but these quantities cannot be prescribed independently.

4.2 General Solution in the Time-Domain

Fourier superposition may now be used to derive an integral formula for the solution $\varphi(\mathbf{x},t)$ of the wave equation (43). Application of the *convolution theorem*

$$\int_{-\infty}^{\infty} f_1(\omega) f_2(\omega) \mathrm{e}^{-i\omega t} d\omega = \frac{1}{2\pi}\int_{-\infty}^{\infty} f_1(\tau) f_2(t-\tau) d\tau \tag{49}$$

to (48) yields **Kirchhoff's formula**

$$\begin{aligned}\varphi(\mathbf{x},t) = &\oint_{\mathrm{S}} \left\{ \varphi(\mathbf{y},\tau)\frac{\partial G}{\partial y_j}(\mathbf{x},\mathbf{y},t-\tau) - \frac{\partial \varphi}{\partial y_j}(\mathbf{y},\tau) G(\mathbf{x},\mathbf{y},t-\tau)\right\} n_j d\mathrm{S}(\mathbf{y}) d\tau \\ &+ \int G(\mathbf{x},\mathbf{y},t-\tau) f(\mathbf{y},\tau) d^3\mathbf{y} d\tau,\end{aligned} \tag{50}$$

where the retarded time integration with respect to τ is taken over $(-\infty,\infty)$, and $G(\mathbf{x},\mathbf{y},t-\tau)$ is an outgoing solution of (40) and is related to $G(\mathbf{x},\mathbf{y};\omega)$ by (41).

The special case in which G is the free space Green's function (42) should be noted:

$$\begin{aligned}\varphi(\mathbf{x},t) = &-\frac{1}{4\pi}\oint_S \frac{v_n(\mathbf{y}, t-|\mathbf{x}-\mathbf{y}|/c_o)}{|\mathbf{x}-\mathbf{y}|} d\mathrm{S}(\mathbf{y}) \\ &-\frac{1}{4\pi}\frac{\partial}{\partial x_j}\oint_S \frac{\varphi(\mathbf{y}, t-|\mathbf{x}-\mathbf{y}|/c_o)}{|\mathbf{x}-\mathbf{y}|} n_j d\mathrm{S}(\mathbf{y}) \\ &+\frac{1}{4\pi}\int \frac{f(\mathbf{y}, t-|\mathbf{x}-\mathbf{y}|/c_o)}{|\mathbf{x}-\mathbf{y}|} d^3\mathbf{y}.\end{aligned} \tag{51}$$

In this formula we have set $v_n = \partial\varphi/\partial y_n =$ normal velocity on S. The surface integrals respectively represent the sound produced by distributions of monopole and dipole sources on S, of strengths $v_n(\mathbf{x},t)$ and $\varphi(\mathbf{x},t)$ per unit area. These sources cannot be prescribed independently.

5 Compact Green's Functions

The compact Green's function provides a formal and intuitive procedure for calculating the leading order monopole and dipole terms in the multipole expansion of the sound produced by sources near a solid body.

5.1 Time-Harmonic Problems

Consider the particular Green's function $G(\mathbf{x},\mathbf{y};\omega)$ determined by the Helmholtz equation (47) that has *vanishing normal derivative* on the surface S of an *acoustically compact* body of diameter ℓ. This equation is to be solved as a function of $\mathbf{y}$ for fixed value of $\mathbf{x}$, which we take to be in the *acoustic far field* of S.

The determination of $G(\mathbf{x},\mathbf{y};\omega)$ can be posed as a scattering problem in which the spherical wave (37) (assumed to be generated by a point source at $\mathbf{x}$, and which is just equal to the free space Green's function) is incident on S. If the coordinate origin is within S and $\mathbf{y}$ is close to S, the compactness condition $\kappa_o\ell \ll 1$ permits $G(\mathbf{x},\mathbf{y};\omega)$ to be expanded in the form

$$G(\mathbf{x},\mathbf{y};\omega) = \frac{-\mathrm{e}^{i\kappa_o|\mathbf{x}|}}{4\pi|\mathbf{x}|}\left(1 - \frac{i\kappa_o x_i}{|\mathbf{x}|}\Big(y_i - \varphi_i^*(\mathbf{y})\Big) + \sum_{n\geq 2}(\kappa_o\ell)^n\Phi_n\left(\frac{\mathbf{x}}{|\mathbf{x}|},\mathbf{y}\right)\right),$$
$$\mathbf{y}\sim \mathrm{O}(\ell),\ |\mathbf{x}|\to\infty. \qquad (52)$$

The first term in the large braces represents the incident wave (37) evaluated at $\mathbf{y}=\mathbf{0}$. The next term is $\mathrm{O}(\kappa_o\ell)$ and includes a component $-i\kappa_o x_i y_i/|\mathbf{x}|$ from the incident wave plus a correction $-i\kappa_o x_i\varphi_i^*(\mathbf{y})/|\mathbf{x}|$ due to S. To this order of approximation $Y_i(\mathbf{y}) \equiv y_i - \varphi_i^*(\mathbf{y})$ is a solution of Laplace's equation satisfying $\partial Y_i/\partial y_n = 0$ on S. Since $\varphi_i^*(\mathbf{y})$ must decay with distance from S, this implies that φ_i^* is simply the velocity potential of the incompressible motion that would be produced by translational motion of S as a *rigid body* at unit speed in the i-direction. The remaining terms in (52) are of order $(\kappa_o\ell)^2$ or smaller. When they are neglected the resulting approximation for G can be used to determine the monopole and dipole terms in the multipole expansion of the solution of a fluid-structure interaction problem.

The potentials φ_i^* are uniquely defined by the shape of the body and satisfy

$$\frac{\partial\varphi_i^*}{\partial y_n}(\mathbf{y}) = n_i \text{ on S.} \qquad (53)$$

When the body is rigid they also determine the *added mass* tensor M_{ij}, which is symmetric and given by

$$\mathrm{M}_{ij} = -\rho_o\oint_S \varphi_i^*(\mathbf{y})n_j d\mathrm{S}(\mathbf{y}) \qquad (54)$$
$$= -\rho_o\oint_S \varphi_i^*\partial\varphi_j^*/\partial y_n d\mathrm{S}(\mathbf{y}) \equiv -\rho_o\oint_S \varphi_j^*\partial\varphi_i^*/\partial y_n d\mathrm{S}(\mathbf{y})$$

A body in translational accelerated motion at velocity $\mathbf{U}(t)$ in an ideal, incompressible fluid experiences a reaction force from the fluid equal to $-\partial(\mathrm{M}_{ij}U_j)/\partial t$ in the i-direction. An external force F_i acting through the center of mass of a body of mass m therefore produces accelerated motion determined by $\partial(mU_i + \mathrm{M_{ij}}U_j)/\partial t = F_i$.

Example: Sound generation by a compact, rigid body executing small amplitude translational oscillations (Fig. 6).

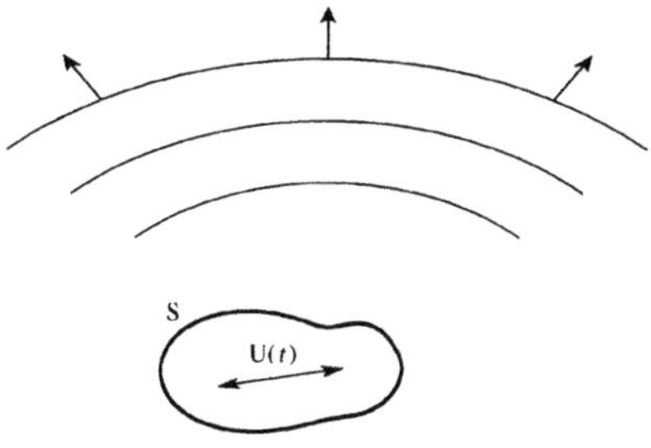

Fig. 6. Sound generation by a rigid body executing small amplitude translational oscillations.

Let the translational velocity be $\mathbf{U}(t)$ and consider first the contribution to the sound from the Fourier component $\mathbf{U}(\omega)\mathrm{e}^{-i\omega t}$. Use Green's function (52) in the general solution (48), with $f = 0$, and $v_n = \mathbf{U}.\mathbf{n}$ on S. Then

$$\varphi(\mathbf{x},\omega) = \frac{-U_j(\omega)\mathrm{e}^{i\kappa_o|\mathbf{x}|}}{4\pi|\mathbf{x}|}\left(\oint_S n_j d\mathrm{S}(\mathbf{y}) - \frac{i\kappa_o x_i}{|\mathbf{x}|}\oint_S Y_i n_j d\mathrm{S}(\mathbf{y}) + \mathrm{O}(\ell^2(\kappa_o\ell)^2)\right),$$
$$Y_i = y_i - \varphi_i^*(\mathbf{y}).$$

The first integral vanishes identically; the second is equal to $(m_o\delta_{ij} + \mathrm{M}_{ij})/\rho_o$, where m_o is the mass of fluid displaced by the body. When the body is acoustically compact $U_j(\omega) \neq 0$ only for $\kappa_o\ell \ll 1$. Hence, multiplying by $\mathrm{e}^{-i\omega t}$ and integrating over $-\infty < \omega < \infty$

$$\varphi(\mathbf{x},t) \approx \frac{-x_i}{4\pi\rho_o c_o|\mathbf{x}|^2}\frac{\partial}{\partial t}\left(m_o U_i + \mathrm{M}_{ij}U_j\right)(t - |\mathbf{x}|/c_o), \quad |\mathbf{x}| \to \infty. \tag{55}$$

This is a dipole field of strength $[m_o U_i(t) + \mathrm{M}_{ij}U_j(t)]/\rho_o$.

Example: $\varphi_i^*(\mathbf{y}) = -R^3 y_i/2|\mathbf{y}|^3$ for a rigid sphere of radius R, and the added mass tensor $\mathrm{M}_{ij} = \frac{2}{3}\pi R^3\rho_o\delta_{ij} \equiv \frac{1}{2}m_o\delta_{ij}$. The sound generated by low frequency, small amplitude translational oscillations at velocity $\mathbf{U}(t)$ is

$$\varphi(\mathbf{x},t) \approx -\frac{R^3\cos\theta}{2c_o|\mathbf{x}|}\frac{\partial U}{\partial t}(t - |\mathbf{x}|/c_o), \quad |\mathbf{x}| \to \infty,$$

where θ is the angle between $\mathbf{U}$ and the observer direction $\mathbf{x}/|\mathbf{x}|$.

5.2 Time-Domain Problems

Consider the general solution (50) of the wave equation when S is *stationary* and *rigid.* Choose Green's function to have vanishing normal derivative on S. Then, using the expansion (52) to calculate $G(\mathbf{x}, \mathbf{y}, t-\tau) = \frac{-1}{2\pi}\int_{-\infty}^{\infty} G(\mathbf{x}, \mathbf{y}; \omega) \mathrm{e}^{-i\omega t} d\omega$, we find

$$\varphi(\mathbf{x}, t) \approx \frac{1}{4\pi|\mathbf{x}|}\bigg(\int f(\mathbf{y}, t-|\mathbf{x}|/c_o) d^3\mathbf{y} + \frac{x_i}{c_o|\mathbf{x}|}\frac{\partial}{\partial t}\int Y_i(\mathbf{y}) f(\mathbf{y}, t-|\mathbf{x}|/c_o) d^3\mathbf{y}$$

$$+\sum_{n\geq 2}\frac{(i\ell)^n}{c_o^n}\frac{\partial^n}{\partial t^n}\int \Phi_n(\mathbf{x}/|\mathbf{x}|, \mathbf{y}) f(\mathbf{y}, t-|\mathbf{x}|/c_o) d^3\mathbf{y}\bigg), \quad |\mathbf{x}| \to \infty \quad (56)$$

The term of order n in the series $\sim \mathrm{O}\big((\omega\ell/c_o)^n\big)$. When the body is compact it is usually sufficient to retain only the first two terms in the large brackets, the monopole and dipole. This is equivalent to approximating Green's function by

$$\begin{aligned} G(\mathbf{x}, \mathbf{y}, t-\tau) &\approx \frac{1}{4\pi|\mathbf{x}|}\left(\delta(t-\tau-|\mathbf{x}|/c_o) + \frac{\mathbf{x}\cdot\mathbf{Y}}{c_o|\mathbf{x}|}\delta'(t-\tau-|\mathbf{x}|/c_o)\right) \\ &\approx \frac{1}{4\pi|\mathbf{x}|}\delta\Big(t-\tau-(|\mathbf{x}|-\mathbf{x}\cdot\mathbf{Y}/|\mathbf{x}|)/c_o\Big) \\ &\approx \frac{1}{4\pi\,|\mathbf{x}-\mathbf{Y}|}\delta\Big(t-\tau-|\mathbf{x}-\mathbf{Y}|/c_o\Big), \quad |\mathbf{x}| \to \infty. \end{aligned} \quad (57)$$

This result can be made symmetric, in accordance with the reciprocal theorem of acoustics, by replacing $\mathbf{x}$ by $\mathbf{X} \equiv \mathbf{x} - \varphi^*(\mathbf{x})$. We then define the **Compact Green's function for a rigid body with surface S:**

$$G(\mathbf{x}, \mathbf{y}, t-\tau) = \frac{1}{4\pi\,|\mathbf{X}-\mathbf{Y}|}\delta\Big(t-\tau-|\mathbf{X}-\mathbf{Y}|/c_o\Big), \quad (58)$$

where $X_i = x_i - \varphi_i^*(\mathbf{x})$, $Y_i = y_i - \varphi_i^*(\mathbf{y})$, and φ_i^* is the velocity potential of incompressible flow that would be produced by rigid body motion of S at unit speed in the i-direction, so that $\partial G(\mathbf{x}, \mathbf{y}, t-\tau)/\partial x_n = \partial G(\mathbf{x}, \mathbf{y}, t-\tau)/\partial y_n = 0$ on S.

In the frequency domain the compact approximation is

$$G(\mathbf{x}, \mathbf{y}; \omega) = \frac{-\mathrm{e}^{i\kappa_o|\mathbf{X}-\mathbf{Y}|}}{4\pi|\mathbf{X}-\mathbf{Y}|}. \quad (59)$$

In these generalized and symmetric formulae (58) and (59), the coordinate origin may be taken at an arbitrary point in space. However, it is still necessary that *either* $\mathbf{x}$ or $\mathbf{y}$ should lie in the far field of the body. When *both* $\mathbf{x}$ and $\mathbf{y}$ are in the far field predictions made with the compact Green's function are the same as when the body is *absent*, because for distant sources the amplitude of the sound scattered by a compact rigid object is $O((\kappa_o \ell)^2)$ smaller than the incident sound, i.e. is of *quadrupole* intensity.

Example: In incompressible fluid $c_o \to \infty$. The potential of motion produced by an arbitrary distribution of sources interacting with a rigid body then satisfies the Poisson equation

$$\nabla^2 \varphi = q(\mathbf{x}, t). \tag{60}$$

The monopole and dipole components of the solution can be calculated from the limiting form of the compact Green's function

$$G(\mathbf{x}, \mathbf{y}, t-\tau) = \frac{-\delta(t-\tau)}{4\pi|\mathbf{X} - \mathbf{Y}|} \tag{61}$$

provided that at least one of the points $\mathbf{x}$ or $\mathbf{y}$ is in the *hydrodynamic* far field of the body.

Example: Green's function for a wedge Green's function for the rigid wedge of Fig. 7 (which is infinite in the x_3-direction) can be expanded in the form

$$G(\mathbf{x}, \mathbf{y}; \omega) = \frac{-i}{4} \int_{-\infty}^{\infty} \sum_{n=0}^{\infty} \sigma_n \Big(\mathrm{H}_\nu^{(1)}(\gamma r_o) \mathrm{J}_\nu(\gamma r) \mathrm{H}(r_o - r)$$
$$+ \mathrm{H}_\nu^{(1)}(\gamma r) \mathrm{J}_\nu(\gamma r_o) \mathrm{H}(r - r_o) \Big) \cos \nu(\theta - \pi) \cos \nu(\theta_o - \pi) \mathrm{e}^{ik_3(x_3 - y_3)} dk_3,$$

in terms of the cylindrical polar coordinates

$$\mathbf{x} = (r\cos\theta, r\sin\theta, x_3), \quad \mathbf{y} = (r_o \cos\theta_o, r_o \sin\theta_o, y_3),$$

where $\mathrm{H}_\nu^{(1)}$, J_ν are Hankel and Bessel functions, $\sigma_0 = 1/(2\pi - \chi)$, $\sigma_n = 2\sigma_0$ $(n \geq 1)$; $\nu = n\pi/(2\pi - \chi)$; $\gamma = \sqrt{\kappa_o^2 - k_3^2}$, and $\mathrm{sgn}(\gamma) = \mathrm{sgn}(\omega)$ when γ is real, or γ is positive imaginary.

Taking the limit as $|\mathbf{x} - y_3 \mathbf{i}_3| \to \infty$ (where $\mathbf{i}_3$ is a unit vector in the x_3-direction), using the stationary phase approximation (by expanding the integrand about $k_3 = \kappa_o \cos\psi$, where ψ is the angle between $\mathbf{x} - \mathbf{i}_3 y_3$ and the edge of the wedge)

$$\int_{-\infty}^{\infty} \mathrm{H}_\nu^{(1)}(\gamma r) \mathrm{J}_\nu(\gamma r_o) \mathrm{e}^{ik_3(x_3 - y_3)} dk_3$$

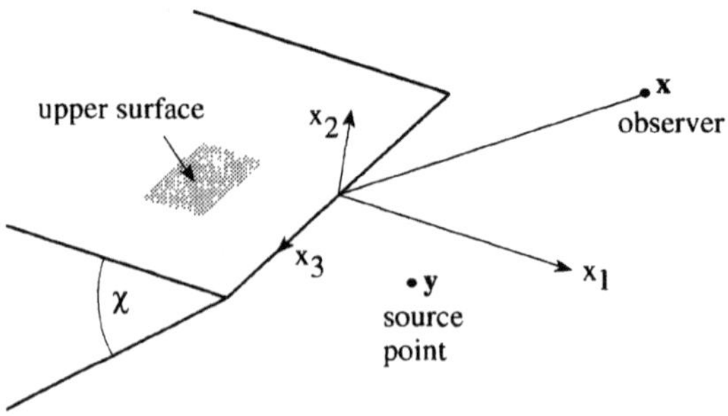

Fig. 7. Coordinate system for a wedge.

$$\approx \sqrt{\frac{2}{\pi\kappa_o r \sin\psi}} \mathrm{J}_\nu(\kappa_o r_o \sin\psi)\mathrm{e}^{-i(\nu+\frac{1}{2})\frac{\pi}{2}} \int_0^\pi \mathrm{e}^{i\kappa_o|\mathbf{x}-y_3\mathbf{i}_3|\cos(\psi-\psi')}\kappa_o \sin\psi'\, d\psi'$$

$$\approx \sqrt{\frac{2\kappa_o \sin\psi}{\pi r}} \mathrm{J}_\nu(\kappa_o r_o \sin\psi)\mathrm{e}^{-i(\nu+\frac{1}{2})\frac{\pi}{2}} \int_{-\infty}^\infty \mathrm{e}^{i\kappa_o|\mathbf{x}-y_3\mathbf{i}_3|(1-\xi^2/2)}\, d\xi$$

$$= \frac{-2i}{|\mathbf{x}-y_3\mathbf{i}_3|} \mathrm{J}_\nu(\kappa_o r_o \sin\psi)\mathrm{e}^{i(\kappa_o|\mathbf{x}-y_3\mathbf{i}_3|-\nu\pi/2)}$$

$$G(\mathbf{x},\mathbf{y};\omega) \approx \frac{-1}{2|\mathbf{x}-y_3\mathbf{i}_3|}\sum_{n=0}^\infty \sigma_n \mathrm{J}_\nu(\kappa_o r_o \sin\psi) \times \cos\nu(\theta-\pi)\cos\nu(\theta_o-\pi)\mathrm{e}^{i(\kappa_o|\mathbf{x}-y_3\mathbf{i}_3|-\nu\pi/2)}, \quad |\mathbf{x}-y_3\mathbf{i}_3| \to \infty. \tag{62}$$

For this geometry we can define a compact Green's function for a source *whose distance from the edge* is small compared to the acoustic wavelength, i.e. $\kappa_o r_o \ll 1$. Using the formula $\mathrm{J}_\nu(z) \approx (\frac{1}{2}z)^\nu/\nu!$ for $|z| \ll 1$, and taking the wedge angle $\chi = 0$ we obtain in this limit the compact Green's function for the *half-plane* $x_1 < 0$, $x_2 = 0$ (Fig. 8):

$$G(\mathbf{x},\mathbf{y};\omega) = G_0(\mathbf{x},\mathbf{y};\omega) + G_1(\mathbf{x},\mathbf{y};\omega) + \ldots, \tag{63}$$

where, for $|\mathbf{x}-y_3\mathbf{i}_3| \to \infty$ and $\kappa_o\sqrt{y_1^2+y_2^2} \ll 1$,

$$G_0(\mathbf{x},\mathbf{y};\omega) = \frac{-1}{4\pi|\mathbf{x}-y_3\mathbf{i}_3|}\mathrm{e}^{i\kappa_o|\mathbf{x}-y_3\mathbf{i}_3|},$$

$$G_1(\mathbf{x},\mathbf{y};\omega) = \frac{-1}{\pi\sqrt{2\pi i}}\frac{\sqrt{\kappa_o}\varphi^*(\mathbf{x})\varphi^*(\mathbf{y})}{|\mathbf{x}-y_3\mathbf{i}_3|^{3/2}}\mathrm{e}^{i\kappa_o|\mathbf{x}-y_3\mathbf{i}_3|}. \tag{64}$$

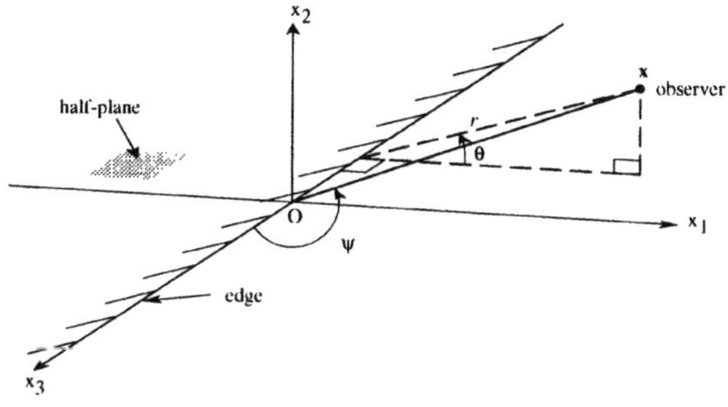

Fig. 8. Coordinates for the half-plane.

and

$$\varphi^*(\mathbf{x}) = \sqrt{r}\sin(\theta/2) \tag{65}$$

is a velocity potential of incompressible flow around the edge expressed in terms of polar coordinates $(x_1, x_2) = r(\cos\theta, \sin\theta)$. The component G_0 of G is the radiation from a point source at $\mathbf{y}$ when scattering is neglected; G_1 is the first correction due to presence of the half-plane.

Example: Green's function for two-dimensional sources near the edge of a half-plane.

When the source has no dependence on the coordinate x_3 parallel to the edge the sound radiates in two-dimensions and the corresponding compact Green's function is obtained by integrating (64) over $-\infty < y_3 < \infty$, using the method of stationary phase for $\kappa_o\sqrt{x_1^2 + x_2^2} \to \infty$. Multiplying the result by $(-1/2\pi)\mathrm{e}^{-i\omega(t-\tau)}$ and integrating over $-\infty < \omega < \infty$, we deduce that the time-domain component G_1 of G has the simple representation

$$G_1(\mathbf{x}, \mathbf{y}, t - \tau) \approx \frac{\varphi^*(\mathbf{x})\varphi^*(\mathbf{y})}{\pi|\mathbf{x}|}\delta(t - \tau - |\mathbf{x}|/c_o), \quad |\mathbf{x}| \to \infty, \tag{66}$$

where $\mathbf{x} = (x_1, x_2)$, $\mathbf{y} = (y_1, y_2)$ in two-dimensions. The two-dimensional counterpart of G_0 is independent of $\mathbf{y}$ and is therefore of use only in applications involving monopole sources, which are uninfluenced by the edge to first order.

Example: Compact Green's function for sources near the entrance of a long duct of uniform cross-sectional area $\mathcal{A}$.

This has applications to the calculation of the compression wave generated by a train entering a long tunnel. $G(\mathbf{x}, \mathbf{y}; t - \tau)$ is required for source points $\mathbf{y}$ in the vicinity of the mouth of the duct (the *tunnel portal*) and for observer positions $\mathbf{x}$ in the 'far field' of the mouth within the duct (Fig. 9).

Take the coordinate origin in the entrance plane of the duct, and assume the duct extends parallel to x_1-axis to $x_1 = -\infty$. In the reciprocal problem

(a)

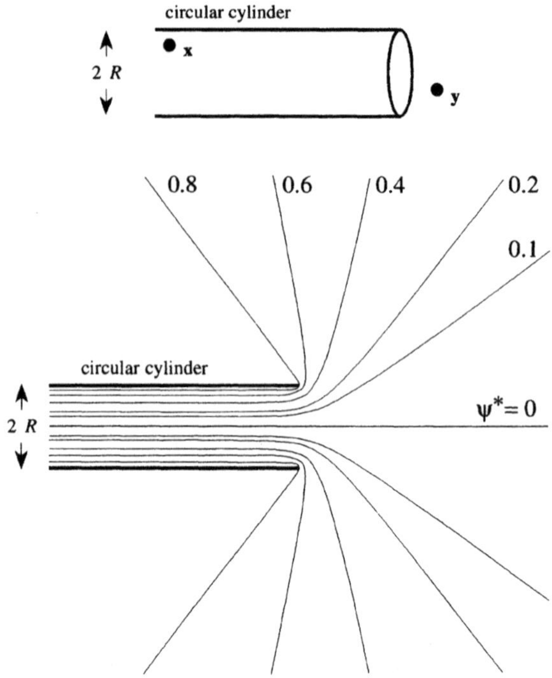

(b)

Fig. 9. (a) Source and observer positions for an unflanged circular cylindrical duct. (b) Stream surfaces $\psi^*(\mathbf{x})$ = constant corresponding to the velocity potential $\varphi^*(\mathbf{x})$ of uniform flow from the duct; the labels indicate the fraction of the total volume flux between the stream surface and the axis of symmetry.

a point source is placed at $\mathbf{x}$ and $G(\mathbf{y},\mathbf{x};\omega)$ is determined as a function of $\mathbf{y}$ near the mouth. Only plane waves can propagate in the duct in the compact approximation, when $\kappa_o\sqrt{\mathcal{A}}$ is small. The potential due to the source at $\mathbf{x}$ is equal to $e^{i\kappa_o|y_1-x_1|}/2i\kappa_o\mathcal{A}$ when $|x_1-y_1| \gg \sqrt{\mathcal{A}}$ in a cylinder that is infinite in both directions. To the right of the source ($y_1 > x_1$), but at distances $\gg \sqrt{\mathcal{A}}$ from the mouth, we therefore write

$$G(\mathbf{y},\mathbf{x};\omega) = \frac{e^{-i\kappa_o x_1}}{2i\kappa_o\mathcal{A}}\{e^{i\kappa_o y_1} + \mathcal{R}e^{-i\kappa_o y_1}\}, \tag{67}$$

where $\mathcal{R}$ determines the amplitude and phase of the wave reflected from the open end. Near the end, where $\kappa_o y_1 \ll 1$

$$G(\mathbf{y},\mathbf{x};\omega) \approx \frac{e^{-i\kappa_o x_1}}{2i\kappa_o\mathcal{A}}\{1+\mathcal{R}+i\kappa_o y_1(1-\mathcal{R})\}, \quad \sqrt{\mathcal{A}} \ll |y_1| \ll 1/\kappa_o. \tag{68}$$

The irrotational motion in the neighbourhood of the mouth may be regarded as incompressible, and when small effects of low frequency radiation from the open end are ignored we can write

$$G(\mathbf{y},\mathbf{x};\omega) \approx \mathcal{T}\varphi^*(\mathbf{y}), \tag{69}$$

where the velocity potential $\varphi^*(\mathbf{y})$ is assumed to satisfy

$$\begin{aligned} \varphi^*(\mathbf{x}) &\approx x_1 - \ell' && \text{when } |x_1| \gg \sqrt{\mathcal{A}} \text{ within the duct,} \\ &\approx -\mathcal{A}/4\pi|\mathbf{x}| && \text{when } |\mathbf{x}| \gg \sqrt{\mathcal{A}} \text{ outside the duct,} \end{aligned} \tag{70}$$

in which ℓ' is the 'end correction' of the duct opening ($\approx 0.61R$ for an unflanged circular cylinder of radius $R = \sqrt{\mathcal{A}/\pi}$). Thus, within the duct at $|y_1| \gg \sqrt{\mathcal{A}}$, $G(\mathbf{y},\mathbf{x};\omega) \approx \mathcal{T}(y_1 - \ell)$, and this must agree with (68). Equating corresponding terms and neglecting small quantities $\sim O(\kappa_o^2\mathcal{A})$ relative to unity, we find

$$\mathcal{R} \approx -e^{2i\kappa_o\ell}, \quad \mathcal{T} \approx \frac{e^{-i\kappa_o(x_1-\ell)}}{\mathcal{A}}, \tag{71}$$

and therefore that

$$\begin{aligned} G(\mathbf{y},\mathbf{x};\omega) &\approx \frac{\varphi^*(\mathbf{y})}{\mathcal{A}} e^{-i\kappa_o(x_1-\ell)}, \text{ near the mouth and outside the duct} \\ &\approx \frac{e^{-i\kappa_o(x_1-\ell)}}{2i\kappa_o\mathcal{A}} \left[e^{i\kappa_o(y_1-\ell)} - e^{-i\kappa_o(y_1-\ell)}\right], \\ &\qquad \text{for } |y_1| \gg \sqrt{\mathcal{A}} \text{ in the duct.} \end{aligned} \tag{72}$$

Now $\varphi^*(\mathbf{y}) \sim O(\sqrt{\mathcal{A}})$ in the vicinity of the tunnel mouth, and decreases in magnitude to zero like $\mathcal{A}/|\mathbf{y}|$ with distance from the mouth outside the tunnel. Hence, when $\kappa_o\sqrt{\mathcal{A}} \ll 1$ the following representation provides a uniform approximation to both of the formulae in (72)

$$G(\mathbf{y},\mathbf{x};\omega) \equiv G(\mathbf{x},\mathbf{y};\omega) \approx \frac{e^{-i\kappa_o(x_1-\ell)}}{2i\kappa_o\mathcal{A}} \left[e^{i\kappa_o\varphi^*(\mathbf{y})} - e^{-i\kappa_o\varphi^*(\mathbf{y})}\right]. \tag{73}$$

This is applicable provided the point $\mathbf{x}$ is within the duct far from the mouth. It is generalized by replacing the factor $(x_1 - \ell)$ in the first exponent by $\varphi^*(\mathbf{x})$ and re-writing (73) in the form

$$G(\mathbf{x},\mathbf{y};\omega) \approx \frac{1}{2i\kappa_o\mathcal{A}} \left[e^{i\kappa_o|\varphi^*(\mathbf{x})-\varphi^*(\mathbf{y})|} - e^{-i\kappa_o[\varphi^*(\mathbf{x})+\varphi^*(\mathbf{y})]}\right]. \tag{74}$$

This is valid for arbitrary source and observer positions provided at least one of them is within the cylinder at a distance from the mouth greatly exceeding $\sqrt{\mathcal{A}}$.

Equation (41) now gives

$$G(\mathbf{x},\mathbf{y};t-\tau) \approx \frac{c_o}{2\mathcal{A}}\left\{\mathrm{H}\left[t-\tau-\frac{|\varphi^*(\mathbf{x})-\varphi^*(\mathbf{y})|}{c_o}\right]\right.$$
$$\left. -\mathrm{H}\left[t-\tau+\frac{(\varphi^*(\mathbf{x})+\varphi^*(\mathbf{y}))}{c_o}\right]\right\}, \tag{75}$$

where $\mathrm{H}(\xi)$ is the Heaviside step function ($=0,\ 1$ according as $\xi \lessgtr 0$).

6 Influence of Solid Bodies on the Generation of Aerodynamic Sound

6.1 Integral Representation of Aerodynamic Sound

Most practical problems of sound generation by flow involve moving boundaries, moving sources interacting with such boundaries, or turbulence in shear layers separating a quiescent medium from a high speed flow. To apply Lighthill's equation (5) in these circumstances closed control surfaces S are introduced. These may coincide with the surface of a moving solid, or mark a convenient interface between fluid regions of widely differing mean properties. We then seek the solution of the equation by imposing suitable boundary conditions on S, either by first performing subsidiary calculations to determine, say, the pressure and velocity on S or, when S coincides with the surface of a solid, by application of suitable impedance conditions.

It is usually assumed that the ambient medium is in a mean state of rest at large distances from the bodies and source region, and an integral representation of the acoustic pressure $p \approx c_o^2(\rho-\rho_o)$ analogous to the Kirchhoff integral formula (51) is formed in terms of the free space Green's function (42). The most general representation of this kind is the *Ffowcs Williams - Hawkings* equation, and is applicable for a control surface S in arbitrary motion. In this lecture, however, we shall consider only the limiting case of stationary surfaces S, which is called *Curle's* formula:

$$p(\mathbf{x},t) = \frac{\partial^2}{\partial x_i \partial x_j}\int_{\mathrm{V}}\left[T_{ij}\right]\frac{d^3\mathbf{y}}{4\pi|\mathbf{x}-\mathbf{y}|} - \frac{\partial}{\partial x_i}\oint_{\mathrm{S}}\left[\rho v_i v_j + p'_{ij}\right]\frac{n_j d\mathrm{S}(\mathbf{y})}{4\pi|\mathbf{x}-\mathbf{y}|}$$
$$+\frac{\partial}{\partial t}\oint_{\mathrm{S}}\left[\rho v_j\right]\frac{n_j d\mathrm{S}(\mathbf{y})}{4\pi|\mathbf{x}-\mathbf{y}|}, \tag{76}$$

where $p'_{ij} = (p-p_o)\delta_{ij} - \sigma_{ij}$, the surface normal n_j is directed into the fluid, and the square brackets denote evaluation at the retarded time $t-|\mathbf{x}-\mathbf{y}|/c_o$.

The volume integral is over the fluid, and represents the sound produced by the Reynolds stress fluctuations in that region; it is formally identical with the corresponding result for the quadrupole sound generated in an unbounded flow. The contribution to $p(\mathbf{x},t)$ from this term is of order $(\ell/|\mathbf{x}|)\rho_o v^2 M^2$ at low Mach numbers, when the mean fluid density is uniform. The surface integrals respectively represent the radiation from dipole and monopole distributions on S. Evidently, when S consists of an arbitrary, fixed control surface drawn in the fluid, whose interior is filled with fluid in low Mach number motion, the net contribution from the monopoles and dipoles must actually reduce to a quadrupole. This indicates that extreme care must be exercised when making order of magnitude estimates of the acoustic contributions from surface integrals of this kind. Indeed, for an acoustically compact surface of dimension ℓ, the dipole strength might be naively estimated to be of order $\rho_o v^2 \ell^2$; since $\partial/\partial x_i \sim (x_i/c_o|\mathbf{x}|)\partial/\partial t \sim M/\ell$ when $|x| \to \infty$, it would then follow that the far field acoustic pressure is of order $(\ell/|\mathbf{x}|)\rho_o v^2 M$. This is $O(1/M)$ ($\gg 1$) larger than the quadrupoles when M is small. This estimate is clearly wrong in general, but happens to be correct when S coincides with the surface of a compact rigid body (Fig. 10).

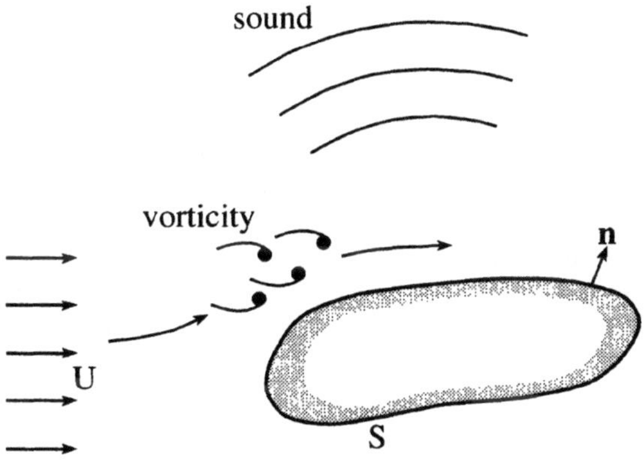

Fig. 10. Sound production by turbulence near a stationary, rigid body.

The more general Ffowcs Williams - Hawkings equation is applicable to moving surfaces, and is frequently applied, for example, to predict the radiation from rapidly rotating rotor blades. Blade motion introduces additional dipole and quadrupole sources associated with surface acceleration and retarded time effects of fluid volume displaced by the blade. At high rotation speeds the blade chords are not usually compact, and predictions typically rely on numerical integrations of the near field Navier-Stokes equations.

6.2 Curle's Theory Applied to Compact Bodies

For a *stationary* rigid body Curle's equation (76) identifies the additional sound caused by the presence of the body with surface dipoles of strength $p'_{ij}n_j$ per unit area, i.e. with the *force* exerted on the fluid per unit area of S.

In typical applications the length scale ℓ of the fluid velocity near S is of the same order as the dimensions of the body. The hydrodynamic time scale $\sim \ell/v$, and the motion in the neighborhood of the body is acoustically compact provided $c_o\ell/v \gg \ell$, i.e. when the Mach number v/c_o is small. Retarded time variations on S may then be neglected, so that when the coordinate origin is taken in the neighborhood of S, the leading order contribution from the surface integrals of (76) is the dipole field

$$p(\mathbf{x},t) \approx \frac{x_i}{4\pi c_o|\mathbf{x}|^2}\frac{\partial F_i}{\partial t}(t - |\mathbf{x}|/c_o), \tag{77}$$

where $F_i(t) = \oint_S p'_{ij}(\mathbf{y},t)n_j dS(\mathbf{y})$ is the net force exerted on the fluid by the rigid body.

The order of magnitude of $F \sim \rho_o v^2 \ell^2$, and $\partial/\partial t \sim v/\ell$. The dipole generated acoustic pressure is therefore of order $(\ell/|\mathbf{x}|)\rho_o v^2 M$ $(M = v/c_o)$, with corresponding sound power $\Pi_a \sim \ell^2 \rho_o v^3 M^3$ and acoustic efficiency $\Pi_a/(\ell^2\rho_o v^3) \sim O(M^3)$. For the direct quadrupole radiation (in a fluid of uniform mean density), the acoustic pressure $\sim (\ell/|\mathbf{x}|)\rho_o v^2 M^2$, the same as in the absence of the body, and the efficiency is $\sim O(M^5)$. The radiation is accordingly dominated by the dipole when M is small, and as $M \to 0$ the acoustic power exceeds the quadrupole power by a factor $\sim 1/M^2 \gg 1$. Precisely how small M should be for this to be true depends on the details of the flow.

The fluid motion close to a compact body may be assumed to be described by the equations of incompressible flow. For a stationary body the net surface force $\mathbf{F}(t)$ is given in terms of the vorticity $\boldsymbol{\omega}(\mathbf{x},t)$ and velocity $\mathbf{v}(\mathbf{x},t)$ by

$$F_i = -\rho_o \int_V \nabla X_i \cdot \boldsymbol{\omega} \wedge \mathbf{v} d^3\mathbf{x} + \eta \oint_S \boldsymbol{\omega} \wedge \nabla X_i \cdot \mathbf{n} dS, \tag{78}$$

where η is the shear coefficient of viscosity, and X_i is defined as in (57) by

$$X_i = x_i - \varphi_i^*(\mathbf{x}), \tag{79}$$

$\varphi_i^*(\mathbf{x})$ being the velocity potential of the incompressible flow that would be produced in an ideal fluid by motion of S at unit speed in the i-direction. The two integrals in (78) respectively represent the contributions to the force from (i) the vector sum of the normal stresses induced on S by the vorticity, and (ii) the viscous skin friction. The latter can usually be neglected when the Reynolds number $v\ell/\nu$ is large $(\nu = \eta/\rho_o)$.

The dipole radiation (77) may therefore be cast in the form

$$p(\mathbf{x},t) \approx \frac{-\rho_o x_i}{4\pi c_o|\mathbf{x}|^2}\frac{\partial}{\partial t}\left\{\int [(\boldsymbol{\omega}\wedge\mathbf{v})\cdot\nabla Y_i]\,d^3\mathbf{y} - \nu\oint_S [\boldsymbol{\omega}\wedge\nabla Y_i\cdot\mathbf{n}]\,dS(\mathbf{y})\right\}$$

$$\approx \frac{-\rho_o x_i}{4\pi c_o|\mathbf{x}|^2}\frac{\partial}{\partial t}\int (\boldsymbol{\omega}\wedge\mathbf{v})(\mathbf{y},t-|\mathbf{x}|/c_o)\cdot\nabla Y_i(\mathbf{y})d^3\mathbf{y}, \quad |\mathbf{x}|\to\infty, \tag{80}$$

where [ ] denotes evaluation at $t-|\mathbf{x}|/c_o$, $Y_i = y_i - \varphi_i^*(\mathbf{y})$, and the second line corresponds to the high Reynolds numbers limit in which the 'skin friction' contribution is neglected.

6.3 Vortex Sound

This result is closely related to the vortex theory of aerodynamic sound (§2.3), and we now give a brief derivation of (80) from equation (24), which is applicable for flows of very small Mach number and uniform mean density.

To solve (24) for the problem depicted in Fig. 10, in the presence of a rigid, acoustically compact body S we make use of the general formula (50), choosing the Green's function to have vanishing normal derivative on S, thereby obtaining

$$B(\mathbf{x},t) \approx -\oint_S G(\mathbf{x},\mathbf{y},t-\tau)\frac{\partial B}{\partial y_j}(\mathbf{y},\tau)n_j dS(\mathbf{y})d\tau + \int G(\mathbf{x},\mathbf{y},t-\tau)\frac{\partial}{\partial y_j}(\boldsymbol{\omega}\wedge\mathbf{v})_j(\mathbf{y},\tau)d^3\mathbf{y}d\tau. \tag{81}$$

The volume integral is transformed by use of the divergence theorem:

$$\int G\mathrm{div}(\boldsymbol{\omega}\wedge\mathbf{v})d^3\mathbf{y} = -\oint_S G(\boldsymbol{\omega}\wedge\mathbf{v})_j n_j dS - \int(\boldsymbol{\omega}\wedge\mathbf{v}).\nabla G d^3\mathbf{y}.$$

When the motion near S is regarded as incompressible, Crocco's equation (15) (in which $\nabla s \equiv \mathbf{0}$) may be approximated by $\partial\mathbf{v}/\partial t + \nabla B = -\boldsymbol{\omega}\wedge\mathbf{v} - \nu\,\mathrm{curl}\boldsymbol{\omega}$, and (81) then becomes

$$B(\mathbf{x},t) \approx -\int(\boldsymbol{\omega}\wedge\mathbf{v})_j(\mathbf{y},\tau)\frac{\partial G}{\partial y_j}(\mathbf{x},\mathbf{y},t-\tau)d^3\mathbf{y}d\tau + \nu\oint_S \boldsymbol{\omega}(\mathbf{y},\tau)\wedge\frac{\partial G}{\partial \mathbf{y}}(\mathbf{x},\mathbf{y},t-\tau).d\mathbf{S}(\mathbf{y})d\tau + \oint_S G(\mathbf{x},\mathbf{y},t-\tau)\frac{\partial v_j}{\partial\tau}(\mathbf{y},\tau)n_j dS(\mathbf{y})d\tau, \tag{82}$$

wherein the last term vanishes when S is rigid.

In the acoustic far field $B(\mathbf{x},t) \approx p(\mathbf{x},t)/\rho_o$. Thus, when the compact Green's function (58) is used we find, correct to dipole order,

$$p(\mathbf{x},t) \approx \frac{-\rho_o x_i}{4\pi c_o|\mathbf{x}|^2}\frac{\partial}{\partial t}\left\{\int [(\boldsymbol{\omega}\wedge\mathbf{v})\cdot\nabla Y_i]\,d^3\mathbf{y} - \nu\oint_S [\boldsymbol{\omega}\wedge\nabla Y_i\cdot\mathbf{n}]\,dS(\mathbf{y})\right\}$$

$$\approx \frac{-\rho_o x_i}{4\pi c_o|\mathbf{x}|^2}\frac{\partial}{\partial t}\int \Big[(\boldsymbol{\omega}\wedge\mathbf{v})\cdot\nabla Y_i\Big]d^3\mathbf{y}, \quad |\mathbf{x}|\to\infty, \quad \frac{v\ell}{\nu}\gg 1, \tag{83}$$

where [ ] denotes evaluation at $t-|\mathbf{x}|/c_o$; in the second line the skin friction contribution has been discarded for high Reynolds number flow.

Equation (83) is identical with the prediction (80) based on Curle's equation (76).

6.4 Influence of Vortex Shedding

The neglect of viscous boundary forces is a satisfactory approximation when the Reynolds number (Re) is large, provided $\rho_o v_i v_j$ (or the vorticity) is known everywhere. Indeed, for turbulence of length scale ℓ and frequency v/ℓ, the depth of viscous penetration from a stationary boundary $\sim \mathrm{O}(\sqrt{\nu\ell/v})$, which is always much smaller than the dimension ℓ of the turbulent flow. However, an impinging turbulence eddy, or the mean flow, can induce 'separation' at a sharp edge, where a substantial quantity of new vorticity appears in the flow. Accurate estimates the sound produced in these circumstances must include contributions from the shed vorticity. It is still not usually necessary to include the viscous surface stresses, provided Re $\gg 1$.

The importance of sound generation by shed vorticity can be established by considering the idealized *two-dimensional* flow depicted in Fig. 11a. A *line* vortex of circulation Γ is parallel to the edge ($x_3 = 0$) of a semi-infinite, rigid plate ($x_1 < 0,\ x_2 = 0$). If $\mathbf{x}_o(t) = (x_{o1}, x_{o2})$ is the position of the vortex at time t, then in the absence of vortex shedding $\boldsymbol{\omega} = \Gamma\mathbf{i}_3\delta(\mathbf{x}-\mathbf{x}_o(t))$ and the sound generated by the vortex-edge interaction is given by the first integral on the right of equation (82). In the acoustic far field $B(\mathbf{x},t) = p(\mathbf{x},t)/\rho_o$, and the appropriate two-dimensional Green's function at very low Mach numbers (such that the vortex may always be considered to be well within an acoustic wavelength from the edge) is given by equation (66). Then

$$p(\mathbf{x},t) \approx -\frac{\rho_o\sin(\theta/2)}{\pi\sqrt{|\mathbf{x}|}}\int\left[\frac{\partial\varphi^*}{\partial\mathbf{y}}\cdot\boldsymbol{\omega}\wedge\frac{d\mathbf{x}_o}{dt}\right]d^2\mathbf{y}$$

$$= -\frac{\rho_o\Gamma\sin(\theta/2)}{\pi\sqrt{|\mathbf{x}|}}\left[\mathbf{i}_3\cdot\frac{d\mathbf{x}_o}{dt}\wedge\nabla\varphi^*(\mathbf{x}_o)\right] \tag{84}$$

$$= \frac{\rho_o\Gamma\sin(\theta/2)}{\pi\sqrt{|\mathbf{x}|}}\left[\frac{d\mathbf{x}_o}{dt}\cdot\nabla\psi^*(\mathbf{x}_o)\right] \equiv \frac{\rho_o\Gamma\sin(\theta/2)}{\pi\sqrt{|\mathbf{x}|}}\left[\frac{D\psi^*}{Dt}\right], \quad |\mathbf{x}|\to\infty,$$

(a)

(b)

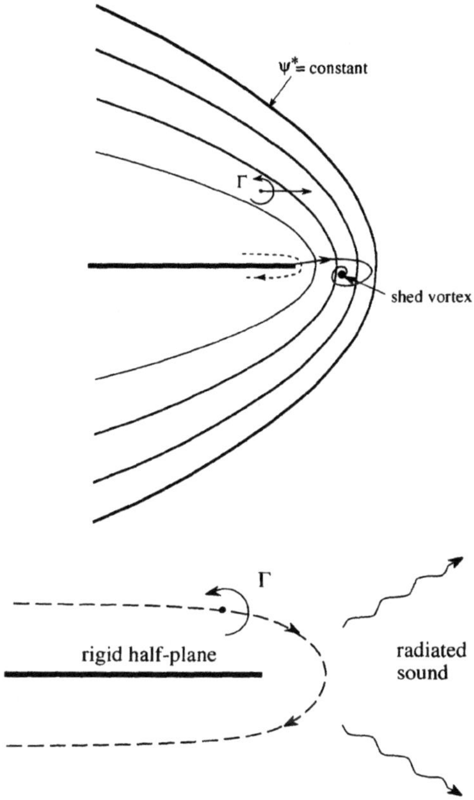

Fig. 11. Motion of a line vortex near an edge

where $[D\psi^*/Dt]$ is evaluated at the retarded position of the vortex and $\mathbf{x} = r(\cos\theta, \sin\theta)$ (as in Fig. 8). The function $\psi^*(\mathbf{x}) = -\sqrt{r}\cos(\frac{1}{2}\theta)$ is the imaginary part of the complex potential $w = \varphi^* + i\psi^* \equiv -i\sqrt{z}$ of ideal flow around the edge, where $z = x_1 + ix_2$ (see §5.2). $\psi^* =$ constant on each of the parabolic streamlines of the potential flow φ^* around the edge. A vortex that translates along one of these streamlines would be 'silent', and the actual edge-noise depends on the *rate at which the trajectory of the vortex cuts across the streamlines of this ideal edge-flow.*

The following argument now supplies a qualitative picture of the influence of vortex shedding on sound generation. Let the circulation Γ be in the sense indicated Fig. 11, so that fluid near the plate is induced to flow in a clockwise direction around the edge, as indicated by the dashed curve in Fig. 11a, and is impeded by viscous stresses. When the Reynolds number is large, separation occurs near the edge and vorticity of opposite sign is shed into the flow. Let us assume for simplicity that the shed vorticity rolls-up into a concentrated core of strength Γ_s. Equation (84) gives the total radiation in the form

$$p(\mathbf{x},t) \approx \frac{\rho_o \sin(\theta/2)}{\pi\sqrt{|\mathbf{x}|}}\left(\Gamma\left[\frac{D\psi^*}{Dt}\right]_\Gamma + \Gamma_s\left[\frac{D\psi^*}{Dt}\right]_{\Gamma_s}\right), \quad |\mathbf{x}| \to \infty,$$

where the derivatives are evaluated at the retarded positions of Γ and Γ_s respectively. Both vortices translate across the curves $\psi^* =$ constant in the direction of decreasing ψ^*, and the derivatives therefore have the same sign. Hence, because Γ and Γ_s have opposite signs, *sound produced by the shed vortex will tend to cancel the edge-generated sound attributable to the incident vortex Γ alone.* A conclusion of this kind is valid for a wide range of fluid-structure interactions, although not necessarily without qualification in cases where the back-reaction of the shed vorticity on the flow produces an overall increase in the amplitude of the motion (e.g., in wind instruments and other resonant systems).

Example: The method of conformal mapping can be used to show that in an ideal fluid, at rest at infinity, the vortex Γ in Fig. 11b moves along a path given by the polar equation

$$r_o = \ell \sec(\frac{1}{2}\theta_o) \quad (-\pi < \theta_o < \pi),$$

where ℓ is the distance of closest approach to the edge of the plate (at $\theta = 0$), provided the Mach number $M = \Gamma/8\pi\ell c_o \ll 1$. If the vortex is at $\mathbf{x}_o = (-\infty, 2\ell)$ at $t = -\infty$, and crosses the x_1-axis at $t = 0$, one finds

$$r_o = \ell\sqrt{1 + (\Gamma t/8\pi\ell^2)^2}, \quad \theta_o = 2\arctan(-\Gamma t/8\pi\ell^2).$$

and that, when vorticity production at the edge is ignored, the acoustic pressure is

$$p(r,\theta,t) \approx \frac{\rho_o\Gamma^2}{(4\pi\ell)^2}\left(\frac{\ell}{|\mathbf{x}|}\right)^{\frac{1}{2}} \sin(\frac{1}{2}\theta)\left[\frac{\Gamma t/8\pi\ell^2}{[1 + (\Gamma t/8\pi\ell^2)^2]^{5/4}}\right]_{t-|\mathbf{x}|/c_o},$$
$$r \equiv |\mathbf{x}| \to \infty.$$

This is plotted as the dotted pressure signature in Fig. 12. When r is large the total radiated energy is

$$E_o = \int_{-\infty}^{\infty} dt \int_0^{2\pi} (p^2(r,\theta,t)/\rho_o c_o)\, r d\theta = \frac{32\pi}{3}\rho_o U^2 M \ell^2, \qquad (85)$$

where $M = U/c_o$, $U = \Gamma/8\pi\ell$.

Example: A quantitative estimate of the effect of vortex shedding on the solution of the previous example is illustrated in Fig. 12. The strength of the incident vortex is now denoted by Γ_o, and the shed vorticity is modelled

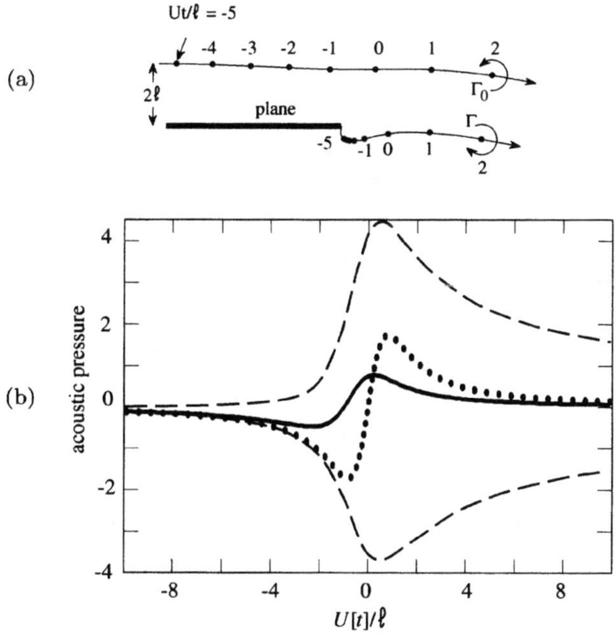

Fig. 12. (a) Vortex trajectories and (b) acoustic pressures.

by a line vortex of variable strength $\Gamma(t)$. The so-called *Brown and Michael* equation is used to determine $\Gamma(t)$ and to calculate the trajectories of the vortices (see Howe 1998 for details). The vortex locations are indicated at different times. The solid curve in Fig. 12b is the net acoustic pressure signature $p/[\rho_o U^2 \sqrt{\ell/r}\sin(\theta/2)]$, where $U = \Gamma_o/8\pi\ell$; the broken curves are the separate contributions from the two vortices, which are effectively equal and opposite at positive retarded times, when the amplitude of the sound becomes very much less than that calculated in the previous example in the absence of shedding (shown dotted in the figure). Very little sound is generated by the shed vorticity prior to the arrival of the incident vortex at the edge, and the acoustic pressure then coincides with that in the absence of shedding.

Numerical evaluation of the integral (85) for the radiated energy E with shedding reveals that $E/E_o \approx 0.18$ (~ -7.5dB), where E_o is the value in the absence of shedding.

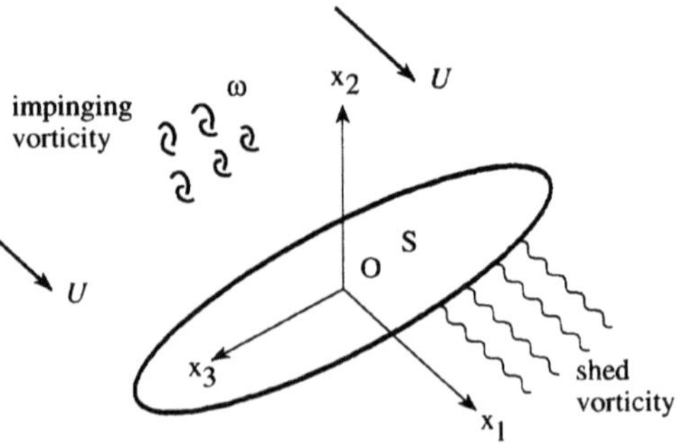

Fig. 13. Spinning vortices.

7 Vortex-Airfoil Interaction Noise

Consider an inhomogeneous field of vorticity, a 'gust', convected in high Reynolds number, homentropic flow past a stationary rigid airfoil (Fig. 13). The undisturbed flow has speed U in the x_1-direction, where the origin is at a convenient point within the airfoil, with x_3 along the span and x_2 vertically 'upwards'. The Mach number $M = U/c_o$ is taken to be sufficiently small that convection of sound by the flow can be ignored.

The aerodynamic sound sources include vorticity in the gust together with any shed from the airfoil, either in response to excitation by the gust, or as 'tip' vortices responsible for the mean lift. The problem is linearized by assuming that $u \ll U$, where $\operatorname{curl} \mathbf{u} = \boldsymbol{\omega}$, i.e. by requiring the gust velocity, and the perturbation velocities caused by airfoil thickness, twist, camber, and the angle of attack, to be small. When the source term $\operatorname{div}(\boldsymbol{\omega} \wedge \mathbf{v})$ is expanded about the undisturbed mean flow, only the gust vorticity and additional vorticity shed when the gust encounters the airfoil contribute to the acoustic radiation to first order. In other words, thickness, twist, camber and angle of attack may ignored, and the airfoil regarded as a rigid lamina in the plane $x_2 = 0$. In this approximation, quadrupoles are neglected, and the vorticity convects as a *frozen* pattern of vortex filaments at the undisturbed mean stream velocity $\mathbf{U} = (U, 0, 0)$. In particular, the wake vorticity is confined to a vortex sheet downstream of the trailing edge.

When convection of sound by the flow is neglected, the linearized equation describing the production of sound becomes

$$\left(\frac{1}{c_o^2}\frac{\partial^2}{\partial t^2} - \nabla^2\right) B = \operatorname{div}(\boldsymbol{\omega} \wedge \mathbf{U}), \tag{86}$$

with solution,

$$B(\mathbf{x}, t) = \frac{p(\mathbf{x}, t)}{\rho_o} \approx -\int (\boldsymbol{\omega} \wedge \mathbf{U})(\mathbf{y}, \tau) \cdot \frac{\partial G}{\partial \mathbf{y}}(\mathbf{x}, \mathbf{y}, t - \tau) d^3\mathbf{y} d\tau, \quad |\mathbf{x}| \to \infty, \tag{87}$$

where $\partial G/\partial y_2 = 0$ on both sides ($y_2 = \pm 0$) of the projection of the airfoil planform onto the y_1y_3-plane. At sufficiently small Mach numbers the airfoil chord will be acoustically compact, and G may be approximated by the compact Green's function.

7.1 Blade-Vortex Interactions in Two-Dimensions

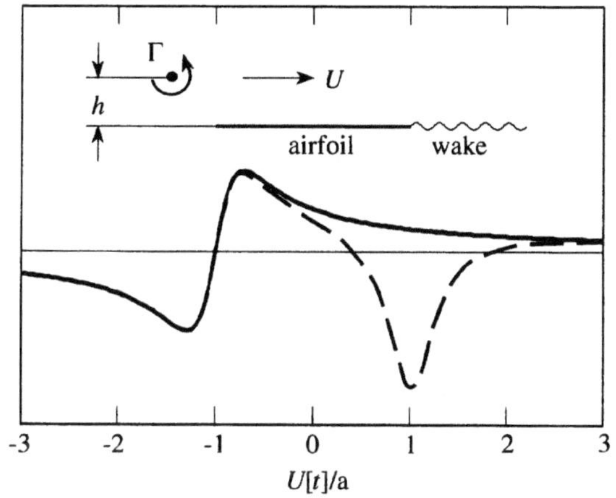

Fig. 14. Blade-vortex interaction in two-dimensions.

Consider a two dimensional airfoil of chord $2a$, and suppose the flow is independent of the spanwise coordinate x_3. The airfoil is replaced by a flat, rigid strip at zero angle of attack (Fig. 14), which occupies $-a < x_1 < a$, $x_2 = 0$.

When $M \ll 1$ the wavelength of the sound is much larger than the chord, and (87) may be evaluated by using the compact Green's function (58), where,

$$Y_1 = y_1, \quad Y_2 = \mathrm{Re}\left(-i\sqrt{z^2 - a^2}\right), \quad Y_3 = y_3; \qquad z = y_1 + iy_2. \qquad (88)$$

In two-dimensions $\boldsymbol{\omega} = (0, 0, \omega_3(x_1, x_2, t))$, $\boldsymbol{\omega} \wedge \mathbf{U} = (0, \omega_3 U, 0)$, and (87) gives

$$p(\mathbf{x}, t) \approx \frac{-\rho_o x_2}{4\pi c_o} \frac{\partial}{\partial t} \int (\nabla Y_2 \cdot \boldsymbol{\omega} \wedge \mathbf{U})\,(y_1, y_2, t - |\mathbf{x} - y_3\mathbf{i}_3|/c_o) \frac{d^3\mathbf{y}}{|\mathbf{x} - y_3\mathbf{i}_3|^2},$$
$$r \equiv \sqrt{x_1^2 + x_2^2} \to \infty, \quad (89)$$

where the dependence of the retarded time on y_3 has been retained since the source extends over the infinite interval $-\infty < y_3 < \infty$. This formula simplifies in the frequency domain. When the y_3-integration is performed by the method of stationary phase for $\kappa_o r \to \infty$, i.e. by means of the approximation

$$\int_{-\infty}^{\infty} \frac{\mathrm{e}^{i\kappa_o|\mathbf{x}-y_3\mathbf{i}_3|}}{|\mathbf{x}-y_3\mathbf{i}_3|^2}\,dy_3 = \int_{-\infty}^{\infty} \frac{\mathrm{e}^{i\kappa_o\sqrt{r^2+\lambda^2}}}{r^2+\lambda^2}\,d\lambda \approx \frac{\mathrm{e}^{i\kappa_o r}}{r^2}\int_{-\infty}^{\infty} \mathrm{e}^{i\kappa_o\lambda^2/2r}\,d\lambda$$
$$= \frac{1}{r}\sqrt{\frac{2\pi i}{\kappa_o r}}\mathrm{e}^{i\kappa_o r}$$

we find

$$p(r,\Theta,\omega) \approx \frac{\cos\Theta}{2}\sqrt{\frac{\kappa_o}{2\pi i r}}F_2(\omega)\mathrm{e}^{i\kappa_o r}, \quad \kappa_o r \to \infty,$$

where $\Theta = \arccos(x_2/r)$ and $F_2 \equiv -\rho_o \int \nabla Y_2 \cdot \boldsymbol{\omega} \wedge \mathbf{U} dy_1 dy_2$ is the force exerted on the fluid per unit span of the airfoil when the motion is regarded as incompressible (see (78)). This is a dipole pressure field, with peak radiation in directions normal to the airfoil. Transforming back to the time-domain

$$p(r,\Theta,t) \approx \frac{\cos\Theta}{2\sqrt{2\pi i c_o r}}\int_{-\infty}^{\infty} \sqrt{\omega}F_2(\omega)\mathrm{e}^{-i\omega(t-r/c_o)}d\omega, \quad r \to \infty. \tag{90}$$

The path of integration passes just above the real axis (§3.1), so that $\sqrt{\omega} = i|\omega|^{1/2}$ when $\omega < 0$.

Let $\Omega(\mathbf{x},t)\mathbf{i}_3 = \Omega(x_1 - Ut, x_2)\mathbf{i}_3$ denote the vorticity in the incident gust, convecting with the mean flow. In the frequency domain, the x_2-component of velocity induced by the gust in the plane $x_2 = 0$ of the airfoil (the 'upwash') can therefore be written $u_2(\omega)\mathrm{e}^{i\omega x_1/U}$. According to *thin airfoil theory* the force F_2 is given in terms of u_2 by,

$$F_2(\omega) = -2\pi a \rho_o u_2(\omega) U \mathcal{S}(\omega a/U), \tag{91}$$

where,

$$\mathcal{S}(x) = \frac{2}{\pi x\left(\mathrm{H}_0^{(1)}(x) + i\mathrm{H}_1^{(1)}(x)\right)} \tag{92}$$

is called the *Sears function.*

Let the gust consist of a line vortex of circulation Γ convecting along the line $x_2 = h > 0$ (Fig. 14). When time is measured from the instant the vortex crosses the mid-chord of the airfoil, we have

$$\Omega(x_1,x_2,t) = \Gamma\delta(x_1 - Ut)\delta(x_2 - h), \quad \Omega(x_1,x_2,\omega) = \frac{\Gamma}{2\pi U}\delta(x_2 - h)\mathrm{e}^{i\omega x_1/U},$$

and the upwash velocity $u_2(\omega) = (-i\Gamma/4\pi U)\mathrm{sgn}(\omega)\exp(-|\omega h/U|)$. The sound generated by the vortex-airfoil interaction can now be calculated from (90) and (91). The integral in (90) cannot be evaluated in closed form, but the strong interaction case $h \ll a$, where the vortex passes close to the airfoil, is approximated asymptotically by

$$p(r,\Theta,t) \approx \frac{\rho_o \Gamma U \cos\Theta}{4\pi}\sqrt{\frac{M}{ra}}\left[\frac{(Ut+a)a/h^2}{1+(Ut+a)^2/h^2}\right]_{t-r/c_o}, \quad r \to \infty.$$

The pressure signature (illustrated for $h/a = 0.2$ by the solid curve in Fig. 14) is generated predominantly as the vortex passes the leading edge of the airfoil at $t = -a/U$, and has characteristic frequency $\omega \sim U/h$. The dominance of sound generation at the leading edge is typical of all interactions occurring at high *reduced frequencies* $\omega a/U$ ($\sim a/h \gg 1$), and occurs for the reason foreshadowed in §6.4: at high Reynolds number the induced velocity near the trailing edge is very large and the inertia of fluid approaching the edge causes the flow to separate and shed new vorticity into the wake of the airfoil. When the reduced frequency is large the sound generated by the wake is just equal and opposite to that produced by the incident vortex interacting with the edge. This conclusion can also be interpreted in terms of the surface pressure fluctuations. According to linear theory, a gust convecting at the mean stream velocity U produces no fluctuations in pressure except when the upwash interacts with the edges of the airfoil. At high reduced frequencies the surface pressure generated when the vortex is upstream of the trailing edge is very small and remains small as it convects over the edge because the shed vorticity cancels the upwash that would otherwise be generated. The linear theory wake consists of a vortex sheet downstream of the edge, whose elements convect at the mean velocity U. Thin airfoil theory determines the wake vorticity by imposing the *Kutta condition* that the pressure (and velocity) should be finite at the edge. Thus, surface forces responsible for the sound tend to be concentrated near the leading edge. Observe that linear theory does not permit the corresponding singular pressure at the leading edge be removed in this way because vorticity shed there is swept over the rigid surface of the airfoil, where its effects are cancelled by image vortices in the airfoil.

The component p_w of the sound pressure attributable to the wake can be found indirectly by first calculating the surface force

$$F_2^{\Omega} = -\rho_o \int \nabla Y_2 \cdot \Omega \mathbf{i}_3 \wedge \mathbf{U} dy_1 dy_2,$$

produced by the incident vortex Γ alone. The wake-induced force $F_2^w = F_2 - F_2^{\Omega}$, where $F_2(\omega)$ is given by (91) and (92), and when this is substituted for F_2 in (90) we find

$$p_w(r,\Theta,t) \approx \frac{\rho_o \Gamma U \cos\Theta}{4\pi}\sqrt{\frac{M}{ra}}\left[\frac{a/h}{1+(Ut-a)^2/h^2}\right]_{t-r/c_o}, \quad r \to \infty.$$

Most of the shedding occurs when the vortex is close to the trailing edge when $h \ll a$, and the sound produced by the wake is radiated when the vortex is close to the edge ($t \approx a/U$). If the contribution from the wake is excluded the pressure signature (now caused by a *potential flow* interaction of Γ with the airfoil) assumes the form indicated by the broken curve in Fig. 14.

7.2 Three-Dimensional Interactions

When a *frozen* three dimensional gust $\boldsymbol{\Omega}(\mathbf{x} - \mathbf{U}t) \equiv \boldsymbol{\Omega}(x_1 - Ut, x_2, x_3)$ is swept past the airfoil of Fig. 13 at low Mach number, the acoustic pressure is given by

$$p(\mathbf{x}, t) \approx \frac{-\rho_o U \cos\Theta}{4\pi c_o|\mathbf{x}|}\frac{\partial}{\partial t}\int \left[\omega_3 \frac{\partial Y_2}{\partial y_2} - \omega_2 \frac{\partial Y_2}{\partial y_3}\right]_{t-|\mathbf{x}|/c_o} d^3\mathbf{y}, \quad |\mathbf{x}| \to \infty, \quad (93)$$

provided the chord is also compact, where $\Theta = \arccos(x_2/|\mathbf{x}|)$ is the angle between the radiation direction and the normal to the airfoil, and the origin is taken in the airfoil within the interaction region. If the aspect ratio of the airfoil is large, then $\partial Y_2/\partial y_3 \ll \partial Y_2/\partial y_2$, except in the tip regions of the airfoil.

Let the interaction occur at an inboard location where the chord may be regarded as constant, with both the leading and trailing edges at right angles to the mean flow. The planform in interaction region is then locally the same as the two-dimensional airfoil of Fig. 14, and Y_2 can be approximated by (88) where $2a$ is the local chord, so that (93) becomes

$$\begin{aligned} p(\mathbf{x}, t) &\approx \frac{\cos\Theta}{4\pi c_o|\mathbf{x}|}\frac{dF_2}{dt}(t - |\mathbf{x}|/c_o), \quad |\mathbf{x}| \to \infty, \\ F_2(t) &= -\rho_o U \int \omega_3(\mathbf{y}, t)\frac{\partial Y_2}{\partial y_2}(\mathbf{y}) d^3\mathbf{y}, \end{aligned} \quad (94)$$

where $-F_2$ is the unsteady lift force on the airfoil during the interaction when the motion is regarded as incompressible, and the vorticity ω_3 includes contributions from the impinging gust together with any shed into the vortex sheet wake.

To calculate F_2 by the method of thin airfoil theory the incident gust is represented by the Fourier integral

$$\boldsymbol{\Omega}(\mathbf{x} - \mathbf{U}t) = \frac{1}{U}\int_{-\infty}^{\infty} \boldsymbol{\Omega}(\omega/U, k_2, k_3) e^{i(\mathbf{k}.\mathbf{x} - \omega t)} d\omega dk_2 dk_3, \quad k_1 = \omega/U, \quad (95)$$

where $\boldsymbol{\Omega}(\mathbf{k})$ is the spatial Fourier transform of the gust vorticity at $t = 0$ (defined as in (33) with $n = 3$). A similar representation is easily obtained for the solenoidal induced velocity $\mathbf{u}(\mathbf{x} - \mathbf{U}t)$ in the absence of the airfoil, by using the relation $\mathbf{u}(\mathbf{k}) = i\mathbf{k} \wedge \boldsymbol{\Omega}(\mathbf{k})/k^2$. The upwash on the airfoil is then

$$u_2(x_3, \omega) e^{i\omega x_1/U} = \frac{i e^{i\omega x_1/U}}{U}\int_{-\infty}^{\infty} \frac{(\mathbf{k} \wedge \boldsymbol{\Omega}(\mathbf{k}))_2 e^{ik_3 x_3}}{k_1^2 + k_2^2 + k_3^2} dk_2 dk_3. \quad (96)$$

By analogy with (91) the force per unit span due to each upwash velocity component $u_2(k_3,\omega)\exp[i(\omega x_1/U + k_3x_3)]$ of $u_2(x_3,\omega)\mathrm{e}^{i\omega x_1/U}$ can be written

$$-2\pi a\rho_o u_2(k_3,\omega)U\hat{\mathcal{S}}(\omega a/U,k_3)\mathrm{e}^{ik_3x_3},$$

where $\hat{\mathcal{S}}(\omega a/U,k_3)$ is the generalization of the Sears function (92) to oblique gusts, such that $\mathcal{S}(\omega a/U)=\hat{\mathcal{S}}(\omega a/U,0)$. This force can be evaluated for each Fourier component of (96), and the net force obtained by integration with respect to x_3 over the span. The integration yields $\delta(k_3)$ when the length scale of the gust is small compared to the airfoil span, so that

$$F_2(\omega)=4\pi^2 i\rho_o aU\omega\mathcal{S}(\omega a/U)\int_{-\infty}^{\infty}\frac{\Omega_3(\omega/U,k_2,0)}{\omega^2+U^2k_2^2}\,dk_2. \tag{97}$$

The first of equations (94) accordingly gives the radiation in the form

$$p(\mathbf{x},t)\approx\frac{2i\rho_oU^2\cos\Theta}{c_o|\mathbf{x}|}\frac{\partial}{\partial t}\iint_{-\infty}^{\infty}\frac{\Omega_3(\omega/U,k_2,0)\mathrm{e}^{-i\omega[t]}\,dk_2d\omega}{(\omega^2+U^2k_2^2)\left(\mathrm{H}_0^{(1)}(\omega a/U)+i\mathrm{H}_1^{(1)}(\omega a/U)\right)},\quad |\mathbf{x}|\to\infty, \tag{98}$$

where $[t]=t-|\mathbf{x}|/c_o$.

This represents the radiation produced by the vorticity in both the gust and wake. Similar formulae can be derived for their separate contributions. The component of F_2 produced by the gust alone is obtained by substituting the representation (95) of the gust vorticity into the second of equations (94). The first equation then yields the following formula for the sound generated by the *potential interaction* of the gust and airfoil:

$$p_\Omega(\mathbf{x},t)\approx\frac{\pi\rho_oaU\cos\Theta}{c_o|\mathbf{x}|}\frac{\partial}{\partial t}\iint_{-\infty}^{\infty}\frac{\omega\Omega_3(\omega/U,k_2,0)\mathrm{J}_1(\omega a/U)\mathrm{e}^{-i\omega[t]}\,dk_2d\omega}{\omega^2+U^2k_2^2},\quad |\mathbf{x}|\to\infty. \tag{99}$$

The wake generated acoustic pressure is $p(\mathbf{x},t)-p_\Omega(\mathbf{x},t)$.

7.3 Blade-Vortex Interactions in Three Dimensions

Consider a gust consisting of a rectilinear vortex of circulation Γ orientated with its axis in the direction of the unit vector $\mathbf{n}$. Choose the origin on the midchord of the airfoil such that the axis of the vortex passes through the origin at time $t=0$ (Fig. 15). Let the axial vorticity be

$$\boldsymbol{\Omega}(\mathbf{x})=\frac{\Gamma\mathbf{n}\mathrm{e}^{-(r_\perp/R)^2}}{\pi R^2}, \tag{100}$$

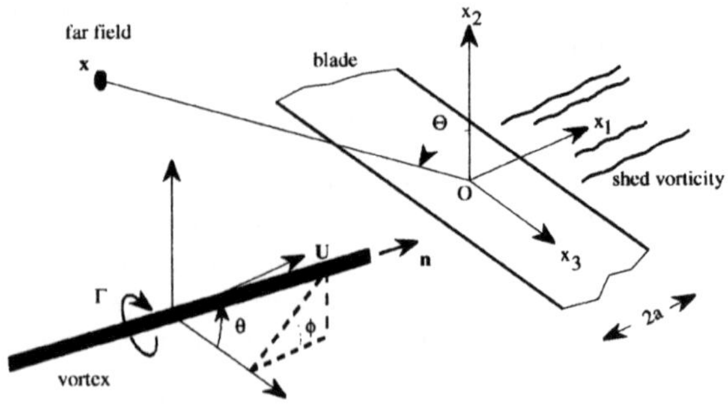

Fig. 15. Blade-vortex interaction.

where $r_\perp$ denotes perpendicular distance from the axis, and R is the nominal radius of the vortex core, which is assumed to be small compared to the chord $2a$. Then

$$\boldsymbol{\Omega}(\mathbf{k}) = \frac{\Gamma \mathbf{n}}{4\pi^2}\delta(\mathbf{n}.\mathbf{k})\mathrm{e}^{-(kR/2)^2},$$

and (98) yields the acoustic pressure in the form

$$p(\mathbf{x},t) \approx \frac{\rho_o U^2 \Gamma |\sin\phi|\cos\Theta}{2\pi^2 c_o \tan\theta |\mathbf{x}|}\int_{-\infty}^{\infty} \frac{\exp\{-(\omega R/2U\sin\phi)^2 - i\omega[t]\}\,d\omega}{\omega\left(\mathrm{H}_0^{(1)}(\omega a/U) + i\mathrm{H}_1^{(1)}(\omega a/U)\right)}, \quad |\mathbf{x}| \to \infty, \tag{101}$$

where the polar angles θ, ϕ define the unit vector $\mathbf{n}$ by

$$\mathbf{n} = (\sin\theta\cos\phi, \sin\theta\sin\phi, \cos\theta), \quad 0 < \theta < \pi,\ 0 < \phi < 2\pi.$$

$2R/|\sin\phi|$ is the length in the mean flow direction of the projected area of the circular cross-section of the core onto the airfoil. Most of the sound is produced when this area is swept over the leading edge of the airfoil, and is characterized by frequencies $\omega \sim U|\sin\phi|/R$, so that the acoustic signature is more sharply peaked when $\phi \approx \pi/2$ or $3\pi/2$. Thus, when $|\sin\phi| \sim \mathrm{O}(1)$ and $R \ll a$, the dominant frequencies satisfy $\omega a/U \gg 1$, and the value of the integral in (101) can be approximated by replacing the Hankel functions by their large argument asymptotic approximations. This procedure supplies

$$p(\mathbf{x},t) \approx \frac{\rho_o U^2 \Gamma \cos\Theta}{8c_o|\mathbf{x}|}\frac{|\sin\phi|^{\frac{3}{2}}(a/\pi R)^{\frac{1}{2}}}{\tan\theta}\Im(\alpha_+), \quad |\mathbf{x}| \to \infty, \tag{102}$$

where $\Im$ and arguments $\alpha_\pm$ are defined by

$$\begin{aligned}\Im(\alpha) &= |\alpha|^{\frac{1}{2}}\left(\mathrm{I}_{-\frac{1}{4}}(\alpha^2/8) + \mathrm{sgn}(\alpha)\mathrm{I}_{\frac{1}{4}}(\alpha^2/8)\right)\mathrm{e}^{-\alpha^2/8},\\ \alpha_\pm &= (2a|\sin\phi|/R)\{1 \pm U[t]/a\},\end{aligned} \tag{103}$$

and $\mathrm{I}_{\pm\frac{1}{4}}$ are modified Bessel functions of the first kind. The potential flow interaction acoustic pressure p_Ω is similarly determined from equation (99):

$$p_\Omega(\mathbf{x},t) \approx \frac{\rho_o U^2 \Gamma \cos\Theta}{8c_o|\mathbf{x}|} \frac{|\sin\phi|^{\frac{3}{2}}(a/\pi R)^{\frac{1}{2}}}{\tan\theta} \Big(\Im(\alpha_+) - \Im(\alpha_-) \Big), \quad |\mathbf{x}| \to \infty, \tag{104}$$

$\Im(\alpha)$ is non-negative, and exhibits a unique maximum (≈ 1.938) at $\alpha = 0$. Equation (102) therefore describes the generation of a single acoustic pulse when the vortex is severed by the leading edge of the blade, as illustrated by the solid curve in Fig. 16a for $R/a = 0.1$. The sign of the pulse is determined by the angle θ between the vortex axis and the span; no sound is generated when $\theta = \pi/2$, when the upwash vanishes. The broken curve is the potential flow interaction pressure p_Ω, which is an odd function of the retarded time $[t]$. The large negative peak in p_Ω, from the potential flow interaction with the trailing edge, is canceled by an equal and opposite contribution generated by the wake.

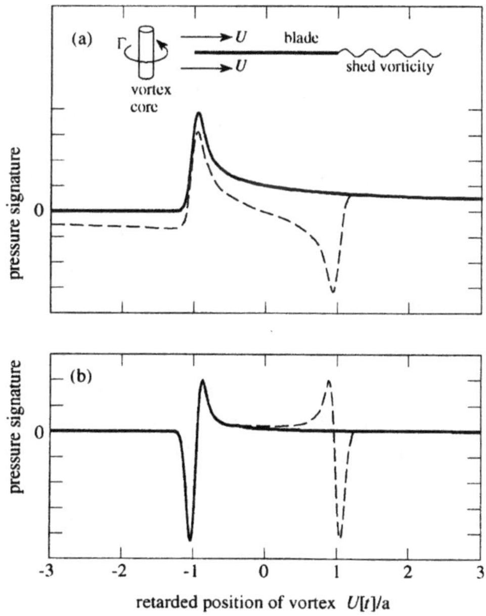

Fig. 16. Blade-vortex interaction acoustic pressure: (a) axial vorticity, (b) azimuthal vorticity.

In practice the axial velocity in the core of the vortex differs from that in the ambient, irrotational flow, giving rise to an 'axial velocity defect'. This occurs, for example, in tip vortices shed from helicopter rotor blades, and is associated with an azimuthal component of vorticity, so that the vortex lines spiral around the vortex axis. The jet-like flow in the core produces a localized upwash when the vortex is cut by the airfoil. Let the maximum velocity defect V occur on the vortex axis, and the variation across the core be modeled by $\mathbf{v} = V\mathbf{n}\exp[-(r_\perp/R)^2]$, so that the axial volume flux is $\pi R^2 V$. The Fourier transform of the azimuthal vorticity is

$$\boldsymbol{\Omega}(\mathbf{k}) = (iVR^2/4\pi)(\mathbf{k}\wedge\mathbf{n})\delta(\mathbf{n}.\mathbf{k})\exp[-(kR/2)^2].$$

When $R \ll a$, the corresponding acoustic pressure $p(\mathbf{x},t)$ and the potential flow interaction acoustic pressure are given by

$$p(\mathbf{x},t) \approx \frac{-\rho_o U^2 V\sqrt{\pi a R}\cos\Theta}{4c_o|\mathbf{x}|}\sin\phi|\sin\phi|^{\frac{1}{2}}\Im'(\alpha_+),$$

$$p_\Omega(\mathbf{x},t) \approx \frac{-\rho_o U^2 V\sqrt{\pi a R}\cos\Theta}{4c_o|\mathbf{x}|}\sin\phi|\sin\phi|^{\frac{1}{2}}\left(\Im'(\alpha_+)+\Im'(\alpha_-)\right), \quad (105)$$

$$|\mathbf{x}|\to\infty,$$

where $\Im'(\alpha) = d\Im(\alpha)/d\alpha$ (which assumes the maximum and minimum values 3.585, -1.647 respectively at $\alpha = -0.78,\ 2.47$). These pressures do not depend on θ, the angle between the spanwise direction and the vortex axis, because the acoustic source is now equivalent to two equal and opposite monopoles on the upper and lower surfaces of the airfoil whose strengths are proportional to the volume flux in the core, which is independent of the vortex orientation (except for the cases $\theta = 0,\ \pi$, which are not covered by this approximation). The shapes of the pressure signatures are shown in Fig. 16b.

Experimental arrangements of the kind shown in Fig. 17 have been used to investigate blade vortex interaction noise. An airfoil A is set at a finite angle of attack to a low Mach number mean stream at velocity U_o in an open jet wind tunnel. The tip vortex from the airfoil is periodically 'chopped' by a rotor at a downstream station B whose plane of rotation is perpendicular to the mean flow. Observations are made of the blade-vortex interaction noise by filtering from the measured acoustic pressure the (typically much stronger) dipole radiation generated by the accelerating mean pressure loading of the rotor (see Howe 1998 for further details).

8 Vorticity Production by Sound

Vorticity is produced by fluid motion over a solid surface, the rate of production being greatest in regions where the pressure and velocity in the primary flow change rapidly, such as at corners and sharp edges. The kinetic energy of the vortex motion is derived from the primary flow, and when this is a sound wave incident, say, on a sharp edge, vorticity diffuses from the edge and causes

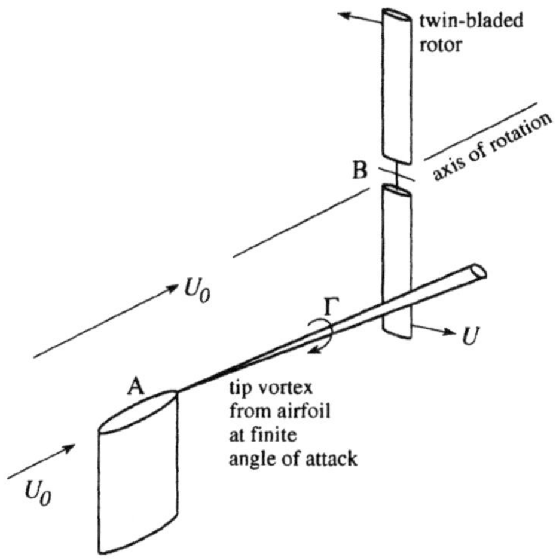

Fig. 17. Experimental arrangement for studying blade-vortex interaction noise.

the sound to be damped. When the fluid is stationary dissipation of sound is caused by nonlinear convection of vorticity from the edge and subsequent viscous damping, both of which are generally weak. Damping is significantly increased, however, in the presence of mean flow. At high Reynolds numbers viscous effects are important only very close to the edge, where vorticity is produced; shed vorticity is swept away by the flow, and its kinetic energy is permanently lost to the sound.

8.1 Interaction of Sound with a Solid Surface

Consider the situation of Fig. 18, where sound is diffracted by a rigid pipe in the presence of a low subsonic, homentropic flow produced by a fan within the pipe. Let $\mathbf{v}_\omega(\mathbf{x}, t)$ denote the velocity induced by the vorticity $\boldsymbol{\omega}(\mathbf{x}, t)$ as if the flow were incompressible. In the absence of solid boundaries $\mathbf{v}_\omega$ is given

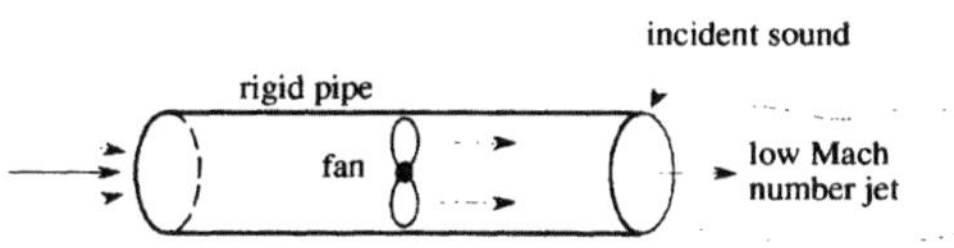

Fig. 18. Sound incident on a pipe with low Mach number exhaust flow.

by the Biot-Savart induction formula

$$\mathbf{v}_\omega(\mathbf{x},t) = \operatorname{curl}\int \frac{\boldsymbol{\omega}(\mathbf{y},t)d^3\mathbf{y}}{4\pi|\mathbf{x}-\mathbf{y}|}.$$

This must be augmented by the velocity field of an image system of the vorticity in the solid boundaries; this can always be done to ensure that $\operatorname{div}\mathbf{v}_\omega = 0$, and that the normal component of $\mathbf{v}_\omega$ vanishes on the instantaneous boundary positions. The potential component of the motion produced by the fan is expressed in terms of a velocity potential ψ which satisfies Laplace's equation $\nabla^2\psi = 0$, with normal velocity $\partial\psi/\partial x_n$ on a bounding surface equal to that of the surface. The total fluid velocity $\mathbf{v}$ may then be cast in the form

$$\mathbf{v} = \mathbf{v}_\omega + \nabla\psi + \nabla\varphi, \tag{106}$$

where the residual velocity $\nabla\varphi$ accounts for the influence of compressibility. This includes contributions from incident and diffracted sound waves, and from compressible motions caused by the boundary motions and the production of aerodynamic sound.

Now $\rho - \rho_o \sim \mathrm{O}(M^2)\rho_o$, so that when the Mach number is small the total kinetic energy of the fluid is approximately

$$\begin{aligned} E_T &= \frac{1}{2}\rho_o\int\Big(\mathbf{v}_\omega + \nabla(\psi+\varphi)\Big)^2 d^3\mathbf{x} \\ &\equiv \frac{1}{2}\rho_o\int \mathbf{v}_\omega^2\, d^3\mathbf{x} + \frac{1}{2}\rho_o\int[\nabla(\psi+\varphi)]^2 d^3\mathbf{x} \end{aligned} \tag{107}$$

where the integrations are over the fluid. The split into the final two terms follows from the divergence theorem and the condition that $\mathbf{v}_\omega\cdot\mathbf{n} = 0$ on the solid boundaries.

The kinetic energy is therefore partitioned between the kinetic energy $E = \frac{1}{2}\rho_o\int \mathbf{v}_\omega^2\, d^3\mathbf{x}$ of the vortical motion, and that of the acoustic field and the incompressible, potential flow caused by boundary motion. The latter is localized to the neighborhood of the fan, and represents energy temporarily stored in the fluid, in that ψ vanishes when the boundary motion ceases. However, continuous operation of the fan generates vortical kinetic energy in the jet exhausting from the pipe which would ultimately become unbounded in the absence of dissipative processes, whereas the energy associated with ψ is always finite and localized.

The motion of the fan and the interaction of sound with the boundaries generates vorticity which produces changes in the vortical kinetic energy E. To calculate this change the momentum equation for homentropic flow is written

$$\rho_o\frac{\partial\mathbf{v}_\omega}{\partial t} + \nabla\left(\rho_o\left[B + \frac{\partial\psi}{\partial t} + \frac{\partial\varphi}{\partial t}\right]\right) = -\rho_o\boldsymbol{\omega}\wedge\mathbf{v} - \eta\operatorname{curl}\boldsymbol{\omega}, \tag{108}$$

where (for $M^2 \ll 1$) $B = p/\rho_o + \frac{1}{2}v^2$ is the total enthalpy, and dissipation terms proportional to div$\mathbf{v}$ (including that associated with the bulk viscosity) are neglected, since viscous effects are important only near boundaries where the motion may be regarded as incompressible. An energy equation is formed by taking the scalar product of (108) with $\mathbf{v}_\omega$. Noting that div $\mathbf{v}_\omega = 0$, we find

$$\frac{\partial}{\partial t}\left(\frac{1}{2}\rho_o \mathbf{v}_\omega^2\right) + \text{div}\left(\rho_o \mathbf{v}_\omega\left[B + \frac{\partial \psi}{\partial t} + \frac{\partial \varphi}{\partial t}\right]\right) = \rho_o \boldsymbol{\omega}\wedge\mathbf{v}\cdot(\nabla\psi + \nabla\varphi) - \eta \mathbf{v}_\omega \cdot \text{curl}\boldsymbol{\omega}. \tag{109}$$

The argument of the divergence is the flux of vortical kinetic energy when convection by sound and unsteady surface motions is excluded. In irrotational regions it reduces to the energy flux vector $-\rho_o \mathbf{v}_\omega \partial\Phi/\partial t$, where Φ is the velocity potential less the contributions from surface motions and sound. The first term on the right is a kinetic energy *source* for the vortex field arising from the coupling of the vorticity with the surface motions and the sound.

By integrating over the fluid, and using the relation

$$\int \partial f/\partial t \, d^3\mathbf{x} = \partial/\partial t \int f \, d^3\mathbf{x} + \oint_{\mathrm{S}} f \nabla\psi \cdot d\mathbf{S}$$

where S is a solid boundary with surface element $d\mathbf{S} = \mathbf{n}dS$ directed *into* the fluid, we obtain

$$\frac{\partial E}{\partial t} + 2\eta e_{ij} e_{ij} = \rho_o \int \boldsymbol{\omega}\wedge\mathbf{v}\cdot(\nabla\psi + \nabla\varphi) d^3\mathbf{x} - \frac{1}{2}\rho_o \oint_{\mathrm{S}} \mathbf{v}_\omega^2 \nabla\psi \cdot d\mathbf{S}$$
$$+\eta \oint_{\mathrm{S}} \Big(\boldsymbol{\omega}\wedge\mathbf{v}_\omega - 2(\mathbf{u}_{\mathrm{S}}\cdot\nabla)\mathbf{v}\Big)\cdot d\mathbf{S}, \tag{110}$$

where $\mathbf{u}_{\mathrm{S}}(\mathbf{x}, t)$ is the velocity of the moving boundary, and e_{ij} is the rate of strain tensor $\frac{1}{2}(\partial v_i/\partial x_j + \partial v_j/\partial x_i)$. The left hand side is the sum of the power $\Pi_\omega = \partial E/\partial t$ entering the vortical flow and that dissipated by frictional forces. At high Reynolds numbers the viscous dissipation occurs mainly in regions close to boundaries where the length scales of the motion are very small, and in very small scales of the motion in the body of the fluid. In typical applications the energy supplied by the source (first term on the right) therefore passes to the vortex motions, and is not immediately dissipated by viscous action. The final surface integral is the energy flux associated with the viscous drag; this is usually small compared to the other source terms when the Reynolds number is large, and is frequently neglected.

When the boundaries are stationary (fan not operating in Fig. 18), $\psi = 0$ and vorticity production occurs solely as a result of incident sound waves. It follows that when surface losses are neglected, $\Pi_\omega \sim \mathrm{O}(\epsilon^3)$, where ϵ is a measure of the acoustic amplitude. When, however, boundary motions are sufficient to create a substantial mean flow (the jet in Fig. 18), both $\mathbf{v}$ and the vorticity $\boldsymbol{\omega}$ can have large mean components that are independent of the incident sound. In this case Π_ω is formally $\mathrm{O}(\epsilon^2)$, which suggests that the mean flow can increase considerably the rate of production of vortical energy at the expense of the sound. In many applications the dominant interaction of the sound (or that component of the sound under investigation) with the vorticity occurs in regions where $\nabla\psi$ is negligible (for example, near the nozzle exit in Fig. 18), because, according to potential flow theory, ψ decays exponentially fast axial distance from the fan). The energy supplied to the vortical flow near the nozzle exit must then be extracted directly from the sound in accordance with the

Acoustic dissipation formula:

$$\Pi_\omega \approx \rho_o \int \boldsymbol{\omega} \wedge \mathbf{v} \cdot \mathbf{u} d^3\mathbf{x} \tag{111}$$

which determines the *rate of dissipation of acoustic energy*, where the acoustic particle velocity $\nabla\varphi$ is now denoted by $\mathbf{u}$. Π_ω can be negative in systems where self-sustaining oscillations are maintained by 'acoustic feedback', in which the phase of vorticity production enables a steady transfer of energy to the oscillations from a mean flow.

8.2 Interactions at a Trailing Edge

The diffraction problem of Fig. 19a is easily treated analytically. A plane sound wave is incident on a rigid plate occupying $x_1 < 0,\ x_2 = 0$ in the presence of uniform flow at speed U in the x_1-direction. Separation induced by the sound at the edge generates a wake, which in the linearized approximation is a vortex sheet.

Let the incident plane wave be

$$p(\mathbf{x}, \omega) = p_{\mathrm{I}} \mathrm{e}^{i(n_1 x_1 - n_2 x_2)}, \quad p_{\mathrm{I}} = \text{constant}, \quad \omega > 0. \tag{112}$$

The wake is modelled as a vortex sheet of vorticity $\boldsymbol{\omega} = (v_- - v_+)\delta(x_2)\mathbf{i}_3$ ($x_1 > 0$), where $v_\pm$ are the x_1-components of velocity just above and below the sheet. When $M \ll 1$ we find

$$\boldsymbol{\omega}(x_1, \omega) = \frac{-2p_{\mathrm{I}}\sqrt{2M}}{\rho_o U} \cos(\theta/2) \mathrm{e}^{i\kappa x_1} \delta(x_2)\mathbf{i}_3, \quad x_1 > 0, \tag{113}$$

where $\kappa = \omega/U$ is the *hydrodynamic* wavenumber.

Correct to second order in the acoustic amplitude p_{I}, we can set $\mathbf{v} = U\mathbf{i}_1$ in (111) to express the dissipated acoustic power per unit length of trailing edge in the form

$$\Pi_\omega = \rho_o U \int_0^\infty \langle \Omega(x_1, t) u_2(x_1, 0, t) \rangle dx_1, \tag{114}$$

where Ω is the real part of the coefficient of $\delta(x_2)\mathbf{i}_3$ in (113) after restoration of the time factor $\mathrm{e}^{-i\omega t}$, and the angle brackets denote a time average.

(a)
(b)

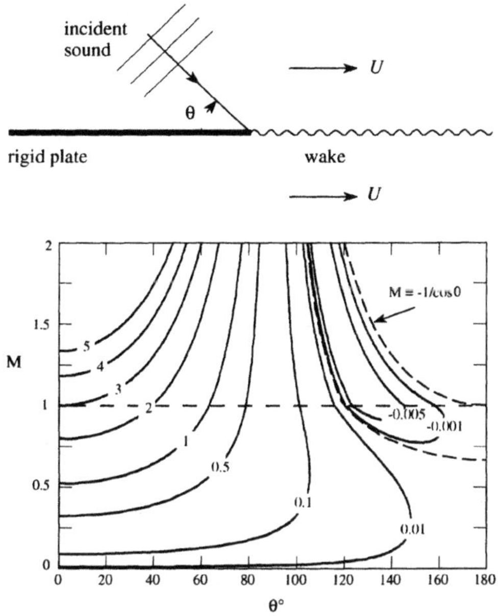

Fig. 19. (a) Sound interacting with a trailing edge.(b) Contours of constant values of $\rho_o \omega \Pi_\omega / 2|p_\text{I}|^2$. Sound incident on a pipe with low Mach number exhaust flow.

To evaluate this integral one must determine the acoustic particle velocity $\mathbf{u} = \nabla \varphi$. The component $u_2(x_1, x_2, t)$ is made up of a diffracted velocity plus a component from the incident sound. It is easily confirmed, however, that the latter makes no contribution to (114), i.e. that the dissipation is controlled by the coupling between the scattered acoustic field and the wake. Moreover, as $M \to 0$ it is only the *singular* part of the acoustic particle velocity u_2 very close to the edge that is important, because the length scale of the vorticity is then infinitesimal on a scale of the acoustic wavelength and the integral is dominated by the contribution from near the edge. In fact, for $M \ll 1$

$$u_2(x_1, 0, t) \approx \mathrm{Re}\left(\frac{-p_\text{I} \cos(\theta/2)}{\rho_o \omega} \sqrt{\frac{2\kappa_o}{\pi x_1}} \mathrm{e}^{i(\kappa x_1 - \omega t)} \right), \quad x_1 > 0, \quad \kappa_o x_1 \ll 1, \tag{115}$$

which exhibits the characteristic potential flow singularity $\sim 1/\sqrt{x_1}$ at the edge. Substituting into (114) we find

$$\Pi_\omega \approx \frac{2|p_\text{I}|^2 M \cos^2(\theta/2)}{\rho_o \omega}, \quad M \ll 1. \tag{116}$$

This is positive definite for all angles of incidence (defined by the angle between the normal to the *phase front* and the x_1-direction), showing that acoustic energy is expended in the formation of the wake, and that the dissipation is greatest for sound waves incident from upstream at θ close to $0°$. At higher Mach numbers (116) becomes

$$\begin{aligned} \Pi_\omega &= \frac{2|p_{\mathrm{I}}|^2}{\rho_o\omega} M\cos^2(\theta/2)(1+M\cos\theta)\left(1+M\cos\theta-\frac{1}{2}M\right), \quad M \le 1, \\ &= \frac{|p_{\mathrm{I}}|^2}{2\rho_o\omega}(1+M\cos\theta)^2(1+2M\cos\theta), \quad M \ge 1. \end{aligned} \tag{117}$$

In the first of these expressions ($M \le 1$), the term $\frac{1}{2}M$ makes a *negative* contribution to the dissipation. It corresponds to the contribution to the acoustic particle velocity $u_2(x_1,0,t)$ in (114) produced by the interaction of the wake with the edge of the plate, i.e. to the *generation of trailing edge noise by the wake.* Consequently, when M exceeds about $\frac{2}{3}$ the dissipation can be negative, resulting in a net transfer of energy from the mean flow to the acoustic field. A more detailed overall picture of the dependence of Π_ω on M and angle of incidence θ is shown in Fig. 19b, which depicts contours of constant $\rho_o\omega\Pi_\omega/2|p_{\mathrm{I}}|^2$. Acoustic energy is dissipated except when (θ, M) lies within the region bounded by the broken curves, on which $\Pi_\omega = 0$. When $M > 1$ the incident wave cannot lie within the top right corner of the figure, bounded by the broken curve on which $\theta = \arccos(-1/M)$; θ cannot exceed this angle because of the convection of sound by the flow.

Note: When $\mathbf{v}$ in (111) can be replaced by the mean velocity and the vorticity assumed to convect at the mean stream velocity, the dissipation formula is valid for *arbitrary* values of the mean flow Mach number, including supersonic! This is because vorticity swept out of the region of interaction with the edge convects at precisely the velocity of the mean stream, and produces neither pressure nor density fluctuations. The dissipated power must therefore be contained in the kinetic energy of a fluid of constant density ρ_o, the rate of increase of which must just equal the rate at which acoustic energy is dissipated.

8.3 Leading Edge Interactions

According to linear theory there can be no exchange of energy between impinging sound and the mean flow at the leading edge of an airfoil in nominally steady potential mean flow (Fig. 20). This is because vorticity generated at the leading edge remains on the airfoil, and its influence is cancelled by that of equal and opposite image vorticity in the rigid surface. In practice the leading edge motion excites *Tollmien-Schlichting* waves in the surface boundary layers or causes flow separation and the appearance of discrete boundary layer structures. An estimate of the effect of such interactions on energy transfers between the flow and the sound can be made by assuming the boundary layers remain thin, and that the *displacement thickness velocity* fluctuation v_δ

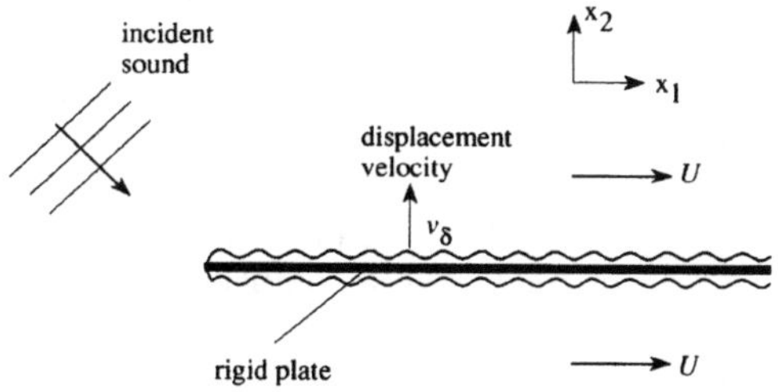

Fig. 20. Interaction of sound with a leading edge.

at the outer edges of the boundary layers is known in terms of the incident sound.

Consider the case in which the mean flow Mach number $M \ll 1$, and take the incident sound in the form (112). The plate occupies the semi-infinite domain $x_1 > 0$, $x_2 = 0$, and the unsteady motion outside the boundary layers is described by the velocity potential $\Phi + \varphi_{\rm I}$, where

$$\varphi_{\rm I}(\mathbf{x}, \omega) = (-ip_{\rm I}/\rho_o\omega)\mathrm{e}^{i(n_1x_1 - n_2x_2)}$$

is the potential of the incident sound. If the boundary layer is thin the displacement boundary condition

$$\partial(\Phi + \varphi_{\rm I})/\partial x_2 = v_\delta^\pm, \quad x_2 = \pm 0, \quad x_1 > 0, \tag{118}$$

can be applied at the upper and lower surfaces of the plate, where $v_\delta^\pm$ is the displacement velocity (in the x_2-direction) for the upper and lower boundary layers.

The simplest hypothesis for $v_\delta^\pm$ is that

$$v_\delta^\pm(x_1, \omega) = v_\pm \mathrm{e}^{i\kappa_c x_1}, \tag{119}$$

where $v_\pm$ are constants and $\kappa_c = \omega/U_c$ is the wavenumber of the 'displacement thickness waves' generated in the boundary layer at the leading edge, and U_c is their phase speed. The internal dynamics of the boundary layer must determine the precise dependence of κ_c on ω.

The values of $v_\pm$ are found by imposing a *leading edge Kutta condition.* The displacement thickness waves cause fluid to be "pumped" around the edge of the plate. By re-writing (118) in the form

$$v_2(x_1, \pm 0, \omega) = \left[\frac{1}{2}(v_+ + v_-) \pm \frac{1}{2}(v_+ - v_-)\right] \mathrm{e}^{i\kappa_c x_1}, \quad x_1 > 0,$$

it can be seen that the pumping action is controlled by the first terms in the square brackets, which produce an anti-symmetric motion with respect to the plane $x_2 = 0$. The normal velocities determined by the second terms in the brackets cannot induce flow about the edge, but generate an exterior potential flow that is finite everywhere. It follows that $v_+ - v_-$ cannot be determined from a leading-edge Kutta condition, and must therefore vanish in the absence of other excitation mechanisms. This is equivalent to setting $v_+ = v_- = v_o$, so that the boundary layer motions on opposite sides of the plate are 180° out of phase, which means that the boundary layer cannot be a net source or sink of fluid. These conditions permit Φ to be uniquely determined in terms of v_o.

For a thin boundary layer the dissipation integral (111) can be expressed in terms of the displacement thickness velocity v_o and the acoustic particle velocity component u_1 parallel to the plate. To see this, note first that $\boldsymbol{\omega}$ vanishes except within the boundary layers on the plate, where the motion may be regarded as incompressible when $M \ll 1$, and where the divergence of the momentum equation for isentropic flow reduces to $\nabla^2 B = -\mathrm{div}(\boldsymbol{\omega}\wedge\mathbf{v})$. Integrate this equation across the boundary layer and suppose the boundary layer thickness is small compared to the length scale of variation of B parallel to the plate. Then, since $\partial B/\partial x_2 = -\partial v_2/\partial t$ just outside the boundary layer, $v_\delta(x_1,\omega) = v_o e^{i\kappa_c x_1}$ where

$$\frac{\partial v_\delta}{\partial t} = \frac{\partial}{\partial x_1}\int_0^\infty (\boldsymbol{\omega}\wedge\mathbf{v})_1 dx_2. \tag{120}$$

The normal component of the acoustic particle velocity vanishes on the plate; therefore, for a thin boundary layer only the tangential velocity u_1 makes a significant contribution to (111), the behavior of which near the leading edge is given by

$$u_1(x_1,\pm 0,\omega) \approx \pm(p_{\mathrm{I}}/\rho_o\omega)\sqrt{2i\kappa_o/\pi x_1}\sin(\theta/2), \quad x_1 > 0, \tag{121}$$

when $M \ll 1$. Then (after restoring the time factor $e^{-i\omega t}$ and taking real parts, etc) (111) yields

$$\Pi_\omega \approx 2\rho_o \int_0^\infty u_1(x_1,0,t)\int_0^\infty (\boldsymbol{\omega}\wedge\mathbf{v})_1 dx_2 dx_1,$$

where the factor of 2 accounts for equal contributions from opposite sides of the plate. However, (120) implies $\int_0^\infty (\boldsymbol{\omega}\wedge\mathbf{v})_1(\mathbf{x},\omega)dx_2 \approx -(\omega v_o/\kappa_c)e^{i\kappa_c x_1}$, and application of the Kutta condition at the leading edge supplies

$$v_o = -(p_{\mathrm{I}}/\rho_o U)\sqrt{2M}\sin(\theta/2),$$

so that

$$\Pi_\omega \approx \frac{-2|p_{\mathrm{I}}|^2 M_c \sin^2(\theta/2)}{\rho_o\omega}, \quad M_c = U_c/c_o \ll 1. \tag{122}$$

Thus, in contrast to the trailing edge problem of §8.2, a leading edge interaction at low Mach numbers results in the transfer of mean flow energy to the acoustic energy. It is a maximum when the impinging sound wave is incident from a downstream direction ($\theta = \pm\pi$); the wave produces a large perturbation of the flow near the edge, which is compensated by the generation of displacement thickness waves of proportionate amplitude. Experiments with laminar boundary layers appear to confirm this direct proportionality between the amplitudes of the incident sound and boundary layer Tollmien-Schlichting waves.

It is not hard to see from the above discussion and §8.2 that the fundamental difference between leading and trailing edges is independent of the structure of the incident sound field. In fact, *any* mechanism that produces a reciprocating flow about an edge will bring about a net transfer of energy to or from the mean flow. This conclusion is important in the study of mean-flow induced resonant oscillations.

8.4 Application to Resonant Oscillations

Vortex shedding and interactions with structures are sources of sound, and can be responsible for the excitation of cavity resonances in ducts, rocket motor combustion chambers, heat exchangers, etc. A resonance is usually strongly excited only at certain mean flow velocities because, as we have seen, sound can also be absorbed during vorticity production, and it is only when certain phase relations are satisfied that production mechanisms become dominant.

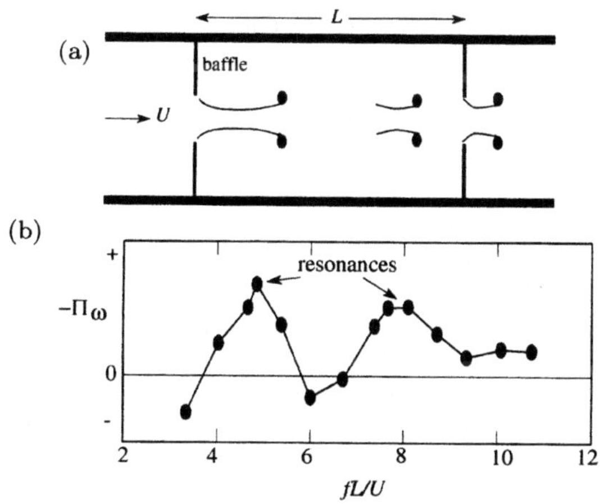

Fig. 21. (a) Baffled section of a duct with flow. (b) Mean power flux (in arbitrary units); adapted from (Hourigan *et al.* 1990).

The dissipation integral (111) can often be used to predict the flow conditions for resonance. Fig. 21a depicts a short baffled section of length L of a long duct of square cross-section conveying air at low Mach number. In an experiment performed by Hourigan *et al.* (1990) a longitudinal acoustic mode (of fixed frequency and wavelength equal approximately to the whole length of the duct) was excited at Strouhal numbers $fL/U \approx 4.7$, 7.7 and 11.0, where U is the mean velocity just upstream of the baffled section. For these conditions, respectively one, two and three vortex pairs were observed within the baffled section, generated by shedding from the upstream baffles. By measuring the amplitude of the acoustic pressure fluctuations at the resonance frequency for a range of mean flow velocities, the following four-step numerical procedure can be used to determine whether the acoustic excitation can be attributed to the unsteady vortex flow in the baffled section of the duct:

(1) The acoustic field is determined by solving the linearized acoustic equations at the observed resonance frequency f. The amplitude of the sound field is then set equal to the measured value.

(2) The calculated acoustic field is incorporated into an incompressible numerical solution of the Navier-Stokes equation (via the no-slip boundary condition and the contribution of the acoustic particle velocity to the convection of vorticity) to determine the velocity and vorticity distributions in the duct. This procedure locks the vortex shedding rate onto the acoustic frequency.

(3) The average power $\langle -\Pi_\omega \rangle$ fed into the acoustic field is calculated using (111).

(4) Steps (1) - (3) are repeated for a range of flow velocities U, and $\langle -\Pi_\omega \rangle$ is plotted as a function of fL/U. Those values of the Strouhal number where $\langle -\Pi_\omega \rangle$ is large and positive are identified with flow conditions at which resonant excitation is possible.

Fig. 21b shows calculated values of $\langle -\Pi_\omega \rangle$ plotted against fL/U. The two indicated resonances occur at Strouhal numbers close to the observed values of 4.7 and 7.7, but the higher resonance at $fL/U \approx 11$ is not well resolved. The principal contribution to the power flux integral (111) occurs when the large scale vortex structures (idealized as vortex pairs in Fig. 21a) encounter the downstream baffles. At resonance the principal components of $\boldsymbol{\omega} \wedge \mathbf{v}$ tend to be locally parallel to the acoustic particle velocity $\mathbf{u}$ and be in just the right phase to maximize the energy flux and ensure the continued maintenance of the acoustic field. Note that, unlike resonance phenomena of the edgetone type, the resonance is not maintained by a feedback loop in which a pressure pulse generated at the downstream baffle triggers the shedding of a new vortex. In the present case the feedback is via the resonance: the motion in the acoustic field that induces vortex shedding from the upstream baffles is generated over many cycles, and has been reflected back-and-forth many times between the ends of the duct.

9 Perforated Screens

The absorption of acoustic energy by vorticity production can be exploited to enhance the effectiveness of acoustic liners and perforated screens: 'jetting' in surface apertures increases the rate of energy conversion, and the increase can be even larger in the presence of mean flow, because vorticity generated by sound at aperture edges is convected away and its energy is permanently lost to the sound. Most practical screens used to 'absorb' sound are perforated with small (usually circular) apertures. The apertures may still be large on the scale of the viscous motions, however, in which case the influence of viscosity is insignificant except near the edges. This is important if enhanced levels of dissipation are to be achieved by mean flow convection. Screens may be of the 'grazing' or 'bias' flow type and are used in mean flow environments, such as in heat exchanger cavities to damp harmful acoustic oscillations.

A *cavity-backed* screen of the type illustrated in Fig. 22a can greatly increase the efficiency with which sound is absorbed by a bias flow screen. The backing may be formed by the wall of a wind tunnel, with fluid either injected or withdrawn from the main stream at the perforated surface, which behaves as an acoustic liner whose bias flow velocity can be adjusted to yield maximal attenuations at desired frequencies. Predictions (see below) of the attenuation are in good agreement with experiments conducted in air (Fig. 22b), and have been used to design practical devices for absorbing intense sound in, for example, jet engine combustion chambers.

9.1 Rayleigh Conductivity

Consider the unsteady flow through an acoustically compact circular aperture of radius R in a rigid wall of thickness h produced by a uniform, time-dependent pressure differential $[p] = p_+(t) - p_-(t)$. The 'upper' and 'lower' faces of the wall are assumed to coincide respectively with $x_2 = \pm\frac{1}{2}h$, and p_+, p_- are the pressures above and below the wall. The motion near the aperture is regarded as incompressible, with volume flux $Q(t)$ through the aperture (in the x_2-direction), and fluctuating quantities are proportional to $\mathrm{e}^{-i\omega t}$. The Rayleigh conductivity K_R of the aperture is defined by

$$K_R = i\omega\rho_o Q/(p_+ - p_-). \tag{123}$$

When there is no mean flow on either side of the wall this is equivalent to

$$K_R = Q/(\varphi_+ - \varphi_-). \tag{124}$$

where $\varphi_\pm$ are the *uniform* velocity potentials above and below the wall, which satisfy $p\pm = -\rho_o\partial\varphi_\pm/\partial t$. Because of the motion the pressure and velocity potential cannot actually be constant close to the aperture; $p_\pm$ and $\varphi_\pm$ are the limiting values of these quantities at large distances ($\gg R$) from

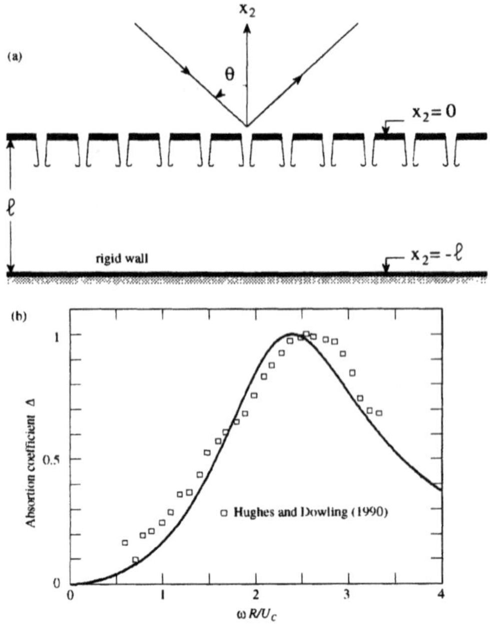

Fig. 22. Sound absorption by a cavity backed, bias flow screen: $R = 0.15$ cm, $\ell = 2$ cm, $U_c = 4.76$ m/sec, $\alpha = 0.024$. Experimental data adapted from Hughes & Dowling (1990).

the aperture. Conductivity has the dimensions of length, and (124) is the analogue of *Ohm's law*, which determines the conductivity of an electrical circuit element as the ratio of the current to the potential difference.

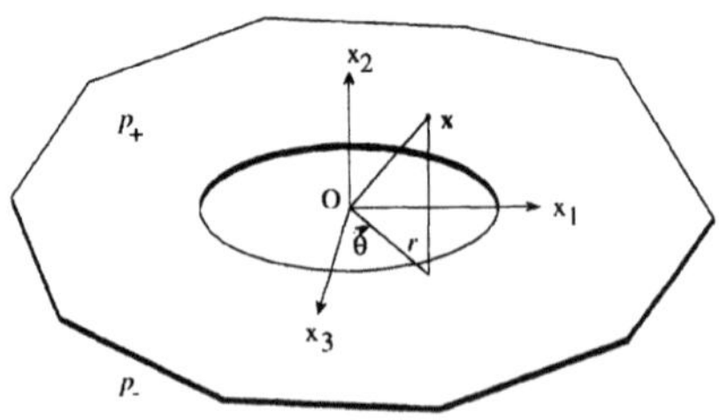

Fig. 23. Circular aperture in a thin wall.

Let us determine K_R for an ideal (inviscid) fluid when $h \to 0$, in the absence of mean flow (Fig. 23). The flow produced by the uniform potential difference $\varphi_+ - \varphi_-$ can be described by a velocity potential Φ; the total potential is then $\varphi = \varphi_\pm + \Phi$, respectively above and below the wall. Φ is a solution of the Laplace equation for incompressible motion, and $\Phi \sim \mp Q/2\pi|x|$ as $x_2 \to \pm\infty$. Symmetry demands that the motion in the aperture must be in the normal direction, so that Φ must be constant there. The general solution (50) applied to incompressible flow (with $f \equiv 0$) can be used to express Φ on either side of the wall in terms of the velocity $v_2(\mathbf{x})$ at the wall and Green's function

$$G(\mathbf{x},\mathbf{y},\omega) = \frac{-1}{4\pi|\mathbf{x}-\mathbf{y}|} + \frac{-1}{4\pi|\mathbf{x}-\mathbf{y}'|}, \quad y' = (y_1, -y_2, y_3), \tag{125}$$

which has vanishing normal derivatives $\partial G/\partial x_2$, $\partial G/\partial y_2$ respectively on x_2 and $y_2 = 0$, i.e.,

$$\varphi(\mathbf{x}) = \varphi_\pm - \frac{\operatorname{sgn}(x_2)}{2\pi}\int_{-\infty}^{\infty}\frac{v_2(y_1,\pm 0,y_3)}{|\mathbf{x}-\mathbf{y}|}\,dy_1dy_3, \quad (y_2 = 0), \tag{126}$$

where the integration is over the upper/lower surfaces $y_2 = \pm 0$ respectively for $x_2 \gtrless 0$. Since $v_2 \equiv 0$ on the rigid portions of the wall, the integration may be restricted to the region S occupied by the aperture. The condition that φ (and the pressure) should be continuous across the aperture now supplies the following integral equation

$$\oint_{\mathrm{S}}\frac{v_2(y_1,0,y_3)}{|\mathbf{x}-\mathbf{y}|}\,dy_1dy_3 = \pi(\varphi_+ - \varphi_-), \quad \text{for } x_2 = y_2 = 0, \quad r = \sqrt{x_1^2 + x_3^2} < R \tag{127}$$

for the aperture velocity v_2, with solution

$$v_2(r) = \frac{(\varphi_+ - \varphi_-)}{\pi\sqrt{R^2 - r^2}}, \quad r < R.$$

Then $Q = \oint_{\mathrm{S}} v_2(x_1,0,x_3)dx_1dx_3 = 2R(\varphi_+ - \varphi_-)$, and the definition (124) yields the **conductivity of a circular aperture in a thin wall:**

$$K_R = 2R. \tag{128}$$

9.2 The Bias Flow Aperture

A steady pressure differential maintained across a wall containing a small aperture produces a nominally steady bias flow. When the Reynolds number $UR/\nu \gg 1$ the motion is uninfluenced by viscosity except near the aperture edge, where the flow separates to form a jet. Vorticity generated at the edge by incident sound is swept away in the jet, and damping occurs by the transfer of acoustic energy to the kinetic energy of the jet, which implies that the Rayleigh conductivity K_R is now a *complex valued* function of frequency.

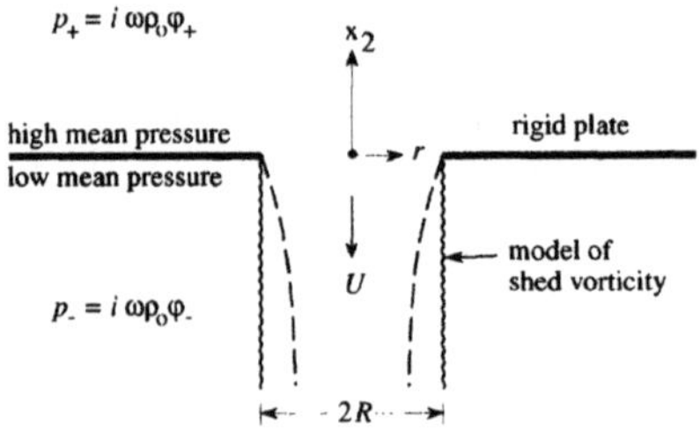

Fig. 24. Aperture with bias flow jet.

An approximate formula for K_R for a circular aperture with a bias flow can be derived using the model shown in Fig. 24. The mean flow is from the upper to the lower surfaces, and the jet velocity U is assumed to have infinitesimal Mach number. According to *free streamline theory*, for an ideal fluid the jet contracts to an asymptotic cross-section equal to about 0.62 times the aperture cross section, in the manner indicated by the dashed profile in the figure; when the flow is turbulent, however, turbulence diffusion will tend to produce an expanding flow. The fluctuations in the flow produced by an applied time harmonic, uniform pressure differential $p_+ - p_-$ may be assumed to be axisymmetric; the additional vorticity generated at the edge of the aperture may be imagined to consist of a succession of vortex rings with infinitesimal cores, which convect at a velocity U_c and form an axisymmetric vortex sheet within the mean shear layer of the jet. To calculate the conductivity the back reaction of these vortex rings on the aperture flow must be determined. This is easily done if the variation in the radius of a ring is neglected, so that the vortex sheet is cylindrical, and equal to the aperture radius R.

Except very near the edges, the motion through the aperture may be regarded as inviscid, incompressible and homentropic, for which case the divergence of the momentum equation reduces to $\nabla^2 B = -\text{div}(\boldsymbol{\omega} \wedge \mathbf{v})$. The vorticity $\boldsymbol{\omega}$ consists of a mean part associated with the undisturbed jet and a fluctuating part produced by the unsteady applied pressure. The dominant unsteady component $(\boldsymbol{\omega} \wedge \mathbf{v})'$ of $(\boldsymbol{\omega} \wedge \mathbf{v})$ is taken to correspond to the elementary vortex rings of radius R translating with velocity U_c parallel to the mean jet, i.e.

$$(\boldsymbol{\omega} \wedge \mathbf{v})'(\mathbf{x}, \omega) = -\gamma_o U_c \delta(r - R)\mathbf{i}_r e^{-i\kappa x_2}, \quad x_2 < 0, \quad r = \sqrt{x_1^2 + x_3^2}, \quad (129)$$

where $\gamma_o \equiv \gamma_o(\omega)$ is the circulation of the perturbation vorticity per unit length of the jet, $\kappa = \omega/U_c$ is the hydrodynamic wavenumber appropriate for convection in the mean flow direction at velocity U_c, and $\mathbf{i}_r$ is a unit vector radially outwards from the jet axis.

For points not on the vortex sheet, the unsteady motion can be described by an axisymmetric velocity potential $\varphi(x_2, r)$. On the side $x_2 < 0$ of the wall, containing the jet, we set

$$\varphi = \varphi_- + \varphi_s + \varphi_w,$$

where $\varphi_- \equiv -ip_-/\rho_o\omega$, $\varphi_w(x_2, r)$ is the velocity potential attributable to the wake vorticity when the aperture is ignored (so that $\partial\varphi_w/\partial x_2 = 0$ at $x_2 = -0$, including the region occupied by the aperture), and $\varphi_s(x_2, r)$ accounts for the influence of the aperture. Similarly, in $x_2 > 0$ we write $\varphi = \varphi_+ + \varphi_s$, where $\varphi_+ \equiv -ip_+/\rho_o\omega$. The potential φ_s is given by the integrated term in (126), in terms of the perturbation normal velocity v_2 in the plane of the aperture. The condition that φ should be continuous across the plane of the aperture then yields the integral equation

$$\oint_S \frac{v_2(y_1, 0, y_3)}{|\mathbf{x} - \mathbf{y}|}\, dy_1 dy_3 = \pi(\varphi_+ - \varphi_-) + \pi\gamma_o R \int_0^\infty \frac{\lambda \mathrm{H}_1^{(1)}(\lambda R)\mathrm{J}_0(\lambda r)d\lambda}{\kappa^2 + \lambda^2},$$
$$r < R, \; x_2 = y_2 = 0, \qquad (130)$$

for $v_2(x_1, 0, x_3) \equiv v_2(0, r)$, where the integrated term on the right hand side is equal to $-\pi\varphi_w(0, r)$.

The solution of this equation is

$$v_2(r) = \frac{\varphi_+ - \varphi_-}{\pi\sqrt{R^2 - r^2}}$$
$$+ \frac{\gamma_o R}{\pi} \int_0^\infty \frac{z\mathrm{H}_1^{(1)}(z)}{z^2 + \kappa^2 R^2} \left(\frac{\cos z}{\sqrt{R^2 - r^2}} + \int_r^R \frac{z \sin(zt/R)\, dt}{\sqrt{t^2 - r^2}} \right) dz, \quad r < R.$$

The strength γ_o of the vortex sheet is obtained by requiring $v_2(0, r)$ to remain finite as $r \to R$. The integrated term in the brace brackets vanishes in this limit, and the remaining terms are finite provided

$$\gamma_o = -(\varphi_+ - \varphi_-) \Big/ R \int_0^\infty \frac{z\mathrm{H}_1^{(1)}(z) \cos z}{z^2 + \kappa^2 R^2}\, dz. \qquad (131)$$

The solution can now be used to express volume flux Q in terms of $\varphi_+ - \varphi_-$, and the definition (124) then gives

$$K_R = 2R(\Gamma_R - i\Delta_R) \qquad (132)$$

where

$$\Gamma_R - i\Delta_R = 1 - \int_0^\infty \frac{\mathrm{H}_1^{(1)}(z) \sin z}{z^2 + \kappa^2 R^2}\, dz \Big/ \int_0^\infty \frac{z\mathrm{H}_1^{(1)}(z) \cos z}{z^2 + \kappa^2 R^2}\, dz$$
$$= 1 + \frac{\frac{\pi}{2}\mathrm{I}_1(\kappa R)\mathrm{e}^{-\kappa R} - i\mathrm{K}_1(\kappa R)\sinh(\kappa R)}{\kappa R[\frac{\pi}{2}\mathrm{I}_1(\kappa R)\mathrm{e}^{-\kappa R} + i\mathrm{K}_1(\kappa R)\cosh(\kappa R)]}. \qquad (133)$$

Γ_R and Δ_R are both positive (for $\omega > 0$), and their dependence on the *Strouhal number* $\kappa R \equiv \omega R/U_c$ is illustrated in Fig. 25. The following asymptotic approximations are easily derived:

$$\begin{aligned} K_R &\approx 2R(1 - i/\kappa R), & \kappa R &\to \infty, \\ &\approx 2R\{\frac{1}{3}(\kappa R)^2 - \frac{i\pi}{4}\kappa R\}, & \kappa R &\to 0. \end{aligned} \tag{134}$$

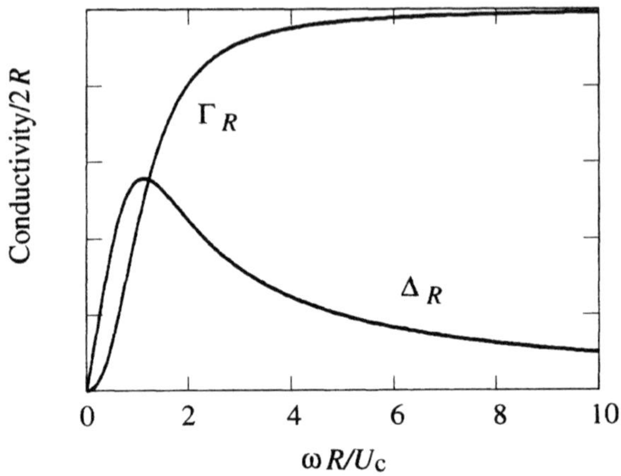

Fig. 25. Conductivity of a bias flow circular aperture.

Thus, at very high frequencies vorticity production has a negligible influence on the fluctuating flow. This is because, when the vorticity length scale $\sim U_c/\omega$ is small, the velocities induced by successive elementary vortex rings largely cancel except in the immediate vicinity of the edge of the aperture. At low Strouhal numbers, however, vorticity of one sign can stretch many aperture diameters downstream and produce a strong effect on the flow; in this case vortex shedding acts to block the aperture. The high frequency asymptotic formula is applicable for κR greater than about 5, but the low Strouhal number approximation is inapplicable unless $\kappa R \leq 0.04$, although the formula for Δ_R may be used for $\kappa R < 0.5$.

The convection velocity. If the bias flow jet emerging from the aperture is initially non-turbulent the vorticity convection velocity U_c may be taken to be approximately equal to the mean velocity in the plane of the aperture. This value is consistent with the approximation in the derivation of K_R, as may be seen from the following argument.

Let p_o denote the steady excess pressure driving the flow through the aperture, V the asymptotic jet velocity far downstream, and A_∞ the asymptotic jet cross-sectional area. Then $V = \sqrt{2p_o/\rho_o}$, and the volume flux $Q = -A_\infty V$ (minus because our convention takes $Q > 0$ in the positive x_2-direction of Fig. 24). The strength of the mean vortex sheet bounding the jet is $\gamma_o = -V$, since the pressure and jet velocity are constant just within the jet. The vortex strength may also be calculated by replacing $\varphi_+ - \varphi_-$ by $p_o/i\omega\rho_o$ in (131) and taking the limit $\omega \to 0$:

$$\gamma_o = -\left(\frac{p_o}{i\omega\rho_o R}\right) \Big/ \left[\frac{\pi}{2}\mathrm{I}_1(\kappa R)\mathrm{e}^{-\kappa R} + i\mathrm{K}_1(\kappa R)\cosh(\kappa R)\right] \to \frac{-p_o}{\rho_o U_c}.$$

Hence $U_c = p_o/\rho_o V = \frac{1}{2}V$. The value of A_∞ is determined by equating the volume flux $Q = -A_\infty V$ to the value given by (123) as $\omega \to 0$, i.e., using (134), to $-\frac{\pi}{2}R^2 p_o/\rho_o U_c$, which gives $A_\infty/\pi R^2 = \frac{1}{2}$. This is identical with the minimum theoretical value obtained by other means (experimentally observed values of the area contraction ratio are in the range 0.61 - 0.64). It follows that the mean jet velocity in the aperture is $\frac{1}{2}V \equiv U_c$.

Energy dissipation. Work is performed by the oscillating pressure load $[p] = p_+ - p_-$ in generating the vorticity. The vortex power Π_ω supplied in this way can be determined by calculating the mean rate $-\oint_S \langle p\mathbf{v}\rangle \cdot d\mathrm{S}$ at which work is performed on a large spherical control surface S of radius $|\mathbf{x}| \gg R$ centered on the aperture (the surface element $d\mathbf{S}$ being radially outwards). On S we can take, after restoring the time factor $\mathrm{e}^{-i\omega t}$,

$$p = \mathrm{Re}\, p_\pm, \quad \text{and} \quad \mathbf{v} \approx \pm\mathrm{Re}\left(\frac{iK_R\mathbf{x}[p]}{2\pi|\mathbf{x}|^3}\rho_o\omega\right), \quad x_2 \gtrless 0,$$

and deduce that

$$\Pi_\omega = \frac{-\mathrm{Im}\{K_R\}|[p]|^2}{2\rho_o\omega} \equiv \frac{R\Delta_R|[p]|^2}{\rho_o\omega}. \tag{135}$$

Thus $\Pi_\omega > 0$ since $\Delta_R > 0$.

9.3 Reflection and Transmission of Sound by a Perforated Screen

The reflection and transmission coefficients for long wavelength sound incident on a homogeneously perforated screen can be expressed in terms of the conductivities of the component apertures, provided the local motion through an aperture may be assumed to be incompressible and identical in structure to the flow in an isolated aperture. The latter condition is fulfilled when the distance between neighboring apertures is much larger than the aperture diameter.

Let the screen occupy $x_2 = 0$ and suppose that there are $\mathcal{N}$ apertures per unit area with equal conductivities K_R. Consider a plane wave (112) incident from 'above' (Fig. 26), with acoustic wavelength very much larger

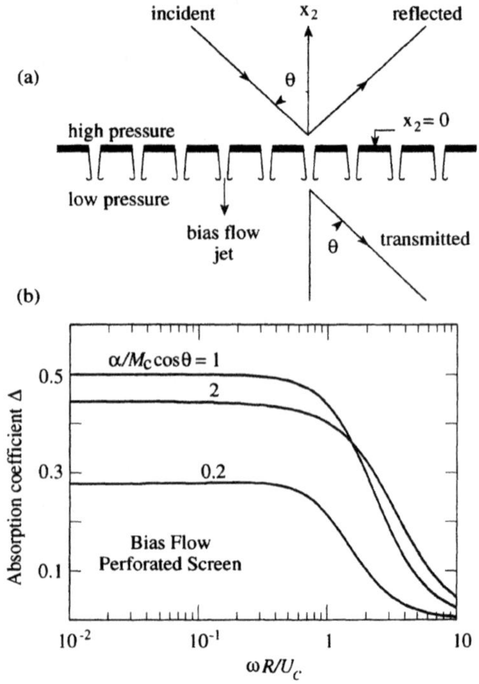

Fig. 26. Sound absorption by a bias flow screen. (a) geometry, (b) absoption coefficient as a function of the Strouhal number.

than the shortest distance d between neighboring apertures. Reflected and transmitted waves are generated such that at distances $|x_2| \gg d$ from the screen the pressure can be written

$$\begin{aligned} p(\mathbf{x}, \omega) &= p_{\mathrm{I}} \left(\mathrm{e}^{-n_2 x_2} + \mathcal{R} \mathrm{e}^{i n_2 x_2} \right) \mathrm{e}^{i n_1 x_1}, \quad x_2 \gg d, \\ &= p_{\mathrm{I}} \mathcal{T} \mathrm{e}^{i(n_1 x_1 - n_2 x_2)}, \quad x_2 \ll -d. \end{aligned} \tag{136}$$

Because the wavelength is large compared to the aperture size and spacing, the velocity potentials $\varphi_\pm$ of the *acoustic field* above and below the wall may be regarded as constant over an aperture, in terms of which the volume flux Q through an aperture is given by equation (124). The total flux per unit area is $\mathcal{N} Q \equiv \mathcal{N} K_R(\varphi_+ - \varphi_-)$. Continuity requires $\partial\varphi_+/\partial x_2 = \partial\varphi_-/\partial x_2 = \mathcal{N} K_R(\varphi_+ - \varphi_-)$ at the screen, which is the same as

$$\partial p_+/\partial x_2 = \partial p_-/\partial x_2 = \mathcal{N} K_R(p_+ - p_-), \quad x_2 = 0, \tag{137}$$

where $p_\pm$ denotes the acoustic pressure above and below the screen, given by (136). The quantity $\mathcal{N} K_R$ is the effective *compliance* of the perforated screen. By substituting from (136) into these equations, it now follows that

$$\mathcal{R} = \frac{\kappa_o \cos\Theta}{\kappa_o \cos\Theta + 2i\mathcal{N}K_R}, \quad \mathcal{T} = \frac{2i\mathcal{N}K_R}{\kappa_o \cos\Theta + 2i\mathcal{N}K_R}, \tag{138}$$

where Θ is the angle of incidence shown in the figure (such that $n_2 = \kappa_o \cos\Theta$).

The absorption coefficient Δ is the fractional acoustic power absorbed by the screen:

$$\Delta \equiv 1 - |\mathcal{R}|^2 - |\mathcal{T}|^2 = \frac{-4\mathcal{N}\kappa_o \cos\Theta \operatorname{Im} K_R}{|\kappa_o \cos\Theta + 2i\mathcal{N}K_R|^2}. \tag{139}$$

When the aperture motion is *ideal*, the conductivity K_R is real, and no energy is dissipated at the screen ($\Delta = 0$).

The bias flow screen: The aperture conductivity determined by (133) is complex, and the absorption coefficient is

$$\Delta = \frac{(8\alpha/\pi M_c)\Delta_R \kappa R \cos\Theta}{|\kappa R \cos\Theta + (4\alpha/\pi M_c)(\Delta_R + i\Gamma_R)|^2} \tag{140}$$

where $\alpha = \mathcal{N}\pi R^2$ is the fractional open area of the screen and $M_c = U_c/c_o$. The dependence of Δ on Strouhal number $\omega R/U_c$ is shown in Fig. 26b for different values of $\alpha/M_c \cos\Theta$. When $\omega R/U_c < 1$ the attenuation becomes maximal for $\alpha \approx M_c \cos\Theta$; as $\omega R/U_c \to 0$ this maximum approaches 0.5, one half of the incident sound being absorbed.

Cavity-backed screen: (Fig. 22) A rigid backing wall can greatly increase the efficiency of sound absorption.

When the acoustic pressure in front of the screen is given by the first line of (136), the pressure in the region $-\ell < x_2 < 0$ between the screen and wall can be taken in the form

$$p(\mathbf{x}, \omega)/p_{\mathrm{I}} = \left(\mathcal{A}\mathrm{e}^{-n_2 x_2} + \mathcal{B}\mathrm{e}^{in_2 x_2}\right)\mathrm{e}^{in_1 x_1},$$

where $\mathcal{A}$ and $\mathcal{B}$ are constants. The rigid wall condition $\partial p/\partial x_2 = 0,\ x_2 = -\ell$, implies that $\mathcal{A}\mathrm{e}^{in_2\ell} = \mathcal{B}\mathrm{e}^{-in_2\ell}$. Two additional equations relating $\mathcal{A}$, $\mathcal{B}$ and the reflection coefficient $\mathcal{R}$ are obtained from the continuity of $\partial p/\partial x_2$ across the screen, and from the relation (137) from which we find

$$\mathcal{R} = \frac{\kappa_o \cos\Theta - i\mathcal{N}K_R\{1 - i\cot(\kappa_o \ell \cos\Theta)\}}{\kappa_o \cos\Theta + i\mathcal{N}K_R\{1 + i\cot(\kappa_o \ell \cos\Theta)\}}. \tag{141}$$

In the absence of flow ($K_R = 2R$) $|\mathcal{R}| = 1$ and there is no absorption of acoustic energy by the screen. In this case, when the frequency is very low ($\kappa_o \ell \ll 1$) the reflection coefficient $\mathcal{R} \to -1$ when $\kappa_o \ell \cos\Theta \to \sqrt{2\mathcal{N}R\ell}$, so that the acoustic pressure vanishes at the surface of the screen. In these circumstances the cavity is excited at a very low frequency resonance whose wavelength is large relative to the depth ℓ. When the sound is at normal incidence ($\Theta = 0$), for example, each aperture may be imagined to be the mouth of a *Helmholtz* resonator of volume $v = \ell/\mathcal{N}$, formed by inserting rigid barriers in the cavity between the wall and screen, whose resonance frequency is given by $\kappa_o \ell = \ell\sqrt{2R/v} \equiv \sqrt{2\mathcal{N}R\ell}$. This suggests that sound satisfying this resonance condition will be strongly absorbed by the bias flow jets, and indeed it can be confirmed by direct calculation of the absorption coefficient $\Delta = 1 - |\mathcal{R}|^2$ from (141) and (133) that it is possible to absorb *all* of the incident acoustic energy of a given frequency by an appropriate choice of the values of R, ℓ and the convection velocity U_c.

An experimental confirmation of this strong absorption is illustrated in Fig. 22b for normally incident sound on a bias flow, cavity backed screen in air for which $R = 0.15$ cm, $\ell = 2$ cm, $U_c = 4.76$ m/sec, and the fractional open area $\alpha = 0.024$. The solid curve is the theoretical prediction of $\Delta = 1 - |\mathcal{R}|^2$; the experimental data are adapted from measurements of Hughes & Dowling (1990), where U_c is identified with the mean flow velocity in the apertures. Theory and experiment exhibit peak absorption in the neighborhood of the Helmholtz resonance frequency where $\kappa_o \ell = \sqrt{2\mathcal{N}R\ell} \equiv \sqrt{2\alpha\ell/\pi R} \approx 0.45$, i.e. at $\omega R/U_c \approx 2.42$ ($\omega/2\pi \approx 1200$ Hz).

10 Compression Wave Generated when a Train Enters a Tunnel

One of the most fascinating applications of classical acoustics is to the 'piston' problem of the generation of a compression wave when a high-speed train enters a tunnel. The train pushes aside the stationary air, most of which flows over the train and out of the tunnel portal, but the build-up of pressure just ahead of the train propagates into the tunnel as a compression wave at the speed of sound (Fig. 27). The wave becomes of environmental concern at train speeds exceeding 200kph, especially in tunnels longer than about 3km with non-ballasted, modern concrete slab tracks, offering little dissipation to the wave, when nonlinear steepening can lead to a shock-like wavefront. The initial thickness of the wavefront $\sim R/M$ where R is the tunnel entrance height and M is the train *Mach* number. The total pressure rise $\Delta p \sim (\mathcal{A}_o/\mathcal{A})\rho_o U^2/(1 - M^2)$, where U is the train speed, ρ_o is the mean air density, and $\mathcal{A}_o$, $\mathcal{A}$ are respectively the cross-sectional areas of the train and tunnel. The ratio $\mathcal{A}_o/\mathcal{A}$ is called the *blockage*, which typically does not exceed about 0.2. At 200kph the initial wave thickness is roughly 40m and Δp is about 0.7% of atmospheric pressure, which is about six or seven times larger than the pressure jump at ground level across the sonic boom from *Concorde*. The new generation of 500kph *Maglev* trains would generate wave amplitudes approaching 5% atmospheric pressure (5kPa) with an initial wave thickness reduced to about 15m if they were permitted to enter a tunnel at the maximum proposed operating speed.

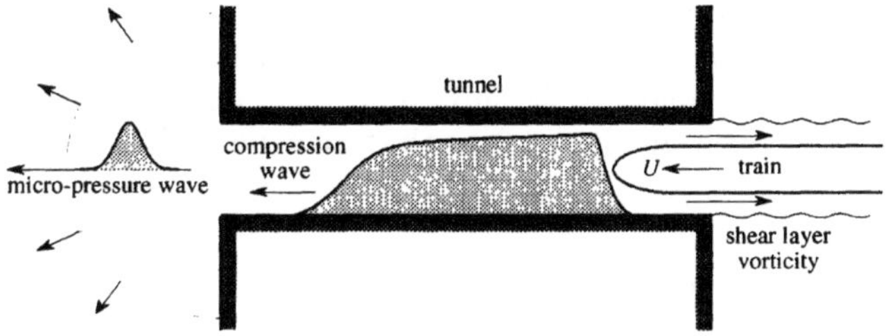

Fig. 27. Schematic illustration of a train entering a tunnel, showing the compression wave, the micro-pressure wave, and the shear layers of the exit flow from the tunnel entrance portal.

Multiple reflections from the ends of the tunnel, from discontinuities such as changes in cross-sectional area, air shafts, side branches, etc., and from the nose and tail of the train can produce complicated interference patterns with locally intense pressure transients. The rapidly varying changes in pressure are frequently a source of serious discomfort to passengers in unsealed rail cars and to maintenance personnel in the tunnel. Nonlinear steepening not only increases the effective frequency of pressure fluctuations experienced during passage of the wave, but is also responsible for the emission from the tunnel of a strong pressure pulse, called the *micro-pressure wave* (see Fig. 27), when the compression wave first arrives at the distant tunnel exit. Elementary acoustic theory shows that at distance r from the tunnel exit the pulse p' is given approximately by the formula

$$p'(\mathbf{x},t) \approx \frac{\alpha \mathcal{A}}{2\pi c_o r}\frac{\partial p}{\partial t}(t - r/c_o), \quad r \gg \sqrt{\mathcal{A}},$$

where α is a constant of order unity, dependent on local conditions near the tunnel exit, and p is the compression wave pressure incident on the tunnel exit evaluated at the exit.

The rattling of windows in neighbouring buildings and severe personal annoyance were first attributed to strong micro-pressure waves in Japan in the 1970's. New high-speed routes typically require 50% or more of a journey to be within a tunnel, and considerable effort continues to be expended (especially in Japan, where long tunnels are common) to understand the mechanisms of wave formation, principally using scale model tests and large scale numerical simulations of the tunnel entry problem.

The application of the equation of vortex sound to study compression wave formation is outlined in this lecture. Linear theory (i.e. correct to first order in $\mathcal{A}_o/\mathcal{A}$) replaces the train by a system of monopole sources distributed over the nose and tail sections, where the cross-sectional area varies; these sources account for most of the pressure rise across the wavefront, and the linearized approximation is found to be adequate when the blockage $\mathcal{A}_o/\mathcal{A}$

is less than about 0.12. The most important second order sources consist of *dipoles* orientated parallel to the train and distributed over the nose and tail, with strength determined by the surface pressure, and 'vortex dipoles' in the free shear layer of the air ejected from the tunnel portal by the entering train.

10.1 The Governing Equations

The dynamics of compression wave generation in Fig. 27 are equivalent to those of the simpler problem depicted in Fig. 28a, in which the tunnel is replaced by a semi-infinite duct. Consider first a tunnel of uniform cross-sectional area $\mathcal{A}$. Let the train be travelling at constant speed U in the negative x-direction, where the origin is at O at ground level in the entrance plane of the tunnel, with the x-axis lying in the vertical plane of symmetry. The pressure, density and speed of sound in the air are denoted respectively by $\bar{p}$, ρ and c. They vary with position and time within the tunnel, and their corresponding undisturbed values are denoted by p_o, ρ_o, c_o

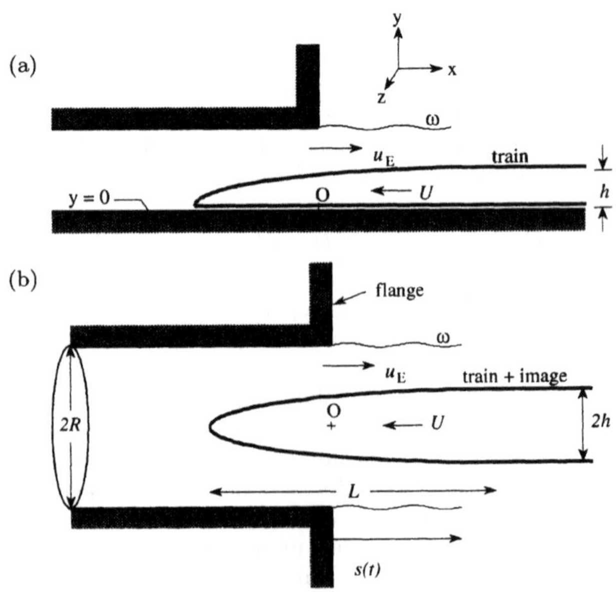

Fig. 28. (a) Train entering tunnel. (b) Dynamically equivalent configuration consisting of the train and its image in the ground plane entering a flanged duct formed by the reflection of the tunnel walls in the ground plane.

The cross-section of the train is assumed to become uniform with constant area $\mathcal{A}_o$ at a distance L from the nose. The uniform maximum height of the train is h, and the aspect ratio of the nose h/L is taken to be sufficiently small, and the train profile sufficiently streamlined, to ensure that flow separation does not occur. In practice the Mach number $M = U/c_o$ does not exceed about 0.4, and the blockage $\mathcal{A}_o/\mathcal{A}$ is typically less than about 0.2. Heat transfer and frictional losses will be neglected during the initial stages of wave formation. The air flow may then be regarded as *homentropic* (so that $\rho \equiv \rho(\bar{p})$) and the compression wave can be calculated from the corresponding vortex sound equation (23)

$$\left(\frac{D}{Dt}\left(\frac{1}{c^2}\frac{D}{Dt}\right) - \frac{1}{\rho}\nabla.(\rho\nabla)\right)B = \frac{1}{\rho}\mathrm{div}(\rho\boldsymbol{\omega}\wedge\mathbf{v}). \tag{142}$$

The air may be regarded as linearly perturbed from its mean state in the compression wave region ahead of the train, and

$$B \approx \frac{p}{\rho_o}, \quad \text{where } p = \bar{p} - p_o. \tag{143}$$

In this simple model the vorticity $\boldsymbol{\omega}$ vanishes everywhere except within the outer shear layer of the exit flow of the air displaced when the train enters the tunnel.

When the actions of frictional ground forces are ignored during entry into the tunnel, the problem of Fig. 28 a is equivalent to calculating the pressure wave generated by the train and the exit flow vorticity together with their *images* in the rigid ground plane ($y = 0$) within a duct consisting of the tunnel and its image in the ground (Fig. 28b). This configuration corresponds to arrangements used in typical model scale experiments, and the remainder of this discussion is framed in terms of this model.

Let $f \equiv f(x + Ut, y, z) = 0$ be a control surface S contained within the fluid that just encloses the moving train, with $f < 0$ inside S (in the region occupied by the train) and $f > 0$ in the exterior fluid region. The surface is fixed relative to the train, and the influence of the train on its surroundings can be represented in terms of monopole and dipole sources distributed on S. The representation is derived by multiplying equation (142) by the Heaviside step function $\mathrm{H} \equiv \mathrm{H}(f)$ ($= 0,\ 1$ according as $f \lessgtr 0$) and rearranging (noting that $Df/Dt = 0$) to obtain

$$\begin{aligned}\left(\frac{D}{Dt}\left(\frac{1}{c^2}\frac{D}{Dt}\right) - \frac{1}{\rho}\nabla\cdot(\rho\nabla)\right)(\mathrm{H}B) &= \frac{1}{\rho}\mathrm{div}(\mathrm{H}\rho\boldsymbol{\omega}\wedge\mathbf{v}) \\ &\quad -\left(\nabla B + \boldsymbol{\omega}\wedge\mathbf{v}\right)\cdot\nabla\mathrm{H} - \frac{1}{\rho}\mathrm{div}\left(\rho B\nabla\mathrm{H}\right).\end{aligned} \tag{144}$$

This equation is analogous to the differential form of the Ffowcs Williams - Hawkings equation (c.f. §6.1); $\mathrm{H}B \equiv B$ in the fluid ($f > 0$), and the two terms on the right hand side involving $\nabla\mathrm{H}$ respectively represent monopole and dipole sources distributed over the moving surface $f(x+Ut,y,z)=0$.

When frictional losses are neglected the momentum equation in the air reduces to $\partial\mathbf{v}/\partial t=-\nabla B-\boldsymbol{\omega}\wedge\mathbf{v}$, and the right of (144) can be written

$$\frac{\partial}{\partial t}(\mathbf{U}\cdot\nabla\mathrm{H})-\mathbf{v}\cdot\nabla\frac{\partial\mathrm{H}}{\partial t}-\frac{1}{\rho}\mathrm{div}\left(\rho B\nabla\mathrm{H}\right)+\frac{1}{\rho}\mathrm{div}(\mathrm{H}\rho\boldsymbol{\omega}\wedge\mathbf{v}), \tag{145}$$

where $\mathbf{U}=(-U,0,0)$.

When source terms of order $M(\mathcal{A}_o/\mathcal{A})^2$ or smaller are neglected, it is permissible to approximate these sources by neglecting the compressibility of the air over S and within the very low Mach number exterior flow from the tunnel portal. If the nonlinear contributions of small changes in density and sound speed to the *propagation* of the compression wave are also ignored, (144) then reduces to

$$\begin{aligned}\left(\frac{1}{c_o^2}\frac{\partial^2}{\partial t^2}-\nabla^2\right)(\mathrm{H}\,B\,)=&\frac{\partial}{\partial t}(\mathbf{U}\cdot\nabla\mathrm{H})+\mathrm{div}\,(\mathbf{v}\,\mathbf{U}\cdot\nabla\mathrm{H})\\&-\mathrm{div}\left\{\left(\frac{p}{\rho_o}+\frac{1}{2}v^2\right)\nabla\mathrm{H}\right\}+\mathrm{div}(\mathrm{H}\boldsymbol{\omega}\wedge\mathbf{v}),\end{aligned} \tag{146}$$

where the relation $\partial\mathrm{H}(f)/\partial t=-\mathbf{U}\cdot\nabla\mathrm{H}(f)$ has been used.

10.2 Linear Theory

The first source on the right of (146) is the only one linear in $\mathcal{A}_o/\mathcal{A}$. When the aspect ratio h/L of the train nose is small the slender body approximation gives

$$\frac{\partial}{\partial t}(\mathbf{U}\cdot\nabla\mathrm{H})(\mathbf{x},t)\approx\frac{\partial}{\partial t}\left(U\frac{\partial\mathcal{A}_\mathrm{T}}{\partial x}(x+Ut)\delta(y)\delta(z-z_\mathrm{T})\right), \tag{147}$$

where $\mathcal{A}_\mathrm{T}(x)$ is the cross-sectional area of the train at distance x *from the tip of the nose*, which is assumed to cross the portal entrance plane at $t=0$. This approximation replaces the monopole distribution by a line source at ground level along the centreline of the path of the train: $-\infty<x<\infty$, $y=0$, $z=z_\mathrm{T}$, where z_T is the offset from the centreline of the tunnel. The source strength is proportional to the rate at which the cross-section changes with distance along the train, and is non-zero only in the vicinity of the train nose (and also the tail).

The corresponding linearized approximation p_1, say, to the compression wave is found by using the compact Green's function (75)

$$\begin{aligned}G(\mathbf{x},\mathbf{x}';t-\tau)\approx\frac{c_o}{2\mathcal{A}}\Bigg\{&\mathrm{H}\left(t-\tau-\frac{|\varphi^*(\mathbf{x})-\varphi^*(\mathbf{x}')|}{c_o}\right)\\&-\mathrm{H}\left(t-\tau+\frac{\varphi^*(\mathbf{x})+\varphi^*(\mathbf{x}')}{c_o}\right)\Bigg\},\end{aligned} \tag{148}$$

where $\varphi^*(\mathbf{x})$ is the velocity potential of a hypothetical uniform incompressible flow *out* of the tunnel portal (see (70)). This approximation for G is applicable for any tunnel whose interior cross-sectional area is ultimately constant and equal to $\mathcal{A}$. It is uniformly valid when regarded as a function of either $\mathbf{x}$ or $\mathbf{x}'$ provided at least one of these points lies within the tunnel at a distance from the portal large compared to the tunnel diameter.

Thus, at points $\mathbf{x}$ within the tunnel, ahead of the train where $B = p/\rho_o$, we have

$$\begin{aligned} p_1 \equiv p_1([t]) &= \rho_o \frac{\partial}{\partial t} \iint_{-\infty}^{\infty} U \frac{\partial \mathcal{A}_T}{\partial x'}(x' + U\tau) G(\mathbf{x}, x', 0, z_T; t-\tau) dx' d\tau \\ &= \frac{\rho_o U c_o}{2\mathcal{A}} \int_{-\infty}^{\infty} \Big\{ \mathcal{A}_T'\Big(x' - M\varphi^*(x', 0, z_T) + U[t]\Big) \\ &\qquad - \mathcal{A}_T'\Big(x' + M\varphi^*(x', 0, z_T) + U[t]\Big)\Big\} dx', \end{aligned} \tag{149}$$

where the prime on $\mathcal{A}_T$ denotes differentiation with respect to the argument, and $[t] = t + (x - \ell')/c_o$ is the effective retarded time. Because nonlinear propagation terms have been ignored, this approximation determines the *initial* form of the compression wave profile, before the onset of nonlinear steepening. It is therefore applicable within the region several tunnel diameters ahead of the train, during and just after tunnel entry.

The main contributions to the integral (150) are from vicinities of the nose and tail of the train, where the cross-sectional area $\mathcal{A}_T$ is changing. The compression wave is generated as the nose enters the tunnel, and may be calculated by temporarily considering a train of semi-infinite length. During the formation of the wave, and in the particular case in which the Mach number is small enough that terms $\sim \mathrm{O}(M^2)$ are negligible, the term $M\varphi^*$ in the arguments of $\mathcal{A}_T'$ in (150) is small, and we then find

$$p_1([t]) \approx \frac{\rho_o U^2}{\mathcal{A}} \int_{-\infty}^{\infty} \frac{\partial \mathcal{A}_T}{\partial x'}(x' + U[t]) \frac{\partial \varphi^*}{\partial x'}(x', 0, z_T) dx', \quad M^2 \ll 1. \tag{150}$$

After the nose has passed into the tunnel, $\partial\varphi^*/\partial x' = 1$ in the region occupied by the nose, and (150) predicts the overall (linear theory) pressure rise to be $\Delta p_1 \approx \rho_o U^2 \mathcal{A}_o/\mathcal{A}$. But the linear theory, asymptotic pressure rise can also be calculated *exactly*, with no restriction on Mach number to be

$$\Delta p_1 = \frac{\rho_o U^2 \mathcal{A}_o}{\mathcal{A}(1 - M^2)}, \tag{151}$$

because this is attained when $\varphi^*(x', 0, z_T) \approx x' - \ell'$ in (150) (see equation (70)). This implies that the approximation (150) can be extended to finite Mach numbers by writing

$$p_1([t]) \approx \frac{\rho_o U^2}{\mathcal{A}(1 - M^2)} \int_{-\infty}^{\infty} \frac{\partial \mathcal{A}_T}{\partial x'}(x' + U[t]) \frac{\partial \varphi^*}{\partial x'}(x', 0, z_T) dx'. \tag{152}$$

This extrapolation of the linear theory to finite values of M turns out to be valid for $M < 0.4$.

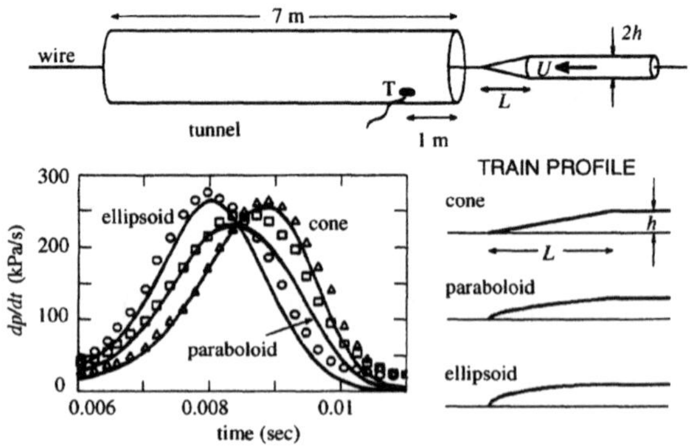

Fig. 29. Measurements at T and predictions (equation (152)) of the compression wave pressure gradient dp/dt for $h/L = 0.2$, $\mathcal{A}_o/\mathcal{A} = 0.116$, $U \approx 230$kph ($M \approx 0.188$).

Figure 29 illustrates schematically an experimental arrangement used by Maeda *et al.* (1993) to investigate the compression wave. Wire-guided, axisymmetric model 'trains' are projected into and along the axis of a 'tunnel' consisting of a 7m long circular cylinder of internal diameter 0.147m. The nose 'aspect ratio' $h/L = 0.2$, the blockage $\mathcal{A}_o/\mathcal{A} = 0.116$, and the projection speed $U \approx 230$kph ($M \approx 0.188$). The data points in Fig. 29 are measurements (made at the point labelled T) of the 'pressure gradient' dp/dt one metre from the entrance for three different train nose profiles (cone, paraboloid and ellipsoid). The solid curves are predictions of equation (152), which underpredicts the maximum pressure gradients by about 8%.

To examine further the validity of linear theory consider a circular cylindrical tunnel of radius R with a *flanged* entrance (Figure 28b), and a train with an ellipsoidal nose profile obtained by rotating the curve

$$y = h\sqrt{(x/L)(2 - x/L)},\ 0 < x < L$$

about the x-axis, so that

$$\frac{\mathcal{A}_T(x)}{\mathcal{A}_o} = \begin{cases} x/L\,(2 - x/L)\,, & 0 < x < L, \\ 1, & x > L, \end{cases} \tag{153}$$

where $\mathcal{A}_o = \pi h^2$. A model of this type with dimensions

$$h = 2.235\text{cm}, \quad L = 11.18\text{cm} \tag{154}$$

with a the tail of identical ellipsoidal shape, and with an overall length of 91.5cm has been used in experiments. The experimental tunnel had length 10m and radius $R = 5$cm (so that the blockage $\mathcal{A}_o/\mathcal{A} \approx 0.2$), and pressure measurements were made at a distance $|x| = 1$m from the tunnel entrance.

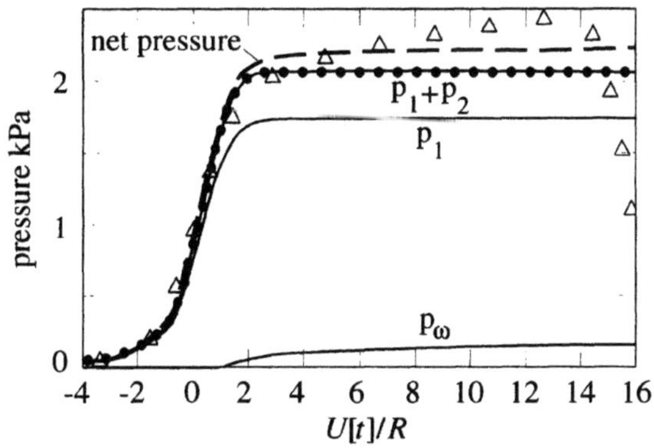

Fig. 30. Illustrating different approximations for the compression wave pressure profile (in kPa) for a train with the ellipsoidal nose profile (153) entering a flanged circular cylindrical tunnel of radius $R = 5$cm at $U = 293$ kph: the solid curves depict the linear theory pressure p_1 (equation (152)), the second order approximation $p_1 + p_2$ of equation (159), and the exit flow vorticity-generated pressure p_ω (equation (164)); the dotted curve (• • •) is the simplified second order approximation (161); the broken-line curve (- - - -) is the net predicted pressure $p_1 + p_2 + p_\omega$ calculated from (161) and (164). The open triangles (△ △ △) are the measured pressure at a distance of 1 m from the tunnel entrance plane.

The curve labelled p_1 in Fig. 30 is the linear theory representation of the wavefront pressure profile calculated from (152), (153) for $U = 293$kph ($M = 0.24$), plotted against the non-dimensional retarded position of the train $U[t]/R$ (the nose of the train being assumed to enter the tunnel at $t = 0$). The open triangles are measured values of the pressure. The initial pressure rise occurs over a time interval $\sim 2R/U$, corresponding to a wavefront thickness $2R/M \sim 41$ cm. The rapid decrease in the measured pressure that occurs for $U[t]/R > 14$ is produced by the rarefaction wave generated when the tail enters the tunnel. The gradual increase in the experimental points after $U[t]/R \sim 2$ is believed to be caused by boundary layer growth on the train and tunnel walls, or possibly by flow separation on either of these surfaces.

10.3 Contribution from the Surface Dipoles

The second and third terms on the right of (146) are dipole sources distributed over the surface of the train. In the slender body approximation the first of these becomes

$$\operatorname{div}(\mathbf{v}\,\mathbf{U}\cdot\nabla\mathrm{H}) \approx \frac{\partial}{\partial x}\left(u_x U \frac{\partial \mathcal{A}_\mathrm{T}}{\partial x}(x+Ut)\delta(y)\delta(z-z_\mathrm{T})\right), \tag{155}$$

where u_x is the x-component of the air velocity $\mathbf{v}$. This source produces a contribution to the compression wave that is

$$\mathrm{O}\left(\rho_\mathrm{o} U^2 \left(\frac{\mathcal{A}_\mathrm{o}}{\mathcal{A}}\right)^2\right)$$

because $u_x \sim U\mathcal{A}_\mathrm{o}/\mathcal{A}$ except in the immediate vicinity of the nose of the train.

The strength of the second dipole is proportional to $p/\rho_\mathrm{o} + \frac{1}{2}v^2$. This can be estimated by means of a quasi-static approximation, using Bernoulli's equation and the *linear theory* uniform pressure $p_1(t)$ just ahead of the train, to be given by

$$\frac{p}{\rho_\mathrm{o}} + \frac{1}{2}v^2 \approx \frac{p_1(t)}{\rho_\mathrm{o}} - u_x U, \tag{156}$$

so that in the slender body approximation

$$\begin{aligned}-\operatorname{div}&\left\{\left(\frac{p}{\rho_\mathrm{o}} + \frac{1}{2}v^2\right)\nabla\mathrm{H}\right\} \\ &\approx \frac{\partial}{\partial x}\left\{\left(\frac{p_1(t)}{\rho_\mathrm{o}} - u_x U\right)\frac{\partial \mathcal{A}_\mathrm{T}}{\partial x}(x+Ut)\delta(y)\delta(z-z_\mathrm{T})\right\}.\end{aligned} \tag{157}$$

Combining (155) and (157), we obtain the

$$\text{net surface dipole} \approx \frac{\partial}{\partial x}\left(\frac{p_1(t)}{\rho_\mathrm{o}}\frac{\partial \mathcal{A}_\mathrm{T}}{\partial x}(x+Ut)\delta(y)\delta(z-z_\mathrm{T})\right). \tag{158}$$

Using the Green's function (148) and neglecting terms of order $M(\mathcal{A}_\mathrm{o}/\mathcal{A})^2$ and smaller, the surface dipole contribution p_2 is found to be given by

$$p_2 \equiv p_2([t]) \approx \frac{p_1([t])}{\mathcal{A}}\int_{-\infty}^{\infty}\frac{\partial \mathcal{A}_\mathrm{T}}{\partial x'}(x'+U[t])\frac{\partial \varphi^*}{\partial x'}(x',0,z_\mathrm{T})dx'. \tag{159}$$

This yields a component of order $\rho_\mathrm{o}U^2(\mathcal{A}_\mathrm{o}/\mathcal{A})^2$ to the overall pressure rise Δp across the compression wavefront.

The solid curve labelled $p_1 + p_2$ in Fig. 30 is the second order approximation $p = p_1 + p_2$ for the train (153), (154), which exhibits a considerably improved agreement with experiment relative to the approximation p_1. A further simplification can be introduced by noting that p_2 is initially very small, which suggests that the integral in (159) may be approximated by its value for large retarded time (when $\partial\varphi^*/\partial x' = 1$), and therefore that

$$p_2 \approx p_1 \frac{\mathcal{A}_o}{\mathcal{A}}. \tag{160}$$

The corresponding prediction of $p_1 + p_2$ across the wavefront is shown dotted in Fig. 30. The difference between the two approximations for $p_1 + p_2$ is insignificant, and the overall pressure wave produced by the monopole and dipole source terms is therefore well represented by

$$p = p_1 + p_2 \approx \frac{\rho_o U^2}{\mathcal{A}(1-M^2)}\left(1 + \frac{\mathcal{A}_o}{\mathcal{A}}\right)\int_{-\infty}^{\infty} \frac{\partial \mathcal{A}_T}{\partial x'}(x' + U[t])\frac{\partial \varphi^*}{\partial x'}(x', 0, z_T)dx'. \tag{161}$$

10.4 Contribution from the Exit Flow Vorticity

When boundary layer growth on the train and on the interior tunnel walls is ignored, the vorticity source $\mathrm{div}(\mathrm{H}\boldsymbol{\omega}\wedge\mathbf{v})$ in equation (146) is non-zero only in the shear layer of the exit flow from the tunnel (see Figure 10.2). Observations indicate that the exit flow shear layer rolls-up and forms a large vortex that is essentially axisymmetric when the train travels along the tunnel centreline. Just after tunnel entry the initial stages of vortex formation may be modelled as in Fig. 28, where the exit flow is assumed to be a uniform jet with speed $u_E(t)$, with vorticity confined to a circular cylindrical vortex sheet of strength u_E convecting in the positive x-direction at speed $\frac{1}{2}u_E$, so that

$$\mathrm{div}(\mathrm{H}\boldsymbol{\omega}\wedge\mathbf{v}) = \begin{cases} \frac{\partial}{\partial r}\left(\frac{1}{2}u_E^2\delta(r-R)\right), & 0 < x < s(t), \\ 0, & \text{elsewhere}, \end{cases} \tag{162}$$

where $s(t)$ is the length of the jet outside the portal, indicated in Fig. 28b, and $r = \sqrt{y^2 + z^2}$ is radial distance from the tunnel axis.

The velocity $u_E(t)$ increases rapidly from zero to its final value of $U\mathcal{A}_o/\mathcal{A}$ during the time $\sim L/U$ in which the nose enters the tunnel. The initial growth of the vortex source strength will be modelled by setting $u_E(t) = U\mathcal{A}_T(Ut)/\mathcal{A}$ for $0 < t < L/U$. The corresponding growth of the vortex generated pressure wave depends on both $u_E(t)$ and $s(t)$; the latter may be approximated by $s(t) = \int_0^t u_E(\tau)d\tau$, $t > 0$. Using Green's function (148), the pressure p_ω generated by the exit flow may then be cast in the form

$$p_\omega([t]) \approx \frac{\rho_o U^2}{2\mathcal{A}}\left(\frac{\mathcal{A}_T(U[t])}{\mathcal{A}}\right)^2 \int_0^{s([t])} 2\pi R\left(\frac{\partial\varphi^*}{\partial r'}\right)_{r'=R} dx'. \tag{163}$$

The integral is equal to the flux through the boundary of the jet (of length s) of the hypothetical irrotational flow from the tunnel portal defined by the velocity potential φ^*. It can therefore be evaluated explicitly in terms of the corresponding Stokes stream function $\psi^*(r', x')$ which satisfies

$$\frac{1}{r'}\frac{\partial \psi^*}{\partial r'} = \frac{\partial \varphi^*}{\partial x'}, \quad \frac{1}{r'}\frac{\partial \psi^*}{\partial x'} = -\frac{\partial \varphi^*}{\partial r'}.$$

On the jet boundary ($r' = R$) $\psi^*(R, x')$ decreases from $\mathcal{A}/2\pi$ at $x' = 0$ to zero at $x' = +\infty$, and we find

$$p_\omega([t]) \approx \frac{\rho_o U^2}{2}\left(\frac{\mathcal{A}_T(U[t])}{\mathcal{A}}\right)^2\left(1 - \frac{2\pi\psi^*\left(R, s([t])\right)}{\mathcal{A}}\right). \tag{164}$$

The same numerical calculation that must be performed to determine the potential φ^* also yields the Stokes stream function ψ^*, thereby permitting straightforward evaluation of the right hand side of (164). The magnitude of p_ω is of the same order as the surface dipole pressure p_2; its growth as the train enters the tunnel is shown in Figure 30.

The broken line curve in Fig. 30 is the predicted overall pressure signature from all three sources:

$$p = p_1 + p_2 + p_\omega. \tag{165}$$

The gradual increase in p_ω as the exit flow extends downstream causes p to continue to increase (but very slowly) after the establishment of the main compression wavefront. The predicted growth rate is smaller than that of the measured pressure, presumably because of the neglect of vorticity sources in the viscous boundary layers on the train and tunnel walls, and within any regions of separated flow.

11 The Flared Portal

11.1 Suppression of the Micro-pressure Wave

The amplitude of the micro-pressure wave that radiates from the far end of the tunnel is proportional to the compression wave pressure gradient dp/dt near the exit. The principal methods used to minimize its subjective and harmful environmental impact are based largely on reducing the initial value of dp/dt at the entrance portal, because wave steepening in a long tunnel tends to be suppressed by relaxation processes when the initial *rise time* of the wavefront is sufficiently large. The maximum pressure gradient varies approximately as U^3, but it also depends critically on tunnel portal geometry and, to a lesser extent, on train nose profile. For a train of given cross-sectional area, nose profile modifications produce only modest increases in the initial rise time. Much larger increases are possible by suitably modifying the tunnel portal geometry. Thus, for adjacent, parallel tunnels the initial wave thickness is increased when the high pressure air produced by an entering train is allowed to vent into the neighboring tunnel through ducting close to the entrance. However, the most impressive increases have been realized by installing a duct-like 'hood' extending 30 - 50 metres ahead of the tunnel entrance. The compression wave forms as the train enters the hood, and its rise time is increased by venting through judiciously positioned windows in the hood walls. For example, a fivefold increase in rise time is achieved by the 49m long hood at the eastern entrance to the Ohirayama tunnel in Japan.

But, hoods tend to be unsightly. A more elegant and aesthetically pleasing, although perhaps more expensive method of increasing the rise time is to *flare* the entrance portal, in the manner illustrated in Fig. 31a. Here we consider a train entering a flanged tunnel of semi-circular cross-section that is *flared* over an axial distance ℓ from the tunnel entrance plane (Fig. 31a). Experiments have been performed to determine the 'optimal' degree of flaring using the axisymmetric model of Fig. 31b, to which the following discussion is confined.

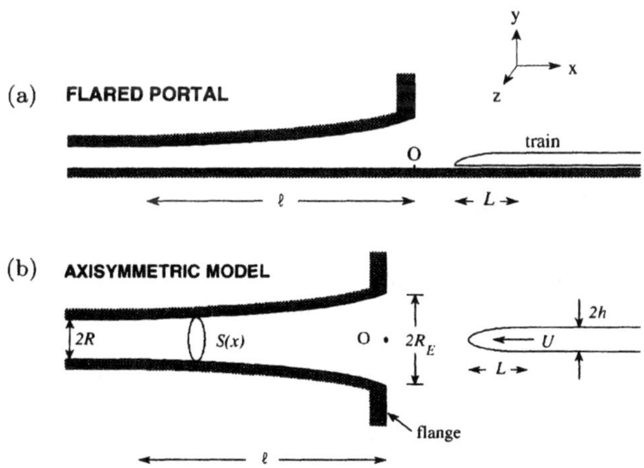

Fig. 31. (a) Train entering a tunnel with a flared portal of length ℓ; (b) axisymmetric flared portal with an infinite flange.

11.2 The Function φ^*

Let the tunnel cross-sectional area $S(x)$ decrease smoothly from $\mathcal{A}_E \equiv \pi R_E^2$ at $x = 0$ to the uniform, constant value $\mathcal{A} = \pi R^2$ for $x \leq -\ell$. Changes in $S(x)$

are assumed to be sufficiently slow that the velocity potential φ^* satisfies the following simplified form of Laplace's equation within the tunnel

$$\frac{1}{S(x)}\frac{\partial}{\partial x}\left(S(x)\frac{\partial \varphi^*}{\partial x}\right) = 0, \quad x < 0, \tag{166}$$

with solution

$$\varphi^*(\mathbf{x}) = \mathcal{A}\int_0^x \frac{d\xi}{S(\xi)} + C, \quad x < 0, \tag{167}$$

where C is a constant.

If the flange is regarded as infinite, the behaviour of $\varphi^*(\mathbf{x})$ outside the tunnel ($x > 0$) is approximately the same as the potential of incompressible flow produced by a 'piston' of area $\mathcal{A}_\mathrm{E}$ in a plane wall at $x = 0$, moving with normal velocity $(\partial\varphi^*/\partial x)_{x=0} = \mathcal{A}/\mathcal{A}_\mathrm{E}$. When $\mathbf{x}$ lies on the axis of symmetry (the x-axis), the condition that $\varphi^*(\mathbf{x}) \to 0$ as $x \to +\infty$ then yields

$$\varphi^*(\mathbf{x}) = -\frac{\mathcal{A}R}{\mathcal{A}_\mathrm{E}}\left[\left(\frac{\mathcal{A}_\mathrm{E}}{\mathcal{A}} + \frac{x^2}{R^2}\right)^{\frac{1}{2}} - \frac{x}{R}\right], \quad x > 0. \tag{168}$$

Equating expressions (167) and (168) at $x = 0$ we find $C = -R\sqrt{\mathcal{A}/\mathcal{A}_\mathrm{E}}$, and the flared portal 'end correction' is then given by

$$\ell' = R\sqrt{\frac{\mathcal{A}}{\mathcal{A}_\mathrm{E}}} + \int_{-\infty}^{0}\left(\frac{\mathcal{A}}{S(\xi)} - 1\right)d\xi. \tag{169}$$

11.3 Optimal Flaring

The flaring is 'optimal' when the pressure increases linearly with distance across the compression wavefront. We can derive the shape of the tunnel area profile $S(x)$ necessary to achieve this by considering the linear theory compression wave p_1 produced by a monopole *point* source entering the tunnel. Equation (150) reveals that this is equivalent to setting

$$\frac{\partial \mathcal{A}_\mathrm{T}}{\partial x'}(x' + U[t]) = \mathcal{A}_\mathrm{o}\delta(x' + U[t]), \tag{170}$$

i.e. to replacing the streamlined train nose profile by a flat-faced or 'snub-nosed' train.

Because the second order approximation (161) for $p_1 + p_2$ differs from the linear theory pressure (150) merely by a multiplicative factor of $(1 + \mathcal{A}_\mathrm{o}/\mathcal{A})$, the linear theory optimal tunnel profile remains optimal at second order. Indeed, substituting from (167) and (170) into (161), we find that

$$p \approx p_1 + p_2 = \frac{\rho_\mathrm{o}U^2}{(1 - M^2)}\left(1 + \frac{\mathcal{A}_\mathrm{o}}{\mathcal{A}}\right)\frac{\mathcal{A}_\mathrm{o}}{S(-U[t])}, \quad \text{for } U[t]/R > 0, \tag{171}$$

after the point source has crossed the entrance plane into the flared section of the tunnel, where most of the pressure rise occurs. During this time $S(-U[t])$ decreases from $\mathcal{A}_E$ to $\mathcal{A}$, and the pressure rises from

$$\frac{\rho_o U^2}{(1-M^2)}\left(1+\frac{\mathcal{A}_o}{\mathcal{A}}\right)\frac{\mathcal{A}_o}{\mathcal{A}_E}$$

at the instant at which the front of the train (the source) enters the tunnel, to its final value of

$$\frac{\rho_o U^2}{(1-M^2)}\left(1+\frac{\mathcal{A}_o}{\mathcal{A}}\right)\frac{\mathcal{A}_o}{\mathcal{A}}$$

when the front enters the uniform section of the tunnel. This increase will be linear when $1/S(-U[t])$ is a linearly increasing function of the retarded time, that is, when the flared section of the tunnel is defined by

$$\frac{S(x)}{\mathcal{A}} = 1\Big/\left[\frac{\mathcal{A}}{\mathcal{A}_E} - \frac{x}{\ell}\left(1-\frac{\mathcal{A}}{\mathcal{A}_E}\right)\right], \quad -\ell < x < 0. \tag{172}$$

It remains to specify the cross-sectional area $S(0) = \mathcal{A}_E$ in the tunnel entrance plane. When the length ℓ of the flared section is prescribed, the value of $\mathcal{A}_E$ must be adjusted to ensure that the pressure varies smoothly at $U[t]/R = 0$, when the train (source) enters the tunnel. This is done by requiring the values of $\partial^2\varphi^*(x,0,0)/\partial x^2$ defined respectively by the approximations (167) and (168) to be equal at $x = 0$, which implies that

$$\left(\frac{\mathcal{A}}{\mathcal{A}_E}\right)^{\frac{3}{2}} = \frac{R}{\ell}\left(1-\frac{\mathcal{A}}{\mathcal{A}_E}\right).$$

Provided that the length ℓ of the flared section exceeds $2R/3\sqrt{3} \approx 0.385R$, we then find

$$\frac{\mathcal{A}_E}{\mathcal{A}} = \left(\frac{\ell}{2R}\right)^{\frac{2}{3}}\left[\left(1+\sqrt{1-\left(\frac{2R}{3\sqrt{3}\ell}\right)^2}\right)^{\frac{1}{3}} + \left(1-\sqrt{1-\left(\frac{2R}{3\sqrt{3}\ell}\right)^2}\right)^{\frac{1}{3}}\right]^2. \tag{173}$$

This gives a minimum value of $\mathcal{A}_E/\mathcal{A} = \frac{4}{3}$ at $\ell/R = 0.385$; when $\ell \gg R$, $\mathcal{A}_E/\mathcal{A} \approx (\ell/R)^{\frac{2}{3}}$. In the experiment described below $\ell/R = 10$, so that the optimal area ratio $\mathcal{A}_E/\mathcal{A} = 5.35$ and $R_E/R = 2.31$. The local tunnel radius is given by $r(x) = R\sqrt{S(x)/\mathcal{A}}$, and the flared sections shown in Fig. 31 are the optimal profile calculated from this formula and equations (172), (173) when $\ell/R = 10$.

11.4 Comparison with Measurements

To first order it would be expected that the predicted linear increase in pressure across the wavefront will also be realized when the 'snub' nosed train is replaced by one of more realistic profile. This is because the linear theory monopole distribution (147) can be replaced to a good approximation by a *point source* at a suitable 'centroid' within the nose of the train. Figure 32 shows an overall view of the experimental arrangement used to verify this conjecture.

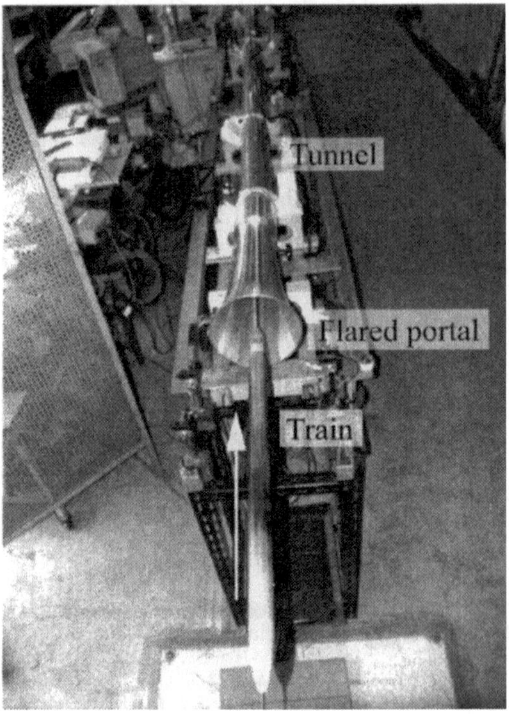

Fig. 32. Unflanged flared portal and cylindrical tunnel with wire-guided axisymmetric model train.

In Fig. 33 the linear theory pressure p_1, determined by (150), (166) and (168), and the second order approximation $p_1 + p_2$ given by (161), (166) and (168), are compared with measurements for a circular cylindrical tunnel fitted with a flanged portal that is optimally flared in accordance with equations (172), (173) with

$$R = 5\text{cm}, \quad \ell = 10R.$$

The open triangles are the pressure measured at distance 1.55m from the portal entrance plane produced by an axisymmetric train of total length 91.5cm. The train is the same as that described in §10.2, both ends of which have ellipsoidal profiles defined as in (153) with semi-axes h, L given by (154), the blockage $\mathcal{A}_o/\mathcal{A} = 0.2$, and the train speed $U = 296$ kph.

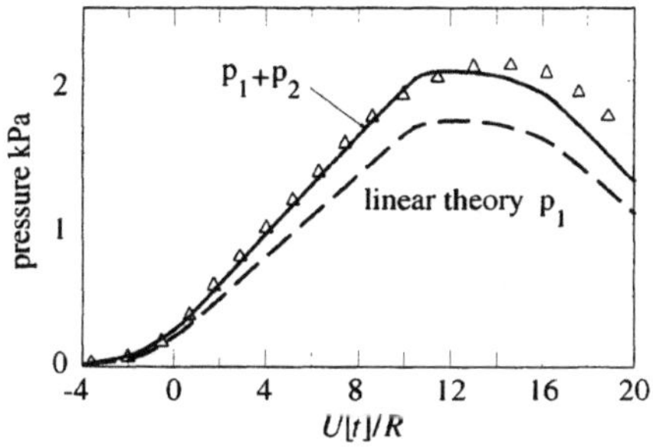

Fig. 33. Compression wave pressure profile (in kPa) generated by a train with ellipsoidal nose and tail profiles defined by (153) entering at $U = 296$ kph a circular cylindrical tunnel of radius $R = 5$cm with an optimally flared portal of length $\ell = 50$cm: $\triangle\ \triangle\ \triangle$, pressure measured at a distance 1.55m from the tunnel entrance plane; - - - -, linear theory pressure predicted by (150), (167), (168); ———, second order approximation $p_1 + p_2$ given by (161), (167), (168).

The nose of the train enters the flared section of the tunnel at $U[t]/R = 0$, following which the pressure rises linearly until just after the nose arrives at the beginning of the uniform section of the tunnel, at $U[t]/R = 10$, so that the compression wave thickness $\sim 10R/M = 207$ cm (representing a fivefold increase over the wave thickness in §2 for the flanged, circular cylindrical portal). The second order approximation (161) gives an excellent rendering of this experimental observation. The subsequent sharp decrease in the pressure is produced by the expansion wave generated when the tail of the train begins to enter the flared section at $U[t]/R \approx 16$. Because of the large expansion in cross-section of the tunnel at the portal, the contribution p_ω to the pressure rise from the exit flow vorticity is reduced by a factor $(\mathcal{A}/\mathcal{A}_{\mathrm{E}})^2 \sim 0.035$, and may be ignored.

Measured and predicted values of the temporal 'pressure gradient' dp/dt are compared in Fig. 34. The value of dp/dt (augmented by nonlinear steepening) governs the strength of the micro-pressure wave radiated from the far end of the tunnel when the compression wave arrives. The initial maximum value of about 310 kPa/s shown in Fig. 34 is about 25% of the corresponding maximum for the pressure profile of Fig. 30 for the unflared circular cylindrical tunnel. The abrupt decrease in the predicted pressure gradient at $U[t]/R = 10$ occurs as the nose of the train enters the uniform section of the tunnel. The measured decrease of dp/dt at these times is much smoother, presumably because various vortex sources discussed above are ignored in the theory; in particular these could well include a large vortex ring-like structure between the train and tunnel wall in the flared section, although no experimental data is currently available to support this conjecture.

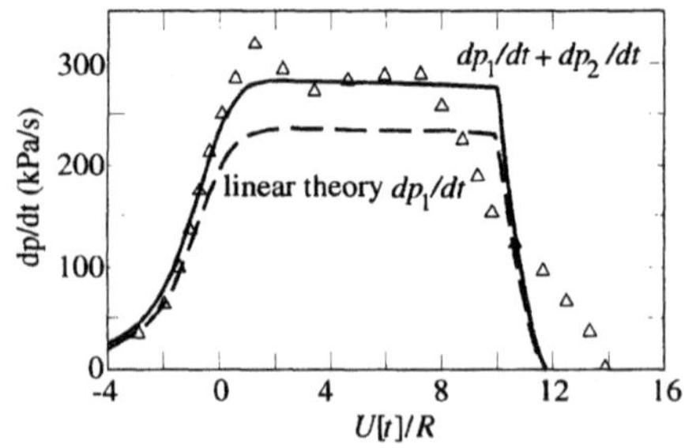

Fig. 34. Compression wave 'pressure gradient' dp/dt (in kPa/s) generated by a train with ellipsoidal nose and tail profiles defined by (153) entering at $U = 296$ kph a circular cylindrical tunnel of radius $R = 5$cm with an optimally flared portal of length $\ell = 50$cm: △ △ △, pressure gradient measured at a distance 1.55m from the tunnel entrance plane; - - - -, linear theory prediction of (150), (167), (168); ———, second order approximation $dp_1/dt + dp_2/dt$ determined by (161), (167), (168).

References

1. Batchelor, G. K. 1967 *An Introduction to Fluid Dynamics.* Cambridge University Press.
2. Blake, W. K. 1986 *Mechanics of flow-induced sound and vibration*, Vol. 2: Complex flow-structure interactions. New York: Academic Press.
3. Cremer, L. and M. Heckl, E. E. Ungar 1988 *Structure-borne Sound* (2nd ed). New York: Springer-Verlag.
4. Crow, S. C. 1970 *Studies in Appl. Math.* **49**, 21 - 44. Aerodynamic sound emission as a singular perturbation problem.
5. Curle, N. 1955 *Proc. Roy. Soc. Lond.* **A231**, 505 - 514. The influence of solid boundaries upon aerodynamic sound.
6. Dowling, A. P. and Ffowcs Williams, J. E. 1983 *Sound and Sources of Sound.* Ellis Horwood.
7. Ffowcs Williams, J. E. and Hawkings, D. L. 1969 Sound generation by turbulence and surfaces in arbitrary motion. *Philosophical Transactions of the Royal Society* **A264**, 321 - 342.
8. Goldstein, M. E. 1976 *Aeroacoustics.* New York: McGraw-Hill.
9. Hinze, J. O. 1975 *Turbulence* (Second edition). New York: McGraw-Hill.
10. Howe, M. S. 1975 Contributions to the theory of aerodynamic sound, with application to excess jet noise and the theory of the flute. *Journal of Fluid Mechanics* **71**, 625 - 673.
11. Howe, M. S. 1989 *J. Fluid Mech.* **206**, 131 - 153. On unsteady surface forces, and sound produced by the normal chopping of a rectilinear vortex.
12. Howe, M. S. 1998 *Acoustics of Fluid-Structure Interactions*, Cambridge University Press.
13. Lamb, Horace *Hydrodynamics* (6th. ed.), 1932. Cambridge University Press (Reprinted 1993).

14. Lesser, M. B. and Crighton, D. G. 1976 *Physical Acoustics* **11**, 69 - 149. Physical acoustics and the method of matched asymptotic expansions.
15. Lighthill, M. J. 1952 *Proc. Roy. Soc. Lond.* **A211**, 564 - 587. On sound generated aerodynamically. Part I: General theory.
16. Lighthill, M. J. 1954 *Proc. Roy. Soc. Lond.* **A222**, 1 - 32. On sound generated aerodynamically. Part II: Turbulence as a source of sound.
17. Lighthill, James 1978 *Waves in Fluids.* Cambridge University Press.
18. Maeda, T., Matsumura, T., Iida, M., Nakatani, K. and Uchida, K. 1993 *Effect of shape of train nose on compression wave generated by train entering tunnel,* Proceedings of the International Conference on Speedup Technology for Railway and Maglev Vehicles, pp 315 - 319 (Yokohama, Japan 22-26 November, 1993).
19. Morse, P. M. and Feshbach, H. 1953 *Methods of Theoretical Physics,* Vols. 1 and 2. New York: McGraw-Hill.
20. Pierce, A. D. 1989 *Acoustics, An introduction to its principles and applications.* American Institute of Physics.
21. Phillips, O. M. 1956 *J. Fluid Mech.* **1**, 607 - 624. The intensity of Aeolian tones.
22. Powell, A. 1960 *J. Acoust. Soc. Am.* **32**, 962 - 990. Aerodynamic noise and the plane boundary.
23. Powell, A. 1961 *Vortex sound.* Rept. No. 61 - 70, Department of Engineering, University of California, Los Angeles.
24. Rayleigh, Lord 1894 *The Theory of Sound,* two volumes. London: Macmillan. (Reprinted by Dover Publications, New York, 1945).
25. Wu, J. C. 1981 *AIAA J.* **19**, 432 - 441. Theory for aerodynamic force and moment in viscous flows.

Sound–Vortex Interaction in Infinite Media

Fernando Lund

Centro para la Investigación Interdisciplinaria Avanzada en Ciencia de los Materiales ***and*** Departamento de Física, Facultad de Ciencias Físicas y Matemáticas, Universidad de Chile, Casilla 487-3, Santiago, Chile

Abstract. This lecture describes the interaction of sound and vorticity in infinite media. The scattering of sound by the vorticity is decribed. Recent results on ultrasound propagation through a disordered flow are given.

1 Introduction

This is a set of lectures on the interaction of sound and vorticity in infinite media, organized in five Sections: Section 2 describes the basic assumptions needed to attack this problem, writes down the corresponding equations, introduces the techniques that are used to tackle them, and derives the formulae for vortex sound. There is an overall small parameter throughout, the Mach number M. Section 3 studies the scattering of sound by vorticity in three dimensions. In anticipation of problems to be studied further down the road, equations are derived to second order in Mach number. The formula relating (linearly) scattered pressure t o (Fourier transform of) vorticity is derived. In two dimensions, but not in three, it is possible to have a distribution of vorticity with non-vanishing net circulation. In this case the tools, notably the Born approximation scheme, that are used in Section 3, do not apply. Moreover, the equations describing the interaction of sound and vorticity in two dimensions are the same equations that describe the interaction of surface waves and vertical vorticity in a! shall ow fluid. It thus turns out to be more rewarding to discuss this latter problem, with its generalization to a not-so-shallow fluid, which lends itself more easily to experimental inquiry. This is done in Section 4, when vorticity is axially symmetric and stationary. Interesting phenomena appear, analogous to the Aharonov-Bohm effect in quantum mechanics. Section 5 studies the propagation of an acoustic wave through a disordered flow using a multiple scattering formalism. It is shown t hat coherent propagation is possible, and formulae are provided for the effective phase velocity and attenuation length. Specific examples are worked out both in two and three dimensions. Finally, two loose ends, a proof of the optical theorem and a discussion of the limits of validity of the Born approximation, are tied up in Section 6. Biliographical notes at the end of each section provide an inroad into the literature.

2 The Basics – Vortex Sound

In order for the issue of sound-vortex interaction to make any sense at all, it is necessary to make a number of assumptions. The main one is that all relevant flow velocities are of low Mach number. It is also necessary to assume that viscosity does not play a significant role. Less significantly from the point of view of the mathematics, but also necessary, the flow is asumed to be homoentropic. Finally, no external sources are considered, only a single component is allowed and, as the title indicates, boundaries are considered to be infinitely far away.

Under these assumptions, the relevant set of equations are

$$\frac{\partial \boldsymbol{v}}{\partial t} + (\boldsymbol{v} \cdot \nabla \boldsymbol{v}) = -\nabla h = -\frac{1}{\rho}\nabla p \tag{1}$$

$$\frac{\partial \rho}{\partial t} + \nabla \cdot (\rho \boldsymbol{v}) = 0 \tag{2}$$

where ρ is density, p is pressure, $\boldsymbol{v}$ is velocity and h is specific enthalpy. Everything is in an Eulerian framework.

The basic idea of the interaction of sound with vorticity is that there are two limits that make physical sense: One is the limit of incompressible flow, in which case density is assumed to be a constant in the above equations. The other is the linear acoustic limit, where all nonlinear terms are dropped from the same equations. Turning on the neglected terms in both cases little by little, ie perturbatively, introduces new phenomena that are identified as being due to the interaction of the previous two.

In order to clarify ideas and to introduce some of the techniques that will be useful later on we first consider the incompressible limit.

2.1 Incompressible Flow

Since density is a constant (strictly speaking it need be a constant only along streamlines), pressure is no longer a dynamical variable and it is called the hydrodynamical pressure, and Eqn. (2) reduces to

$$\nabla \cdot \boldsymbol{v} = 0. \tag{3}$$

Since the velocity field is divergenceless, there exists a vector potential $\boldsymbol{A}$ such that

$$\boldsymbol{v} = \nabla \wedge \boldsymbol{A}. \tag{4}$$

This vector potential is defined up to a scalar field ($\boldsymbol{A}$ and $\boldsymbol{A} + \nabla \Omega$ give rise to the same velocity $\boldsymbol{v}$ for any Ω. This is called a gauge freedom) so that vorticity $\boldsymbol{\omega} = \nabla \wedge \boldsymbol{v}$ satisfies

$$\boldsymbol{\omega} = \nabla \wedge \boldsymbol{v} = \nabla \wedge (\nabla \wedge \boldsymbol{A}) = -\nabla^2 \boldsymbol{A} \tag{5}$$

where we have used, without loss of generality because of the gauge freedom, $\nabla \cdot \boldsymbol{A} = 0$. Because of its definition, vorticity is a solenoidal vector: $\nabla \cdot \boldsymbol{\omega} \equiv 0$. Now, Eqn. (5) is a Poisson equation that can be solved for the vector potential $\boldsymbol{A}$ in terms of the vorticity $\boldsymbol{\omega}$ using Green's functions. This means looking for a solution G to the equation

$$\nabla^2 G = \delta(\boldsymbol{x}, \boldsymbol{x}') \tag{6}$$

and using superposition. Here, of course, it is a great piece of luck that the basic kinematical equation (5) is linear, even though the dynamical equation (1) is not.

Since the Green's function can be understood as the response of the medium to a point source we need boundary conditions; these are that the solution decays at infinity. Here, the fact that we consider unbounded flows makes its first appearance. The Green's function for the Poisson equation is different for different dimensionalities of space. We first consider three dimensions (3D).

When there is invariance under space translations, which means in particular that there are no boundaries, the solution to (6) with vanishing boundary conditions at infinity in 3D is

$$G = -\frac{1}{4\pi}\frac{1}{|\boldsymbol{x} - \boldsymbol{x}'|}. \tag{7}$$

so that the solution to (5) for the vector potential in terms of the vorticity is

$$\boldsymbol{A}(\boldsymbol{x}, t) = \frac{1}{4\pi}\int d^3x' \frac{\boldsymbol{\omega}(\boldsymbol{x}, t)}{|\boldsymbol{x} - \boldsymbol{x}'|} \tag{8}$$

where it is assumed that vorticity is either bounded in space or that it decays sufficiently rapidly for this integral, and all that follow, to make sense. Taking the curl of this equation we obtain the following relation between velocity and vorticity:

$$\boldsymbol{v}(\boldsymbol{x}, t) = -\frac{1}{4\pi}\int d^3x' \frac{(\boldsymbol{x} - \boldsymbol{x}') \wedge \boldsymbol{\omega}(\boldsymbol{x}', t)}{|\boldsymbol{x} - \boldsymbol{x}'|^3} \tag{9}$$

Note that it is implicit here that in an incompressible flow signals travel at infinite speed: a change in the vorticity translates into a change in the velocity at all points in space *at the same time*.

We wish to see what this means in terms of asymtotic behaviour of the velocity field: Let V be the (three dimensional) volume where vorticity is significantly different from zero, and consider distances very far away from this volume: $|\boldsymbol{x}| \gg V^{1/3}$. In this case we have

$$\boldsymbol{v}(\boldsymbol{x}, t) \approx \frac{1}{|\boldsymbol{x}|^3}\left(\boldsymbol{x} \wedge \int d^3x' \left(1 + \frac{3\boldsymbol{x} \cdot \boldsymbol{x}'}{|\boldsymbol{x}|^2}\right)\boldsymbol{\omega}(\boldsymbol{x}', t) - \int d^3x' \boldsymbol{x}' \wedge \boldsymbol{\omega}(\boldsymbol{x}', t)\right). \tag{10}$$

But the three dimensional integral of a solenoidal field identically vanishes. Indeed ($i = 1, 2, 3$),

$$\int d^3x'\omega_i = \int d^3x' \frac{\partial x'_i}{\partial x'_j}\omega_j \tag{11}$$

$$= -\int d^3x' x'_i \partial_j \omega_j \tag{12}$$

$$= 0 \tag{13}$$

where the next to last equality has been obtained integrating by parts (and thus using the boundedness of the vorticity), and the last equality is a consequence of vorticity being solenoidal. Consequently, we have that, in a bounded incompressible flow in three dimensions, velocity falls off at large distances as the inverse cube:

$$\boldsymbol{v}(\boldsymbol{x}, t) \sim \frac{1}{|\boldsymbol{x}|^3} \tag{14}$$

In two dimensions things are a little different. In this case the relation between vector potential $\boldsymbol{A}$ and vorticity $\boldsymbol{\omega}$ is

$$\boldsymbol{A}(\boldsymbol{x}, t) = \frac{1}{2\pi}\int d^2x' \ln|\boldsymbol{x} - \boldsymbol{x}'|\boldsymbol{\omega}(\boldsymbol{x}', t). \tag{15}$$

We now repeat what we did in three dimensions to obtain the far field behaviour of the velocity. There is however, one important difference , since in two dimensions there may exist vortices with non vanishing circulation, for which the integral of vorticity over all of (two dimensional) space will not vanish. We defer this case until Section 4.

$$\boldsymbol{v}(\boldsymbol{x}, t) = \nabla \wedge \boldsymbol{A}(\boldsymbol{x}, t) = \frac{1}{2\pi}\int d^2x' \frac{(\boldsymbol{x} - \boldsymbol{x}') \wedge \boldsymbol{\omega}(\boldsymbol{x}', t)}{|\boldsymbol{x} - \boldsymbol{x}'|^2} \tag{16}$$

$$\approx \frac{1}{|\boldsymbol{x}|^2}\left(\boldsymbol{x} \wedge \int d^2x' \left(1 + \frac{\boldsymbol{x} \cdot \boldsymbol{x}'}{|\boldsymbol{x}|^2}\right)\boldsymbol{\omega}(\boldsymbol{x}', t)\right.$$

$$\left. - \int d^2x' \boldsymbol{x}' \wedge \boldsymbol{\omega}(\boldsymbol{x}', t)\right). \tag{17}$$

$$\sim \frac{1}{|\boldsymbol{x}|^2} \tag{18}$$

when the circulation of the two dimensional vorticity vanishes at infinity.

2.2 Slightly Compressible Flow

Incompressible flow is of course an idealization. All flows imply *some* change in the fluid density. If we allow for small changes $\rho_1(\boldsymbol{x}, t)$ in density so that

$$\rho(\boldsymbol{x}, t) = \rho_0 + \rho_1(\boldsymbol{x}, t)$$

with ρ_0 a constant and $\rho_1 \ll \rho_0$, there will also be small corrections $\boldsymbol{v}'$ to the incompressible velocity $\boldsymbol{u}$ ($\nabla \cdot \boldsymbol{u} = 0$):

$$\boldsymbol{v}(\boldsymbol{x},t) = \boldsymbol{u}(\boldsymbol{x},t) + \boldsymbol{v}'(\boldsymbol{x},t)$$

with $|\boldsymbol{v}'| \ll |\boldsymbol{u}|$, and small corrections p_1 to the hydrodynamic pressure p_0 associated with the incompressible velocity $\boldsymbol{u}$:

$$p(\boldsymbol{x},t) = p_0(\boldsymbol{x},t) + p_1(\boldsymbol{x},t)$$

with $p_1 \ll p_0$.

To leading order in small quantities the dynamical equations then become

$$\frac{\partial \rho_1}{\partial t} + \rho_0 \nabla \cdot \boldsymbol{v}' + \boldsymbol{u} \cdot \nabla \rho_1 = 0 \tag{19}$$

$$\frac{\partial \boldsymbol{v}'}{\partial t} + (\boldsymbol{u} \cdot \nabla)\boldsymbol{v}' + (\boldsymbol{v}' \cdot \nabla)\boldsymbol{u} = -\frac{1}{\rho_0}\nabla p_1 + \frac{\rho_1}{\rho_0^2}\nabla p_0 \tag{20}$$

Now, the imposition that corrections to incompressible velocity have to be small must be qualified with some care: while the density against which small corrections are being compared is certainly nonvanishing, this is not necessarily the case for velocity, particularly in the case we have being considering, when vorticity is bounded in space. Far away from the vorticity, as we have seen, (incompressible) velocity vanishes as the inverse of distance cubed. What happens then to the "small" correction $\boldsymbol{v}'$?

A qualitative answer to the last question can be obtained as follows: in the extreme case of $\boldsymbol{u} = 0$ everywhere, the small density perturbations will be acoustic waves propagating on a fluid at rest with speed of sound c. If $\nabla \wedge \boldsymbol{u} \neq 0$ in a bounded region of space, there will be small density fluctuations within that region, of order

$$\rho_1 \sim \frac{u}{c}$$

and, being waves, they will propagate away from the vorticity distribution, decaying with distance as

$$\frac{1}{c|\boldsymbol{x}|} \quad \text{in 3 dimensions} \tag{21}$$

$$\frac{1}{c\sqrt{|\boldsymbol{x}|}} \quad \text{in 2 dimensions} \tag{22}$$

with velocity following the same law. Comparing (21) with (18) we see that, even in the case of very slightly compressible flow, far away from a vorticity distribution the acoustics will dominate. This is the phenomenon called *vortex sound*, to which we now turn our attention.

The way we shall approach this topic is by thinking of a vorticity distribution as the limit of very many vortex filaments, each of very small circulation, all of them tightly packed together. The reason for doing this is that, in the case of a single vortex filament it is possible to describe the fluid velocity as the gradient of a multivalued scalar field. Indeed, if $\boldsymbol{v}' = \nabla\phi$ and C is a closed curve sorrounding a vortex filament of circulation Γ we have

$$\Gamma = \oint_C \boldsymbol{v}' \cdot d\boldsymbol{\ell} = \oint \nabla\phi \cdot d\boldsymbol{\ell} = \phi_2 - \phi_1 \tag{23}$$

so that it is possible to describe the velocity field generated by a vortex filament as the gradient of a scalar field ϕ as long as ϕ has a discontinuity across some surface bounded by the filament. The surface itself has no physical significance, and measurable quantities cannot depend on it.

To be a little more precise, the vortex filament is described (in three dimensions) by a closed curve $\boldsymbol{X}(\sigma, t)$, where σ is a Lagrangean parameter labeling points along the filament, that bounds a surface $\mathcal{S}$. The tangent to the filament is

$$\boldsymbol{X}' \equiv \frac{\partial \boldsymbol{X}}{\partial \sigma}$$

and the velocity of each point is

$$\dot{\boldsymbol{X}} \equiv \frac{\partial \boldsymbol{X}}{\partial t}$$

As a warm-up, we redo the incompressible case.

2.3 Incompressible Vortex Dynamics

In this case,

$$0 = \nabla \cdot \boldsymbol{v} = \nabla^2 \phi \tag{24}$$

so that the problem at hand is to solve Laplace's equation with boundary conditions: vanishing at infinity, and with a prescribed discontinuity, Γ, across a surface $\mathcal{S}$ bounded by the filament. To go about this, define three new surfaces: $\mathcal{S}_+$ and $\mathcal{S}_-$ that sandwich $\mathcal{S}$ in between them, and $\mathcal{S}_T$, a thin tube that hugs the filament. These three surfaces divide the volume in two parts: Outside, where all functions are regular and single valued, and inside, where there are singularities (at the filament) and discontinuities (across the surface). For convenience, call $\mathcal{S}_\infty$ a closed surface infinitely far away from the filament.

We use Green's theorem in a form convenient for our purposes. Start with the obvious identity

$$\int d^3x' \nabla' \phi(\boldsymbol{x}') \cdot \nabla' G(\boldsymbol{x} - \boldsymbol{x}') = \int d^3x' \nabla' \phi(\boldsymbol{x}') \cdot \nabla' G(\boldsymbol{x} - \boldsymbol{x}') \tag{25}$$

and integrate by parts on both sides, using (6) and (24) to get

$$\phi(\boldsymbol{x}) = \int d^3x' \nabla' \cdot [\phi \nabla' G - (\nabla' \phi) G] \tag{26}$$

and we treat both terms in the right hand side of this equation in turn. Take first

$$\int d^3x'\nabla' \cdot [\phi\nabla' G] = \left[\int_{\mathcal{S}_+} + \int_{\mathcal{S}_-} + \int_{\mathcal{S}_T} + \int_{\mathcal{S}_\infty}\right] dS'_i \phi \nabla'_i G \tag{27}$$

We now make some asumptions: First, that the field ϕ and the Green's function G decay sufficiently fast at infinity that the integral over $\mathcal{S}_\infty$ vanishes. Second, that the tube hugging the vortex filament is thin enough, and that the field ϕ is sufficiently well behaved at short distances, that the integral over $\mathcal{S}_T$ also vanishes, in the limit of zero tube thickness. Both these assumptions must be checked with the solution, if and when one is obtained.

Using the boundary condition that the field ϕ is not single valued but has a discontinuity equal to the vortex circulation Γ when crossing the surface $\mathcal{S}$ that is bounded by the filament, we have, from (27)

$$\int d^3x'\nabla' \cdot [\phi\nabla' G] = \left[\int_{\mathcal{S}_+} + \int_{\mathcal{S}_-}\right] dS'_i \phi \nabla'_i G = \Gamma \int_{\mathcal{S}} dS'_i \nabla'_i G \tag{28}$$

Secondly,

$$\int d^3x'\nabla' \cdot [\nabla'(\phi) G] = \left[\int_{\mathcal{S}_+} + \int_{\mathcal{S}_-} + \int_{\mathcal{S}_T} + \int_{\mathcal{S}_\infty}\right] dS'_i (\nabla'_i \phi) G = 0. \tag{29}$$

To get the last equation, note that the contributions of $\mathcal{S}_+$ and $\mathcal{S}_+$ cancel each other out in (29) since both velocity $\nabla\phi$ and Green's function G are single valued across the surface $\mathcal{S}$, and assume that velocity $\nabla\phi$ is a vector tangent to the tube surrounding the vortex filament at short distances from the filament. Putting all this together we have

$$\phi(\boldsymbol{x}, t) = \Gamma \int dS'_i \nabla'_i G \tag{30}$$

or, using (7),

$$\phi(\boldsymbol{x}, t) = -\frac{\Gamma}{4\pi} \int dS'_i \nabla'_i \left(\frac{1}{|\boldsymbol{x} - \boldsymbol{x}'|}\right) \tag{31}$$

which is the well known result that the solution to Laplace's equation at point $\boldsymbol{x}$ with a discontinuity at a given surface is the solid angle subtended by the surface at point $\boldsymbol{x}$. As it has already been mentioned, the surface $\mathcal{S}$ has no physical meaning, but neither does the scalar potential ϕ. Only its gradient does. So, taking the gradient of (31) and using Stokes' theorem we have

$$\nabla_i \phi(\boldsymbol{x}, t) = \frac{\Gamma}{4\pi} \oint d\sigma \frac{\partial X_k}{\partial \sigma} \epsilon_{ijk} \nabla_j \left(\frac{1}{|\boldsymbol{x} - \boldsymbol{X}(\sigma, t)|}\right) \tag{32}$$

It is a straightforward exercise to check that (31) and (32) satisfy the assumptions that led to them. Taking the limit of continuously distributed vorticity we have the well known Biot-Savart law:

$$\boldsymbol{v}(\boldsymbol{x},t) = \nabla\phi(\boldsymbol{x},t) = \nabla\wedge\left(\frac{1}{4\pi}\int d^3y \frac{\boldsymbol{\omega}(\boldsymbol{y},t)}{|\boldsymbol{x}-\boldsymbol{y}|}\right) \tag{33}$$

2.4 Vortex Sound

What is the point of going through all those multivalued velocity potentials to get, in the end, nothing more than Biot-Savart's law? The point is that the methods and ideas can be generalized from Laplace's equation to the wave equation, and the result is the formula for vortex sound first obtained by Möhring, working on early results of Lighthill.

Going back to the equations (19) for slightly compressible flow, letting $\boldsymbol{u} = 0$ and $\boldsymbol{v}' = \nabla\phi$ we get that the scalar potential ϕ satisfies, possibly after a redefinition, the wave equation

$$\nabla^2\phi - \frac{1}{c^2}\frac{\partial^2\phi}{\partial t^2} = 0 \tag{34}$$

with c the adiabatic speed of sound for the fluid at rest, and we can ask the question: find a solution to this equation that is multivalued, with a prescribed discontinuity across a moving surface $\mathcal{S}(t)$ whose boundary is the vortex filament $\boldsymbol{X}(\sigma,t)$. The strategy is the same as in the incompressible case: sandwich the surface $\mathcal{S}$ in between surface $\mathcal{S}_+$ and $\mathcal{S}_-$, and the vortex filament by a thin tube $\mathcal{S}_T$. We need now the Green's function for the wave equation,

$$\nabla^2 G - \frac{1}{c^2}\frac{\partial^2 G}{\partial t^2} = \delta(\boldsymbol{x}-\boldsymbol{x}')\delta(t-t') \tag{35}$$

The surfaces $\mathcal{S}_+$, $\mathcal{S}_-$ and $\mathcal{S}_T$ bound a volume that hugs the vortex filament and the surface of discontinuity $\mathcal{S}$. We now use a version of Green's theorem that involves an integral over the *exterior* of this time dependent volume:

$$\int dt' \int d^3x' \nabla'\phi\cdot\nabla' G = \int dt' \int d^3x' \nabla'\phi\cdot\nabla' G \tag{36}$$

Integrating by parts in both sides, using (34) and (35) not forgetting the fact that a time dependent volume of integration picks up extra surface terms we get

$$\phi(\boldsymbol{x},t) = \Gamma\int dt' \int_{\mathcal{S}(t')} dS_i'\nabla_i' G + \frac{\Gamma}{c^2}\int dt' \int_{\mathcal{S}(t')} dS_i' V_i \frac{\partial G}{\partial t} \tag{37}$$

where $\boldsymbol{V}$ is the velocity of points on the surface $\mathcal{S}$.

So, this exercise gives a multivalued solution to the wave equation. However, this was not an end in itself but a step in the way to find the sound that is generated by a bounded, time dependent, vorticity distribution, and the wave equation we have just solved is a good approximation to the "real" fluid equations that describe the physical system at hand only in regions where nonlinear terms can be neglected. That is, far away from the vortex filament. We must then find the acoustic pressure, that is proportional to the time derivative of the velocity potential, far away from the filament. Taking the time derivative of (37) and using Stoke's theorem leads to

$$\dot{\phi}(\boldsymbol{x},t) = \Gamma \int dt' \oint d\sigma \left(\frac{\partial \boldsymbol{X}}{\partial t'} \wedge \frac{\partial \boldsymbol{X}}{\partial \sigma} \right) \cdot \nabla G \tag{38}$$

The term "far away" has two meanings: one is, at distances large compared to source size. In this case the Green's function approximates to

$$G(\boldsymbol{x}-\boldsymbol{x}';t-t') \approx G(\boldsymbol{x};t-t') - \boldsymbol{x}' \cdot \nabla G(\boldsymbol{x};t-t').$$

The second is, at distances large compared to wavelength. In this case space variations are dominated by variations over lenghtscales comparable to wavelength, and gradients are approximated by time derivatives, thus:

$$\nabla G \approx -\frac{\hat{x}}{c}\frac{\partial G}{\partial t}$$

Substitution of the last two expressions into (38) leads to

$$\dot{\phi}(\boldsymbol{x},t) = -\Gamma \int dt' \oint d\sigma \left(\frac{\partial \boldsymbol{X}}{\partial t'} \wedge \frac{\partial \boldsymbol{X}}{\partial \sigma} \right) \cdot \left(\frac{\hat{x}}{c}\frac{\partial G}{\partial t} + \frac{\hat{x}(\hat{x}\cdot \boldsymbol{X})}{c^2}\frac{\partial^2 G}{\partial t^2} \right) \tag{39}$$

and using the fact that the Green's function for the wave equation in infinite space is

$$G(\boldsymbol{x}-\boldsymbol{x}';t-t') = -\frac{1}{4\pi}\frac{1}{|\boldsymbol{x}-\boldsymbol{x}'|}\delta(t-t'-c^{-1}|\boldsymbol{x}-\boldsymbol{x}'|) \tag{40}$$

Eqn. (39) turns into

$$\dot{\phi}(\boldsymbol{x},t) \approx \frac{\Gamma}{4\pi c|\boldsymbol{x}|}\hat{x}\cdot \left[\frac{\partial}{\partial t}\oint d\sigma \dot{\boldsymbol{X}} \wedge \boldsymbol{X}' + \frac{1}{c}\frac{\partial^2}{\partial t^2}\oint d\sigma (\hat{x}\cdot \boldsymbol{X}) \cdot \dot{\boldsymbol{X}} \wedge \boldsymbol{X}' \right]_{\mathrm{ret}} \tag{41}$$

where the right hand side is evaluated at the retarded time

$$t_{\mathrm{ret}} = t - \frac{|\boldsymbol{x}|}{c}.$$

Finally, it has to be remembered that the vortex filament evolves, to leading order, under its own induction as if it were embedded in an incompresible fluid. The continuous version of this statement being that we are looking for the acoustic radiation generated by a bounded body of low Mach number vorticity, so that the vorticity dynamics is, to leading order, incompressible. Under this condition the first term on the right of (41) vanishes and the second can be rewritten to get

$$\dot{\phi}(\boldsymbol{x},t) \approx \frac{-\Gamma}{12\pi c^2|\boldsymbol{x}|}\hat{x}\cdot \left[\frac{\partial^3}{\partial t^3}\oint d\sigma (\hat{x}\cdot \boldsymbol{X})\, (\boldsymbol{X}\wedge \boldsymbol{X}') \right]_{\mathrm{ret}} \tag{42}$$

and, taking the continuous limit we obtain Möhring's formula for the far-field acoustic pressure p generated by a low Mach number unsteady vorticity $\boldsymbol{\omega}$ in a fluid of unperturbed density ρ_0:

$$p(\boldsymbol{x},t) = \frac{\rho_0}{12\pi c^2|\boldsymbol{x}|}\hat{x} \cdot \left[\frac{\partial^3}{\partial t^3}\int d^3y(\hat{x}\cdot \boldsymbol{y})\,(\boldsymbol{y}\wedge\boldsymbol{\omega})\right]_{\rm ret} , \tag{43}$$

where we have used the relation $\dot{\phi} = -p/\rho_0$, valid in the linear region. Noteworthy features of this formula are: it is linear in the vorticity; it involves the quadrupole moment of the vorticity (which, of course, does not have a monopole, and the dipole moment does not contribute to spontaneous vortex sound); and it involves third order time derivatives of vorticity thus greatly favoring high frequencies.

2.5 Bibliographical Notes

The classic text to get the basics on fluid dynamics is, of course, L. D. Landau and E. M. Lifshitz "Fluid Mechanics (Second Edition)", Pergamon (1987). Another useful text, easier to follow for students and also with an emphasis on the physics, is A. L. Fetter and J. D. Walecka "Theoretical Mechanics of Particles and Continua", McGraw Hill (1980). The study of the generation of sound by flows, rather than by time-dependent boundary conditions, was pioneered M. J. Lighthill, "On sound generated aerodynamically I. General theory". *Proc. Roy. Soc London* **A211**, 564-587 (1952); "On sound generated aerodynamically II. Turbulence as a source of sound". *Proc. Roy. Soc London* **A222**, 1-32 (1954). The idea of treating the nonlinear terms as perturbations of three limits: acoustic waves, vortical incompressible flow and heat diffusion, was put forth by B. T. Chu and L. S. Kovasznay, "Nonlinear interactions in a viscous heat conducting compressible gas", *J. Fluid Mech.* **3**, 494-514 (1958). The calculations described in subsection 2.4 can be found in F. Lund "Response of a Vortex Filament to High Frequency Sound", *Phys. Fluids A* **1**, 1521-1531 (1989). Möhring's result is in W. Möhring "On vortex sound at low Mach number", *J. Fluid Mech.* **85**, 685-691 (1978). Treatments of vortex sound, including the one presented in this chapter, are often plagued with uncertainties arising from uncontrolled approximations. The main source of difficulty is that, although the flow itself is of low Mach number and thus, to leading approximation, incompressible, the sound far away from it is, of course, a compressibility effect. The proper way to deal with this problem is to use matched asymptotic expansions, which was done by S. C. Crow, in "Aerodynamic sound emission as a singular perturbation problem", *Stud. App. Math.* **49**, 21-44 (1970). For a more recent treatment see T. Kambe, "Acoustic emissions by vortex motions" *J. Fluid Mech.* **173**, 643-666 (1986).

3 Scattering of Sound by Vorticity

Acoustic waves are described when the fluid equations are linearized for small density fluctuations. And vorticity is the only relevant dynamical variable when the flow is incompressible, evolving (in the nonviscous regime) according to the Euler equation. These two regimes can be considered as extreme limits, different terms being neglected in the equations in each case. What happens in between? Terms that have been neglected can be perturbatively turned on, and the resulting phenomena can be physically understood as the interaction of sound and vorticity. One simple question that can be asked is the following: What happens to a point vortex when it is hit by a sound wave? The answer is provided by Kelvin's theorem.

3.1 Kelvin's Theorem

This theorem holds for an adiabatic fluid, and is one of the few general statements that are valid for flows both compressible and vortical. It states that the circulation of a vortex line that moves with the flow is a constant.

Consider a vortex filament of circulation Γ and draw a circuit C around it. The circulation around this circuit

$$\Gamma = \oint_C \boldsymbol{v} \cdot d\boldsymbol{\ell},$$

is, of course, a function of time, and we wish to show that it does not change when C moves as a material line with the flow. To this end compute

$$\Gamma(t+dt) - \Gamma(t) = \oint_C \boldsymbol{v}(t+dt) \cdot d\boldsymbol{\ell}(t+dt) - \oint_C \boldsymbol{v} \cdot d\boldsymbol{\ell} \tag{44}$$

$$= \oint_C \left(\boldsymbol{v}(t) + \frac{d\boldsymbol{v}}{dt}\right) \cdot (d\boldsymbol{\ell} + (d\boldsymbol{\ell} \cdot \nabla)\boldsymbol{v}dt) - \oint_C \boldsymbol{v} \cdot d\boldsymbol{\ell} \tag{45}$$

$$= \oint \left(\boldsymbol{v} \cdot (d\boldsymbol{\ell} \cdot \nabla)\boldsymbol{v} + \frac{d\boldsymbol{v}}{dt} \cdot d\boldsymbol{\ell}\right) dt \tag{46}$$

$$= \oint d\boldsymbol{\ell} \cdot \nabla \left(\frac{1}{2}\boldsymbol{v}^2 + h\right) dt \tag{47}$$

$$= 0 \tag{48}$$

where Eqn. (1) has been used in obtaining the next to last equality.

A consequence of this theorem is that a point vortex moves with the local velocity of the flow. Indeed, suppose that this was not the case. Draw a small circuit around the point vortex and let the circuit move as a material line with the flow. Initially the circulation around this small circuit is different from zero, as it encloses the vortex. After a while, however, it will be zero since it will no longer enclose it, since the vortex, as opposed to the circuit, was not moving with the flow. This is in contradiction with Kelvin's theorem. The last consequence suggests a physical picture to understand the scattering of sound by vorticity: Let an acoustic wave be incident upon a point vortex. From what has just been said, the vortex will oscillate with the flow associated with the wave. But this will be an unsteady distribution of vorticity that, according to the computations of Section 2, will radiate sound. This picture is qualitatively correct, as least for acoustic waves with wavelength much larger than the size of a vorticity distribution, but it is not possible to blindly apply the formulae of Section 2 for two reasons: one is that in the case at hand the vortex is not moving under its own induction, but rather its motion is induced by an outside agent, the acoustic wave; second, the interaction of the acoustic wave with the vortex is not local, since there is a flow around the vortex whose interaction with the wave there is no reason to neglect. The only way to deal with the above issues is to be systematic. We thus start with the equations

$$\frac{\partial \boldsymbol{v}}{\partial t} + (\boldsymbol{v} \cdot \nabla)\boldsymbol{v} = -\frac{1}{\rho}\nabla P \tag{49}$$

$$\frac{\partial \rho}{\partial t} + \nabla \cdot (\rho \boldsymbol{v}) = 0 \tag{50}$$

where viscosity has been neglected altogether. The flow is described by velocity $\boldsymbol{v}$, pressure P and density ρ. We wish to describe the scattering of sound by a vortical flow. The latter is supposed to be of low Mach number, and the incoming sound is supposed to be weak enough not to perturb the vortical flow. We thus proceed by successive approximations starting with the flow, and dynamical quantities are split in two: those that occur in the abscence of incident sound, plus small corrections due to the incident sound:

$$\begin{pmatrix} P \\ \rho \\ \boldsymbol{v} \end{pmatrix} = \begin{pmatrix} P_0 \\ \rho_0 \\ \boldsymbol{u} \end{pmatrix} + \begin{pmatrix} p_s \\ \rho_s \\ \boldsymbol{v}_s \end{pmatrix} \tag{51}$$

The first column on the right describes the background flow (ie in the abscence of incoming sound), and the second column on the right describes the small corrections due to the incoming sound. The background flow is of low Mach number, which does not mean incompressible. It obeys the order zero equations

$$\frac{\partial \boldsymbol{u}}{\partial t} + (\boldsymbol{u} \cdot \nabla)\boldsymbol{u} = -\frac{1}{\rho_0}\nabla P_0 \tag{52}$$

$$\frac{\partial \rho_0}{\partial t} + \nabla \cdot (\rho_0 \boldsymbol{u}) = 0 \tag{53}$$

The background density ρ_0 and pressure P_0 are in turn decomposed in two: their values with the fluid at rest, and small corrections due to to the low Mach number background flow:

$$\begin{pmatrix} P_0 \\ \rho_0 \end{pmatrix} = \begin{pmatrix} P_0^{(0)} \\ \rho_0^{(0)} \end{pmatrix} + \begin{pmatrix} P_0^{(1)} \\ \rho_0^{(1)} \end{pmatrix} \tag{54}$$

Of course, when the fluid is at rest $\boldsymbol{u} = 0$. We will assume that

$$\frac{\partial \boldsymbol{u}}{\partial t} \sim (\boldsymbol{u} \cdot \nabla)\boldsymbol{u} \tag{55}$$

This is valid in the "near field" of the background flow, but *not* far away from it, where the flow will be dominated by the vortex sound generated by it, and where gradients will be dominated by a length scale comparable to the wavelength of said vortex sound, not by the length scale for spatial variations of the flow as in the near field. This already raises the question, how is the sound scattered by the vortical flow going to be disentangled from the sound spontaneously generated by it? The answer is, by sending in sound of a frequency much higher than any frequency associated with the flow. Note also that the ratio

$$\frac{\rho_0^{(1)}}{\rho_0^{(0)}}$$

is of order Mach number squared. The order one equations, that describe the evolution of an incoming acoustic wave that is incident upon the background vortical flow are

$$\frac{\partial \rho_s}{\partial t} + \nabla \cdot (\rho_0 \boldsymbol{v}_s) + \nabla \cdot (\rho_s \boldsymbol{u}) = 0 \tag{56}$$

$$\frac{\partial \boldsymbol{v}_s}{\partial t} + (\boldsymbol{u} \cdot \nabla)\boldsymbol{v}_s + (\boldsymbol{v}_s \cdot \nabla)\boldsymbol{u} = -\frac{1}{\rho_0}\nabla p_s + \frac{\rho_s}{\rho_0^2}\nabla P_0 \tag{57}$$

Taking the time derivative of the first equation, the divergence of the second, adding up the results and using (52) we get

$$\begin{aligned} \nabla^2 p_s - \frac{\partial^2 \rho_s}{\partial t^2} &= -\partial_i[\partial_j(\rho_0 u_j)v_{si}] - (\partial_i \rho_0)v_{sj}\partial_j u_i - \rho_0 \partial_i[(v_{sj}\partial_j)u_i] \\ &\quad -(\partial_i \rho_0)u_j \partial_j v_{si} - \rho_0 \partial_i[(u_j \partial_j)v_{si}] - \partial_i[\partial_j(\rho_0 v_{sj})u_i] \\ &\quad -\partial_i[\partial_j(\rho_s u_j)u_i] - (\partial_i \rho_s)(u_j \partial_j)u_i - \rho_s \partial_i[(u_j\partial_j)u_i]. \end{aligned} \tag{58}$$

No further approximations are performed in arriving at this result. The background pressure P_0 does not appear explicitely because it cancels out in the intermediate calculations. Further rearranging this equation leads to

$$\nabla^2 p_s - \frac{\partial^2 \rho_s}{\partial t^2} = (I) + (II) + (III) + (IV) \tag{59}$$

where

$$\begin{aligned} (I) &= -2\rho_0 \partial_i \partial_j (u_i v_{sj}) \\ (II) &= -\partial_i \partial_j (\rho_s u_j u_i) \\ (III) &= -2(\partial_i \rho_0)\partial_j (u_j v_{si} + v_{sj} u_i) \\ (IV) &= -2(\partial_i \partial_j \rho_0) v_{sj} u_i \end{aligned}$$

and $\partial_i \equiv \partial/\partial x_i$. It is easy to check that (I) is of order Mach number (M), II of order M^2 and both III and IV of order M^3.

3.2 Scattering to Order M

Keeping only terms of order M in (59) we have

$$\nabla^2 p_s - \frac{\partial^2 p_s}{\partial t^2} = -2\rho_0 \partial_i \partial_j (u_i v_{sj}) \equiv S(\boldsymbol{x}, t) \tag{60}$$

where we have called S the right hand side, suggesting an interpretation in terms of a wave equation with a source term. Recalling the definition of the Green's function for the wave equation (35) we can turn the inhomogeneous differential equation (60) into an integral equation that explicitely includes a particular solution p_s^{inc} representing the incident sound, it will soon be taken to be a plane wave:

$$p_s(\boldsymbol{x}, t) = p_s^{\text{inc}}(\boldsymbol{x}, t) + \int d^3x' dt' G(\boldsymbol{x} - \boldsymbol{x}'; t - t') S(\boldsymbol{x}', t') \tag{61}$$

and using the explicit form (40) for the Green's function in infinite space, together with the representation

$$\delta(t - t' - c_0^{-1}|\boldsymbol{x} - \boldsymbol{x}'|) = \frac{1}{2\pi} \int d\nu e^{-i\nu(t - t' - c_0^{-1}|\boldsymbol{x} - \boldsymbol{x}'|)}$$

we get the following expression for the far field behaviour of the acoustic pressure:

$$p_s(\boldsymbol{x}, t) = p_s^{\text{inc}}(\boldsymbol{x}, t) - \frac{1}{8\pi^2 |\boldsymbol{x}|} \int d^3x' dt' d\nu e^{-i\nu(t-t')} e^{i\nu|\boldsymbol{x}|/c_0} e^{-i\nu\hat{x}\cdot\boldsymbol{x}'} S(\boldsymbol{x}', t') \tag{62}$$

where "far field" means at distances large compared to the size of the background flow so we can use[1]

$$|\boldsymbol{x} - \boldsymbol{x}'| \approx |\boldsymbol{x}| - \hat{x} \cdot \boldsymbol{x}' \tag{63}$$

It is *not* assumed that wavelengths are large compared to the size of the background flow. Indeed, as we shall see, the most interesting physics happens when the wavelength is comparable to the relevant length scales of the flow. The second term on the right of (62) decays like the inverse power of distance for large distances from the background flow. It thus deserves to be called the scattered wave: it is acoustic radiation emanating from the interaction with the incident wave. This scattered wave is of order M compared

[1] It is assumed that the term $\hat{x} \cdot \boldsymbol{x}'$ does not vanish. Better, it is asumed that the regions where it vanishes do not contribute significantly to the integral (62). This is a good approximation here, but it would be a bad approximation if the volume of integration were not a volume but a plane, and we were to look at a direction perpendicular to that plane. This happens in diffraction theory and it is the reason that higher order terms mus t be kept in (63). They lead to Fresnel and Fraunhoffer diffraction

with the incident wave so that it makes sense to solve the integral equation by succesive approximations ("Born approximation"). This means, to first order, replacing the velocity $\boldsymbol{v}_s$ associated with the perturbation due to the presence of the incident wave, by the value given by the incident wave itself, which is a plane wave of frequency ν_0, wave vector $\boldsymbol{k}_0$ and (velocity) amplitude $\boldsymbol{v}_s^0$. This means that the source term is approximated as

$$S(\boldsymbol{x},t) \approx -2\rho_0 \nabla_i \nabla_j \left(u_i(\boldsymbol{x},t) \frac{v_{sj}^0}{2} e^{i(\boldsymbol{k}_0 \cdot \boldsymbol{x} - \nu_0 t)} \right). \tag{64}$$

Replacing this value in 62) and integrating by parts we get

$$\begin{aligned} p_s(\boldsymbol{x},t) &= p_s^{\text{inc}}(\boldsymbol{x},t) \\ &+ \frac{\rho_0 v_{sj}^0}{8\pi^2 |\boldsymbol{x}|} \int d\nu e^{-i\nu t} e^{i\nu|\boldsymbol{x}|/c_0} \int d^3x' dt' e^{i\Delta\nu t'} e^{-i\boldsymbol{q}\cdot\boldsymbol{x}'} k_i k_j u_i(\boldsymbol{x}',t'). \end{aligned} \tag{65}$$

where

$$\boldsymbol{k} \equiv \frac{\omega}{c_0} \hat{x}$$

is the scattered wave vector,

$$\boldsymbol{q} \equiv \boldsymbol{k} - \boldsymbol{k}_0$$

is the difference between scattered and incident wave vectors, and

$$\Delta\nu \equiv \nu - \nu_0$$

is the difference between scattered and incident frequencies. The integration by parts that is performed to get (65) assumes that

$$u_i \underset{|\boldsymbol{x}| \to \infty}{\longrightarrow} 0 \tag{66}$$

That is, the background flow must be bounded in all directions in three dimensions. This seemingly obvious requirement is important to remember when trying to apply these results to experiments. The integral over $\boldsymbol{x}'$ and t' in the right hand side of (65) is recognized as the Fourier transform[2] , in space and time, $\tilde{u}_i$ of the background velocity u_i. We can thus write

$$p_s(\boldsymbol{x},t) = p_s^{\text{inc}}(\boldsymbol{x},t) + \frac{\rho_0 v_{sj}^0}{8\pi^2 |\boldsymbol{x}|} \int d\nu e^{-i\nu t} e^{i\nu|\boldsymbol{x}|/c_0} k_i k_j \tilde{u}_i(\boldsymbol{q},\Delta\nu). \tag{67}$$

[2] We use the following definition of Fourier transform:

$$\tilde{f}(q) = \int_{-\infty}^{\infty} dx e^{-iqx} f(x)$$

For an incompressible flow, velocity and vorticity are linearly related in Fourier space. It is thus possible to express (67) in terms of the Fourier transform of vorticity up to terms of order M. To this end, note that since vorticity is the curl of the velocity,

$$\boldsymbol{\omega}(\boldsymbol{x}, t) = \nabla \wedge \boldsymbol{u}(\boldsymbol{x}, t)$$

in Fourier space this relation translates to

$$\tilde{\boldsymbol{\omega}}(\boldsymbol{q}, \Delta) = i\boldsymbol{q} \wedge \tilde{\boldsymbol{u}}(\boldsymbol{q}, \Delta).$$

Taking the cross product in both sides with $\boldsymbol{q}$ we have

$$\boldsymbol{q} \wedge \tilde{\boldsymbol{\omega}} = -iq^2\tilde{\boldsymbol{u}} + i\boldsymbol{q}(\boldsymbol{q} \cdot \tilde{\boldsymbol{u}}).$$

Since the second term on the right hand side is of order M we have, to leading order,

$$\tilde{\boldsymbol{u}} = \frac{i}{q^2}\boldsymbol{q} \wedge \tilde{\boldsymbol{\omega}}, \tag{68}$$

and Eqn. (67) can be written

$$p_s(\boldsymbol{x}, t) = p_s^{\text{inc}}(\boldsymbol{x}, t) + p_s^{\text{scatt}}(\boldsymbol{x}, t) \tag{69}$$

with

$$p_s^{\text{scatt}}(\boldsymbol{x}, t) = \frac{i\rho_0 v_{sj}^0}{8\pi^2|\boldsymbol{x}|} \int d\nu e^{i\nu t} e^{i\nu|\boldsymbol{x}|/c_0} \frac{k_i k_j \epsilon_{iml} q_l \tilde{\omega}_l}{q^2}. \tag{70}$$

This expression can be given a more compact form writing the pressure in the frequency, rather than time, domain:

$$p_s^{\text{scatt}}(\boldsymbol{x}, t) = \int \frac{d\nu}{2\pi} e^{i\nu t} \tilde{p}_s^{\text{scatt}}(\boldsymbol{x}, \nu).$$

Remembering that $\boldsymbol{q} = \boldsymbol{k} - \boldsymbol{k}_0$, noting $\boldsymbol{v}_s^0 = v_s^0 \hat{x}^0$ and $\boldsymbol{k}^0 = k^0 \hat{x}^0$, we get

$$p_s^{\text{scatt}}(\boldsymbol{x}, \nu) = \frac{i\rho_0 \nu^2 \nu_0}{4\pi c^3 |\boldsymbol{x}| q^2} e^{i\nu|\boldsymbol{x}|/c_0} (\boldsymbol{v}_s^0 \cdot \hat{x})(\hat{x} \wedge \hat{x}^0) \cdot \tilde{\omega}_l(\boldsymbol{q}, \Delta\nu). \tag{71}$$

As it was mentioned before, in order to disentangle the scattered sound from the spontaneously generated sound, it is necessary to scatter acoustic waves with frequencies much higher than any relevant flow frequency. Under this condition, the scattered frequency ν will be very similar to the incident frequency ν_0 and Eqn. (71) can be rewritten as

$$p_s^{\text{scatt}}(\boldsymbol{x}, \nu) = \frac{i\nu p_0^{\text{inc}}}{8\pi c^2 |\boldsymbol{x}|} e^{i\nu|\boldsymbol{x}|/c_0} \frac{\sin\theta\cos\theta}{\cos\theta - 1} \tilde{\omega}_\perp(\boldsymbol{q}, \Delta\nu) \tag{72}$$

where $\tilde{\omega}_\perp$ is the component of vorticity perpendicular to the scattering plane, defined as the plane subtended by $\hat{x}$ and $\hat{x}^0$. The scattering angle θ is defined by

$$\cos\theta = \hat{k} \cdot \hat{k_0}$$

and we have used the fact that, for a plane acoustic wave,

$$\rho_0 v_s^0 = \frac{p_0}{c}$$

Eqn. (72) suggests a divergence in the forward direction, i.e. for small scattering angles. Is this really so? To examine what happens, one must not forget that there is also a hidden angle dependence in the Fourier transform of vorticity through its dependence in $\boldsymbol{q}$, and note that

$$\tilde{\omega}_\perp(\boldsymbol{q}, \Delta) = \int d^3x' e^{-i\boldsymbol{q}\cdot\boldsymbol{x}'} \tilde{\omega}_\perp(\boldsymbol{x}', \Delta) \tag{73}$$

$$\underset{|\boldsymbol{q}| \to 0}{\approx} \int d^3x'(1 - i\boldsymbol{q}\cdot\boldsymbol{x}' - \frac{1}{2}(\boldsymbol{q}\cdot\boldsymbol{x}')^2 + \cdots)\tilde{\omega}_\perp(\boldsymbol{x}', \Delta) \tag{74}$$

where the approximation in the second line of the above expression is valid at small scattering angles θ for which $|\boldsymbol{q}|^2 = 2k_0^2(1 - \cos\theta) \approx k_0^2\theta^2$ is also small. More precisely, the expansion (73) is valid when $|\boldsymbol{q}| \ll V^{-1/3}$ where V is the volume outside of which the vorticity vanishes.

We have already seen in Section 2 that the first integral on the right hand side of (74) vanishes in three dimensions because vorticity is solenoidal. Consequently the coefficient of the term that would appear to be divergent in (2) vanishes and, to leading order in small scattering angles, the scattered pressure is proportional to the dipole moment of the vorticity distribution:

$$p_s^{\rm scatt}(\boldsymbol{x}, \nu) \underset{\theta \to 0}{\sim} \int d^3x'(\hat{y}_0 \cdot \boldsymbol{x}')\omega_\perp(\boldsymbol{x}') \tag{75}$$

Stationary Vorticity Distribution. When the flow that is responsible for the scattering is independent of time,

$$\boldsymbol{u}(\boldsymbol{x}, t) = \boldsymbol{u}(\boldsymbol{x}),$$

the first Born approximation (65) yields, instead of (67), the following expression:

$$p_s(\boldsymbol{x}, t) = p_s^{\rm inc}(\boldsymbol{x}, t) + \frac{\rho_0 v_{sj}^0}{4\pi|\boldsymbol{x}|} e^{-i\nu_0 t} e^{i\nu_0|\boldsymbol{x}|/c_0} k_i k_j \tilde{u}_i(\boldsymbol{q}). \tag{76}$$

and (72) is replaced by

$$p_s^{\rm scatt}(\boldsymbol{x}, t) = \frac{i\nu p_0^{\rm inc}}{8\pi c^2} \frac{e^{-i\nu_0(t - |\boldsymbol{x}|/c_0)}}{|\boldsymbol{x}|} \frac{\sin\theta\cos\theta}{\cos\theta - 1} \tilde{\omega}_\perp(\boldsymbol{q}) \tag{77}$$

Optical Theorem?. When the vorticity distribution is stationary, Eqn. (77) indicates, according to the usual definitions of scattering theory, an expression for the scattering amplitude of sound by vorticity in three dimensions, in the first Born approximation :

$$f(\theta) \equiv \frac{i\nu}{4\pi c^2} \frac{\sin\theta\cos\theta}{(\cos\theta - 1)} \tilde{\omega}_\perp(\boldsymbol{q}) \tag{78}$$

with the usual definitions of differential scattering cross section

$$\frac{d\sigma}{d\Omega} \equiv |f(\theta)|^2$$

and total scattering cross section

$$\sigma_T \equiv \int d\Omega \frac{d\sigma}{d\Omega}.$$

From (76) we see that the forward scattering amplitude $f(0)$ is real. An alternative expression for $f(\theta)$ comes from (76):

$$f(\theta) = \frac{k_0}{2\pi c_0} (\hat{k}_0 \cdot \hat{k})(\hat{k}_0 \cdot \tilde{\boldsymbol{u}}(\boldsymbol{k} - \boldsymbol{k}_0)) \tag{79}$$

so that the total scattering cross section can be written

$$\sigma_T = \frac{k_0^4}{(2\pi)^2 c^2} \int d\Omega_k (\hat{k}_0 \cdot \hat{k})^2 \left| \hat{k}_0 \cdot \tilde{\boldsymbol{u}}(\boldsymbol{k} - \boldsymbol{k}_0) \right|^2 \tag{80}$$

The Optical theorem states that the total scattering cross section is proportional to the forward scattering amplitude:

$$\sigma_T(k_0) = \frac{4\pi}{k_0} \mathrm{Im} f(0). \tag{81}$$

Clearly, this theorem is not satisfied at this level, since the scattering cross section is quadratic in the vorticity while the scattering amplitude is linear. Also, as already noted, the forward scattering amplitude $f(0)$ is real in first Born approximation. Alternatively, to leading order (81) is trivially satisfied in the sense that both sides vanish.

Clearly one has to go to the next order in the Born approximation scheme in order to perform a meaningful check of the validity of the optical theorem. This will be done in Section 6.1.

3.3 Bibliographical Notes

The study of the scattering of sound by turbulence goes back at least to the work of A. M. Oboukhov, "Uber die schallstreuung in der turbulenten strömung", *Dokl. Akad. Nauk. SSSR* **30**, 616-620 (1941). Formula (71) can be found, in a slightly more cumbersome notation, in the works of R. H. Kraichnan, "The scattering of sound in a turbulent medium", *J. Acoust. Soc. Am.* **25**, 1096-1104 (1953), and M. J. Lighthill, "On the energy scattered from the interaction of turbulence with sound or shock waves", *Proc. Camb. Philos. Soc.*, **49**, 531-551 (1953). A considerable body of experimental research using acoustics to probe atmospheric turbulence is referenced in A. S. Monin and A. M. Yaglom, *Statistical fluid mechanics*, vol. 2, The MIT Press, Cambridge (1975).

A study of the scattering of sound by vortical flows, not necessarily turbulent, was undertaken by T. Kambe, "Scattering of sound by a vortex system", *J. Japan Soc. Fluid Mech.* **1**, 149-165 (1982) (in japanese), who noted the similarity with the scattering of X rays by crystals. See also M. S. Howe, "On the scattering of sound by a vortex ring", *J. Sound. Vib.* **87**, 567-571 (1983), and A. L. Fabrikant, "Sound scattering by vortex flows", *Sov. Phys. Acoust.* **29**, 152-155 (1983). A Karman vortex street provides a conveniently periodical flow on which to scatter sound. This experiment has been performed by P. R. Gromov, A. B. Ezerskii and A. L. Fabrikant, "Sound scattering by a vortex wake behind a cylinder", *Sov. Phys. Acoust.* **28**, 452-455 (1982) adn by C. Baudet, S. Ciliberto and J. F. Pinton, *Phys. Rev. Lett.* **67**, 193-195 (1991). The possibility of using ultrasound scattering to probe vortical flows, turbulent or not, under controlled laboratory conditions, was emphasized by F. Lund and C. Rojas, "Ultrasound as a probe of turbulence", *Physica D* **37**, 508-514 (1989). A related experiment is described in M. Oljaca et al., "Ultrasound scattering by a swirling jet", *Phys. Fluids* **10**, 886-898 (1998). Ultrasound was used to characterize vorticity filaments by B. Dernoncourt, J.-F. Pinton and S. Fauve, "Experimental study of vorticity filaments in a turbulent swirling flow" *Physica D* **117** 181-190, (1998).

The work described in this Section applies when the frequency of the scattered sound is high compared to the inverse of any relevant timescales of the background flow that is responsible for the scattering, and there is no restriction on wavelength: it is not necessarily long compared to the dimensions of the base flow, as in the case of the generation of sound by vorticity in Section 2, but it cannot be so short that the Born approximation ceases to be valid, as discussed futher in Section 6. The opposite limit, of frequencies not necessarily high compared to flow frequencies, but with wavelength large compared to dimensoins of background flow, has been studied recently by S. G. Llewellyn Smith and R. Ford, "Three-dimensional acoustic scattering by vortical flows.I. General theory", *Phys. Fluids* **13**, 2876-2889 (2001); "Three-dimensional acoustic scattering by vortical flows.II. Axisymmetric scattering by Hill's spherical vortex", *Phys. Fluids* **13**, 2890-2900 (2001).

Further experimental, numerical and analytical work can be found in M. S. Korman and R. T. Beyer, "Nonlinear scattering of crossed ultrasonic beams in the presence of turbulence in water. Part I: Experiment", *J. Acoust Soc. Am.* **85**, 611-620 (1989), T. Colonius, S. K. Lee and P. Moin, "The scattering of sound waves by a vortex: numerical simulations and analytical solutions", *J. Fluid Mech.* **260**, 271-298 (1994), and in J. Reinschke, W. Möhring and F. Obermeier, "Scattering of sound waves by a cylindrical vortex: A semi-analytical theory", *J. Fluid Mech.* **333**, 273-299 (1997).

4 Two Dimensions

From an algebraic viewpoint, the main difference between wave propagation in two and three dimensions lies in the Green's function. In three dimensions, the response of an homogeneous, infinite medium to a source localized at a point in space and time is an outwardly propagating spherical shell. In two dimensions, it is an outwardly propagating spherical front with an infinitely long (albeit decaying) tail in time. From a physical point of view, there is a very important difference: the integral of vorticity over all space needs not vanish, as in the easily realisable example of a vertical vortex (say, in water) with non vanishing circulation. In the latter case, the Born approximation scheme that was used in Section 3 breaks down and the problem has to be restudied from the beginning.

Another interesting feature of two dimensions is that the full nonlinear fluid equations for nonviscous adiabatic flow are the same as the nonlinear equations for the propagation of surface waves over a shallow fluid (ie a fluid whose depth is small compared to wavelength), and experiments with surface waves provide information (such as wave crest visualization and measurement) that is difficult to obtain in acoustics. On the other hand, surface waves that are easy to realize experimentally have wavelengths at least comparable and usually smaller than the fluid depth. This raises the additional question, how do surface waves interact with vorticity in a fluid of arbitrary depth, the main difference with acoustic waves being that surface waves are in general dispersive.

4.1 Vanishing Circulation

When the vorticity distribution has vanishing circulation everything goes pretty much as in three dimensions, and the end result for the scattered pressure is

$$p_s^{\mathrm{scatt}}(\boldsymbol{x},\nu) = \rho_0 v_s^0 \sqrt{\frac{\pi^3\nu}{2c|\boldsymbol{x}|}} \frac{\sin\theta\cos\theta}{\cos\theta - 1} e^{i\left(\frac{\nu|\boldsymbol{x}|}{c}+\frac{3\pi}{4}\right)} \tilde{\omega}(\boldsymbol{q},\Delta) \tag{82}$$

with definitions exactly as in Section 3. The subscript "$\perp$" is now superfluous since vorticity can only be perpendicular to the scattering plane.

4.2 Surface Waves

Consider an incompressible fluid with velocity $\boldsymbol{V}$, pressure P, (uniform) density ρ in a uniform gravitational field g with equilibrium depth h and a free surface at $h + \eta(x, y, t)$. The dynamical equations are

$$\partial_t \boldsymbol{V} + \boldsymbol{V} \cdot \nabla \boldsymbol{V} = -\frac{1}{\rho} \nabla P - g\hat{z} \tag{83}$$

$$\nabla \cdot \boldsymbol{V} = 0 \tag{84}$$

Boundary conditions at the free surface are that surface elements move with the surface and that pressure has a discontinuity that is exactly compensated by surface tension. This means that we are neglecting whatever is on top of the fluid. This is a good approximation for air over water.

$$\left.\begin{aligned} V_z &= \frac{\partial \eta}{\partial t} + V_\perp \cdot \nabla_\perp \eta \\ P &= -\tau \nabla^2 \eta \end{aligned}\right\} z = h + \eta \tag{85}$$

where τ is the surface tension, $V_\perp$ is the horizontal velocity and $\nabla_\perp$ is the horizontal gradient.

The boundary condition at the bottom is that there be no vertical velocity:

$$V_z = 0 \qquad z = 0 \tag{86}$$

The usual algebra that leads to the properties of surface waves proceeds by linearizing these equations when the fluid is at rest. What we wish to do is to linearize around a vortical flow and we shall take as such flow a vertical, stationary, axially symmetric vortex. Velocity $\boldsymbol{V}$, pressure P and surface elevation η are thus split in two: quantities associated with vertical vorticity and small perturbations due to small surface deformations around that background, thus

$$\begin{pmatrix} \boldsymbol{V} \\ P \\ \eta \end{pmatrix} = \begin{pmatrix} \boldsymbol{U} \\ P_0 \\ \eta_0 \end{pmatrix} + \begin{pmatrix} \boldsymbol{v} \\ p_1 \\ \eta_1 \end{pmatrix} \tag{87}$$

where the first column on the right hand side describes the vertical vortex. Using polar coordinates (r, θ, z) the situation with this vortex only is described by the order zero equations

$$\boldsymbol{U} = U_o(r)\hat{\theta} \tag{88}$$

$$P_0 = \rho g(h - z) + p_0(r, t) \tag{89}$$

$$\frac{U_0^2}{r} = \frac{1}{\rho}\frac{\partial p_0}{\partial r} \tag{90}$$

where the last equation determines p_0 once the background flow U_0 has been prescribed. Substitution of (89) into (85) gives

$$p_0 = \rho g \eta_0 - \tau \nabla_\perp^2 \eta_0 . \tag{91}$$

from which the unperturbed surface deformation η_0 due to the stationary vortex can be determined. The next step is to write the order one equations for $\boldsymbol{v}$, p_1 and η_1. With the notation $\boldsymbol{v} = (\boldsymbol{u}, w)$, i.e. separating out the horizontal and vertical components of order one velocity, they are

$$\left(\frac{\partial}{\partial t} + \boldsymbol{U} \cdot \nabla_\perp\right) \boldsymbol{u} + \boldsymbol{u} \cdot \nabla_\perp \boldsymbol{U} = -\frac{1}{\rho} \nabla_\perp p_1 \tag{92}$$

$$\left(\frac{\partial}{\partial t} + \boldsymbol{U} \cdot \nabla_\perp\right) w = -\frac{1}{\rho} \frac{\partial p_1}{\partial z} \tag{93}$$

$$\nabla_\perp \cdot \boldsymbol{u} + \partial_z w = 0 \tag{94}$$

Similarly, the boundary conditions to order one are

$$\left.\begin{aligned} w &= \left(\tfrac{\partial}{\partial t} + \boldsymbol{U} \cdot \nabla_\perp\right) \eta_1 + \boldsymbol{u} \cdot \nabla_\perp \eta_0 \\ p_1 &= \rho g \eta_1 - \tau \nabla^2 \eta_1 \end{aligned}\right\} z = h + \eta \tag{95}$$

at the free surface. At the bottom, Eqn. (86) together with (93) imply

$$\frac{\partial p_1}{\partial z} = 0 \qquad z = 0. \tag{96}$$

Taking the horizontal divergence of (92), the z- derivative of (93) and using (94), it is easy to get

$$\nabla_\perp^2 p_1 + \frac{\partial^2 p_1}{\partial z^2} = -2\rho (\nabla_a U_b)(\nabla_b u_a). \tag{97}$$

where a, b are indexes labeling the horizontal coordinates.

Now, quantities of order one have a length scale for space variations given by the wavelength (and we shall call k the corresponding wavenumber) of the perturbation, and a corresponding timescale given by the inverse of the frequency ν. These two quantities are related by $\nu = c_\phi k$ where c_ϕ is the phase velocity associated with the surface waves, which we shall assume to be of the same order of magnitude as the group velocity c_g . The background flow is assumed to be stationary with a characteristic size a, say, the radius of the vertical vortex. Thus, the left hand side of (97) is of order $k^2 p_1$, and the right hand side is of order $\rho k U u / a$. The order one equation (92) implies

$$\nu u \sim \frac{1}{\rho} k p_1$$

so that the ratio between the right hand side and the left hand side of (97) is of order

$$\frac{U}{c_\phi} \frac{1}{ka}. \tag{98}$$

By analogy with acoustics we shall call the ratio between the background velocity U and the phase velocity c_ϕ the Mach number: $M = U/c_\phi$, and this will be assumed to be small, $M \ll 1$. Also, wavelengths will be assumed to be small compared to vortex size a: $ka \gg 1$. But, it will be assumed that

$$Mka \sim 1$$

since it is in that regime that interesting phenomena appear. The upshot of this paragraph is that the pressure perturbations p_1 obey Laplace's equation

$$\nabla_\perp^2 p_1 + \frac{\partial^2 p_1}{\partial z^2} = 0, \tag{99}$$

up to terms of order M^2. We now wish to estimate the relative weight of the two terms on the right hand side of (95). Since surface elevations have gradients of order wavenumber,

$$\boldsymbol{U} \cdot \nabla_\perp \eta_1 \sim U_0 k \eta_1 .$$

Also, from (90) and (91),

$$\boldsymbol{u} \cdot \nabla_\perp \eta_0 \sim \frac{u\eta_0}{a} \sim \frac{uU_0^2}{ag} .$$

Consequently

$$\frac{\left|\frac{\partial \eta_1}{\partial t}\right|}{|\boldsymbol{u} \cdot \nabla_\perp \eta_0|} \sim M^2$$

and we shall have, to first order in Mach number,

$$w = \left(\frac{\partial}{\partial t} + \boldsymbol{U} \cdot \nabla_\perp\right) \eta_1 \qquad z = h + \eta . \tag{100}$$

Finally, quantities p_1 and η_1 that are evaluated at the free surface $z = h + \eta$ in (95) and (100) can, to order one in Mach number, be evaluated at $z = h$ since corrections are of order kU_0^2/g . This is $\sim (kh)(U_0^2/c_\phi^2)$ in the case of shallow water and $\sim (U_0^2/c_g^2)$ for deep water, with c_g the group velocity , and thus negligible.

We now look for solutions to Laplace's equation (99) in the form of a power series in the vertical coordinate z. This works for surface perturbations whose wavelength is large compared to fluid depth:

$$p_1(x, y, z, t) = \sum_{m=0}^{\infty} z^m \Pi_m(x, y, t) . \tag{101}$$

The boundary condition at the bottom (96) implies

$$\Pi_1 = 0$$

and inserting the power series (101) in Laplace's equation (99) we get

$$\Pi_{2m+1} = 0 \tag{102}$$

$$\Pi_{2m+2} = \frac{-1}{(2m+2)(2m+1)} \nabla_\perp^2 \Pi_{2m}. \tag{103}$$

Calling $\Pi_0 = \Pi$ for short, we have

$$p_1 = \sum_{m=0}^{\infty} \frac{(-1)^m z^{2m}}{(2m)!} \nabla_\perp^{2m} \Pi \tag{104}$$

and each term in the series is $\sim (kh)^{2m}$ as announced. The whole point of the above exercise is to get an autonomous equation for the surface elevation η_1. To this end, defining the total time derivative $D_t = (\partial/\partial t + \boldsymbol{U} \cdot \nabla_\perp)$ we have, from (100) and (93) evaluated at $z = h$,

$$D_t^2 \eta_1 = D_t w|_{z=h} = -\frac{1}{\rho} \frac{\partial p_1}{\partial z}\bigg|_{z=h} = \frac{1}{\rho}\left(-h\nabla_\perp^2 \Pi + \frac{h^3}{6} \nabla_\perp^4 \Pi + \cdots\right) \tag{105}$$

and, from (95) we have

$$\rho g \eta_1 - \tau \nabla_\perp^2 \eta_1 = p_1|_{z=h} = \Pi - \frac{h^2}{2} \nabla_\perp^2 \Pi + \cdots \tag{106}$$

An autonomous equation for the surface elevation is obtained eliminating Π order by order. The result, to order $(kh)^4$, is

$$gh\nabla_\perp^2 \eta_1 + \left(\frac{1}{3} gh^3 - \frac{\tau h}{\rho}\right) \nabla_\perp^4 \eta_1 - D_t^2 \eta_1 = 0. \tag{107}$$

4.3 Shallow Water

Keeping only the leading order terms, in kh, in Eqn. (107) we have

$$gh\nabla_\perp^2 \eta_1 - D_t^2 \eta_1 = 0 \tag{108}$$

which is the same equation one obtains for the propagation of acoustic waves in two dimensions in the presence of an axially symmetric vortex, under the same approximations: low Mach number and wavelength short compared to vortex radius. In the abscence of the vortex, it is just the wave equation in two dimensions with phase velocity c, equal to group velocity, given by $c^2 = gh$. There is no dispersion. In order to have a definite example to work on we shall take a vortex of radius a with uniform vorticity inside, with circulation Γ at the rim, and vanishing vorticity outside:

$$\boldsymbol{U} = \begin{cases} \frac{\Gamma r}{2\pi a^2} \hat{\theta} & \text{if } r \leq a \\ \frac{\Gamma}{2\pi r} \hat{\theta} & \text{if } r > a \end{cases} \tag{109}$$

We substitute this flow into Eqn. (108) and look for solutions in the form of partial waves of frequency ν:

$$\eta_1(r,\theta,t) = \mathrm{Re}\left(\sum_n \tilde{\eta}_{1n}(r)\mathrm{e}^{i(n\theta-\nu t)}\right), \tag{110}$$

the result being two ordinary differential equations for the partial wave amplitudes $\tilde{\eta}_{1n}(r)$, one inside the vortex, $r \leq a$

$$\left(\frac{d^2}{dr^2} + \frac{1}{r}\frac{d}{dr} - \frac{n^2}{r^2} + k_n^2\right)\tilde{\eta}_{1n} = 0, \qquad k_n \equiv \frac{|\nu - n\Gamma/2a^2|}{c}. \tag{111}$$

and another outside, $r > a$:

$$\left(\frac{d^2}{dr^2} + \frac{1}{r}\frac{d}{dr} - \frac{n^2 + 2n\alpha}{r^2} + k^2\right)\tilde{\eta}_{1n} = 0, \qquad k = \frac{\nu}{c}. \tag{112}$$

with $\alpha = \nu\Gamma/2\pi c^2$ a parameter of order one. The boundary conditions under which these equations must be solved are: regularity at the origin, continuity of surface elevation, as well as of its first derivative, across the vortex rim, $r = a$, and, far away, to have a plane wave incident from the right. A similar, but a little simpler, problem appears in the quantum treatment of a charged particle in a magnetic field and we turn our attention to it.

Aharonov-Bohm Effect. In classical electromagnetism, a magnetic field $\boldsymbol{B}$ can always (since $\nabla \cdot \boldsymbol{B} = 0$) be written as the curl of a vector potential $\boldsymbol{A}$: $\boldsymbol{B} = \nabla \wedge \boldsymbol{A}$. This vector potential, however, appears as a mathematical convenience and does not have any physical significance since forces and energies depend always on the magnetic field $\boldsymbol{B}$.

For example, there may exist situations in which a region of space has a nonvanishing vector potential with a vanishing curl, hence a vanishing magnetic field. Classical electromagnetism says that a charged particle moving through this region of space will be unaffected by the vector potential. In quantum mechanics the situation is completely different since the dynamics of a charged particle moving in a magnetic field is described by a wave equation, Schrödinger's equation, in which it is the vector potential $\boldsymbol{A}$, not the magnetic filed $\boldsymbol{B}$, that appears:

$$\frac{1}{2m}\left(-i\hbar\nabla - q\boldsymbol{A}\right)^2\psi = \frac{\hbar^2k^2}{2m}\psi. \tag{113}$$

This equation, with appropriate boundary conditions, describes the possible (stationary) states, themselves labeled by a wave function $\psi(\boldsymbol{x})$, of a particle of mass m and charge q moving through a vector potential $\boldsymbol{A}$. $\hbar$ is Planck's constant, and k is a parameter related to the particle's energy. In a classic 1959 paper, Aharonov and Bohm considered the case of a charged particle moving in two dimensions, outside of a solenoid. The particle cannot penetrate inside, where the magnetic field is uniform, and outside there is a vector potential

$$\boldsymbol{A}(r,\theta) = \frac{\Phi}{2\pi r}\hat{\theta}$$

where Φ is the magnetic flux contained inside the solenoid. The magnetic field associated with this vector potential is zero. Looking for solutions to Schrödinger's equation (113) in partial waves

$$\psi(r,\theta) = \sum_n \tilde{\psi}_n(r) e^{in\theta} \tag{114}$$

leads to an ordinary differential equation for $\tilde{\psi}_n$ that is quite similar to (112):

$$\left(\frac{d^2}{dr^2} + \frac{1}{r}\frac{d}{dr} - \frac{(n-\alpha)^2}{r^2} + k^2\right)\tilde{\psi}_n = 0, \tag{115}$$

where

$$\alpha = \frac{q\Phi}{2\pi\hbar}.$$

The solutions to Eqn. (115) with impenetrable boundary conditions at the surface of the solenoid, $\tilde{\psi}_n(r=a) = 0$, incoming plane wave from the left plus decaying outgoing waves far away, were studied in a remarkable 1980 paper by Berry and collaborators, who also noted the analogy with water waves in interaction with a vertical vortex. These solutions are Bessel function of order $|n-\alpha|$. To practitioners of scattering theory, the unexpected characteristic of the solution is that it is not possible to find a solution to (115), that satisfies the boundary conditions just mentioned, and that can be written globally as a plane wave plus a scattered wave. Instead, the solution comes out as a superposition of a dislocated plane wave ψ_{AB} plus a scattered wave ψ_R:

$$\psi = \psi_{AB} - \psi_R \tag{116}$$

where

$$\psi_{AB}(r,\theta) = \sum_{n=-\infty}^{\infty} (-i)^{|n-\alpha|} e^{in\theta} J_{|n-\alpha|}(kr) \tag{117}$$

and

$$\psi_R(r,\theta) = \sum_{n=-\infty}^{\infty} (-i)^{|n-\alpha|} \frac{J_{|n-\alpha|}(ka)}{H^+_{|n-\alpha|}(ka)} H^+_{|n-\alpha|}(kr) \tag{118}$$

with J_m the Bessel function of order m and H^+_m Hankel's function of order m with outgoing boundary conditions. If α was zero, (117) would be just the well known expansion of a plane wave in partial waves. What about now? ψ_{AB} turns out to be the sum of two terms: one is a dislocate plane wave, and the other is a scattered wave. Both the dislocation and this portion of the scattering dissapear in the limit $\alpha \to 0$. Plots of the solution for several values of the parameters are given in Figure (2). There is, however, one subtle point: near the forward direction, it is not possible to split the asymptotic (for large distances) solution to the problem into incoming (even dislocated), and scattered wave.

Table 1. Aharonov-Bohm effect in quantum and classical mechanics compared and contrasted.

Quantum mechanics	**Fluid mechanics**
magnetic field $\boldsymbol{B} = \nabla \times \boldsymbol{A}$	vorticity $\boldsymbol{\omega} = \nabla \times \boldsymbol{U}$
vector potential $\boldsymbol{A}$	velocity $\boldsymbol{U}$
magnetic flux Φ	velocity circulation Γ
wave function ψ	surface displacement η
dislocation parameter $\alpha = -q\Phi/2\pi\hbar$	dislocation parameter $\alpha = k\Gamma/2\pi c$
dislocated wave is an exact solution	dislocated wave is approximate solution
phase is not measurable	phase is measurable

Back to Shallow Water. We now go back to shallow water. As explained in Table 1, there is a close parallel between the scattering of electrons by a magnetic field confined to a solenoid and the scattering of surface waves by a vortex. This parallel has, however, its limitations. From a physical point of view, there are two main differences between the two situations: water waves penetrate into the vortex, and their description by Eqn. (112), that is so temptingly similar to Eqn. (115) is only approximate: velocities and wavelengths must be small. On the other hand, the phase of surface waves is readily visualized, something that is not possible with Schrödinger wave functions.

So, surface elevations inside the vortex are provided by solutions to Eqn. (111). They are Bessel functions (to insure regularity at the origin) of order n[3]:

$$\eta_1(r,\theta,t) = \mathrm{Re}\sum_n a_n \frac{J_{|n|}(k_n r)}{J_{|n|}(k_n a)} \mathrm{e}^{i(n\theta-\nu t)} \qquad r < a, \tag{119}$$

and, outside the vortex, surface elevations are given by solutions to Eqn. (112), which is a Bessel equation of order $m \equiv \sqrt{n^2+2n\alpha}$. This is similar,

[3] When $2\nu/\omega$ is an integer, say n_d, the corresponding k_{n_d} vanishes. This case must be dealt with separately. See C. Coste, F. Lund and M. Umeki, "Scattering of dislocated wavefronts by vertical vorticity and the Aharonov-Bohm effect I. Shallow water", *Phys. Rev. E* **60**, 4908-4916 (1999).

but not quite the same, as already noted, as the equation governing the quantum mechanics of charged particles outside a solenoid, which is a Bessel equation of order $|n-\alpha|$. By analogy with this case, we write the surface elevations outside the vortex as

$$\eta_1 = \mathrm{Re}(\eta_{AB} + \eta_R) \tag{120}$$

where

$$\eta_{AB}(r,\theta,t) = \sum_n b_n \frac{J_m(kr)}{J_m(ka)} \mathrm{e}^{i(n\theta-\nu t)}, \qquad m \equiv \sqrt{n^2+2n\alpha}, \tag{121}$$

and

$$\eta_R(r,\theta,t) = \sum_n c_n \frac{H_m^+(kr)}{H_m^+(ka)} \mathrm{e}^{i(n\theta-\nu t)}. \tag{122}$$

The coefficients a_n, b_n, and c_n are determined by the boundary conditions: Two at the vortex rim, continuity of surface elevation and of its first derivative, and one at infinity: incoming plus outgoing scattered waves. Comparing the index $m = \sqrt{n^2+2n\alpha}$ that appears in the solution to the water wave equations outside the vortex, with the index $|n-\alpha|$ that appears in the solution to Schrödinger's equation outside the solenoid, it is possible to see one important difference: The solution ψ given by (117) and (118) is symmetric under $\alpha \to -\alpha$, while this is not so for the surface elevation given by (121) and (118). Various examples are given in Figure (2).

4.4 Deeper Water

There are at least two reason to perform calculations in not-so-shallow water: One is to check whether the results that have been just presented persist in a more experimentally accesible situation, and, surface waves in deeper water are dispersive suggesting new physics might appear. Indeed, going back to Eqn. (107) we see that the coefficient in front of the fourth order gradient,

$$\frac{gh}{3} - \frac{\tau h}{\rho}$$

can be either positive or negative, depending on whether the capillarity length l_c, defined by

$$l_c^2 \equiv \left(\frac{\tau}{g\rho}\right)^2,$$

is bigger or smaller than $h/\sqrt{3}$. This corresponds to the fact that, in the abscence of the vortex, the dispersion relation for surface waves is, to the order of approxiametion we are considering,

$$\nu^2 = ghk^2 + \left(\frac{\tau h}{\rho} - \frac{gh^3}{3}\right)k^4. \tag{123}$$

Although it might appear from (123) that the group velocity may vanish when $\tau/g\rho < h\sqrt{3}$, this is not so in reality since this is only an approximate dispersion relation and the value of k for which this might happen would be outside the domain of validity of the long wavelength approximation. What is valid, though, is that the dispersion curve, frequency ν versus wavevector k, has different convexity in the two cases, and both are experimentally accessible. This means that in one case the group veocity is an increasing function of wavelength, while in the other case it is a decreasing function. Solutions to Eqn. (107) are again sought in the form of partial waves, Eqn. (114) leads to a fourth order ordinary differential equation whose analysis is somewhat laborious and details can be found elsewhere[4].

The boundary conditions are, as before, regularity at the origin and, far away, a superposition of outgoing cylindrical waves plus an incoming plane wave. At the vortex rim, surface displacement and its first three derivatives must be continuous. Outside the vortex, the solution is still written as a sum of a dislocated wave, exactly as in (121) plus a scattered wave. The only difference in the dislocated wave is now that the dislocation para meter is given by

$$\alpha = \frac{\Gamma\nu}{cc_g}$$

where

$$c = \sqrt{gh}\left(1 + \left(\frac{\tau}{\rho g} - \frac{h^2}{3}\right)k^2\right)^{1/2}$$

is the phase velocity, and c_g is the group velocity of the surface waves, in the abscence of the vortex, to first order in the ratio of depth to wavelength. When $l_c > h/\sqrt{3}$ there are no big differences with the shallow water case. However, when $l_c < h/\sqrt{3}$ the scattering is much more symmetric, and spiral wave patterns may appear.

4.5 Experimental Results

Surface waves lend themselves easily, at least more easily than acoustic waves, to visualization, and an experiment has been performed to test the conclusions of the above paragraphs[5]: plane fronted waves are excited at the surface of a 10 cm deep water tank, in which a vortex forms spontaneously when letting the water come out of its bottom through a 6 mm hole. Visualization is achieved by illuminating the water surface from above with a parallel beam, and recording the reflected caustic lines. A comparison of theory and experiments is given in Figure (2).

[4] C. Coste and F. Lund, "Scattering of dislocated wavefronts by vertical vorticity and the Aharonov-Bohm effect II. Dispersive waves", *Phys. Rev. E* **60**, 4917-4925 (1999)

[5] F. Vivanco, F. Melo, C. Coste and F. Lund, "Surface Wave Scattering by a Vertical Vortex and the Symmetry of the Aharonov-Bohm Wave Function", *Phys. Rev. Lett.* **83**, 1966-1969 (1999)

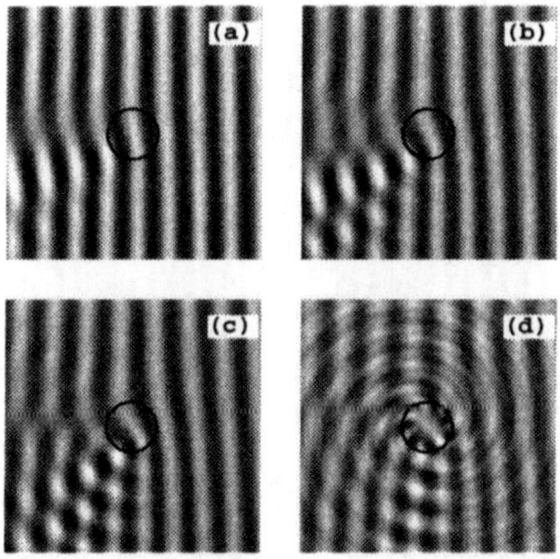

Fig. 1. Density plots of the surface elevation for a surface wave, incident from the right, scattered by a vertical vortex (dark ring). They are obtaine by a partial wave expansion of the solution to the appropriate differential equations, and numerically solving the algebraic conditions that must be satisfied by the coefficients in said expansion, arising from boundary conditions at the origin, at the vortex rim, and at infinity. Panels (a) and (b) have $\alpha = 1$, while (c) and (d) have $\alpha = 2$. On the other hand, (a) and (c) have $l_c > h/\sqrt{3}$, while (b) and (d) have $l_c < h/\sqrt{3}$.

4.6 Bibliographical Notes

The landmark paper of Y. Aharonov and D. Bohm, "Significance of electromagnetic potentials in the quantum theory", *Phys. Rev.* **115**, 485-491 (1959) spawned a large body of research, reviews of which can be found in S. Olariu and I. I. Popescu, "The quantum effects of electromagnetic fluxes", *Rev. Mod. Phys.* **57**, 339-436 (1985), and M. Peshkin and A. Tonomura, *The Aharonov-Bohm Effect*, Lecture Notes in Physics Vol 340 (Springer, Berlin, 1989). It was shown by M. V. Berry, "Quantal phase factors accompanying adiabatic changes", *Proc. R. Soc. London* **A392**, 45-57 (1984) that the Aharonov-Bohm effect was embedded in a more general framework ("Berry's phase"), of wide reaching consequences, that has been reviewed by A. Shapere and F. Wilczek, *Geometric Phases in Physics* (World Scientific, Singapore, 1989).

The fact that Aharonov-Bohm wavefronts would be dislocated was noted by M. V. Berry et al., "Wavefront dislocations in the Aharonov-Bohm effect and its water wave analogue", *Eur. J. Phys.* **1**, 154-162 (1980). They also elucidated the nature of the solution in the forward direction, and proposed, as well as verified, that effects of a similar nature would be observed in the scattering of surface waves by a vortex with an irrotational exterior. The relation between the Aharonov-Bohm effect and the scattering of surface waves by a vortex was studied in some depth, in the limit of wavelengths short compared to vortex core radius, by C. Coste, F. Lund and M. Umeki, "Scattering of dislocated wavefronts by vertical vorticity and the Aharonov-

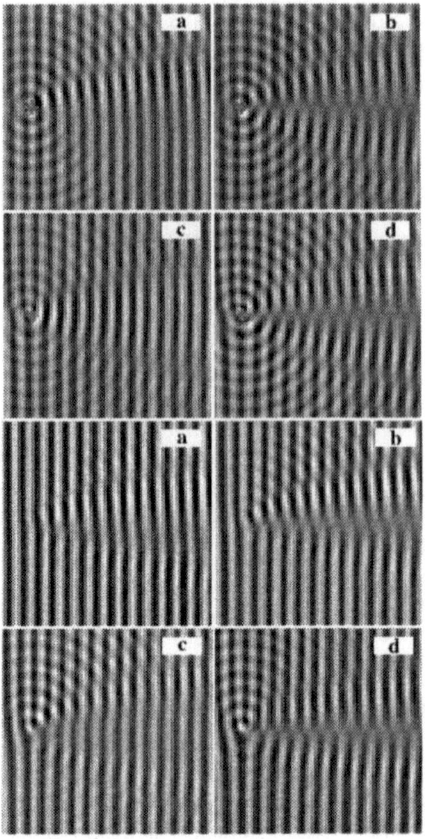

Fig. 2. The left panel presents numerical solutions to the Schrödinger equation for an electron travelling from left to right that is incident upon a magnetic field wholly contained in a solenoid, for four different values of the dislocation parameter α. The upper four correspond to impenetrable boundary conditions at the solenoid's surface. In the lower four the electron is allowed to penetrate the solenoid. Notice the resulting difference in the symmetry of the electron wave function. The right panel presents a comparison of theory and experiment for the scattering of surface waves, travelling from left to right, scattered by a vertical vortex. The upper four are the result of theoretical computations as described in the text, and the lower four correspond to experiment. Parameter values for the theory were taken from the experimental conditions.

Bohm effect I. Shallow water", *Phys. Rev. E* **60**, 4908-4916 (1999) and C. Coste and F. Lund, "Scattering of dislocated wavefronts by vertical vorticity and the Aharonov-Bohm effect II. Dispersive waves", *Phys. Rev. E* **60**, 4917-4925 (1999). The opposite limit, of wavelengths long compared to vortex core radius, was studied by R. Ford and S. G. Llewellyn Smith, "Scattering of acoustic waves by a vortex", *J. Fluid Mech.* **386**, 305-328 (1999). The experiment of Section 4.5 is described in F. Vivanco et al., "Surface Wave Scattering by a Vertical Vortex and the Symmetry of the Aharonov-Bohm Wave Function", *Phys. Rev. Lett.* **83**, 1966-1969 (1999).

5 Multiple Scattering

In Section 3 we studied what happens when a plane acoustic wave is incident upon a bounded body of vorticity. It was found that a scattering problem could be set up, assuming: (i) the vortical flow is of low Mach number; (ii) fluid velocities associated with the acoustic wave are small compared to the target flow, and (iii) acoustic frequencies are high compared to flow frequencies. This scattering problem could be solved in a Born approximation, except in two dimensions when the circulation of the vorticity does not vanish, a topic that was treated in Section 4

A related problem of great practical interest, not just in acoustics, but generally in wave propagation of any kind, is the propagation of a wave through a medium composed of very many scatterers randomly distributed. Radar propagation in clouds, laser propagation in the atmosphere, sound propagation in the ocean and in the atmosphere, electrons in amorphous materials, are but a few examples.

The problem is that the wave will interact many times with many scatterers and the challenge is to extract average parameters that decribe, as simply as possible, what is going on. If the interaction between wave and scatterer is weak, there is a coherent wave that propagates much as a wave would in the abscence of the medium, but with an "effective" phase velocity. This coherent wave will also be attenuated because of the energy that is being scattered away from the direction of propagation. This incoherent energy that is scattered away from the direction of propagation of the coherent wave obeys, in a long wavelength limit, a diffusion equation. In this sense, a random medium changes the nature of wave behaviour from propagative to diffusive at increasing length scales. And it may happen, a topic of much current research, that the corresponding diffusion constant vanishes, in which case it is said that the wave is localized in the sense of Anderson, who first studied this phenomenon for electrons in one dimension. In this Section we study the propagation of an acoustic wave in a medium that is filled with very many vortex filaments randomly distributed. The motivation for this study is due to the fact that vorticity appears to concentrate in thin filaments in many turbulent flows and the question arises, how these structures would affect the propagation of acoustic waves. The strategy is to study first sound propagation over a generic disordered background flow, and then specialize to a background made up of many filaments.

5.1 Equations to Order M^2

We now go back to Eqn. (59), this time keeping terms of order Mach number squared. The reason for doing this is, as will be seen below, that terms of order Mach number give a vanishing contribution to some of the statistical properties we are after.

$$\nabla^2 p_s - \frac{\partial^2 \rho_s}{\partial t^2} = -2\rho_0 \partial_i \partial_j (u_i v_{sj}) - \partial_i \partial_j (\rho_s u_j u_i) \tag{124}$$

An important assumption, that greatly simplifies the analysis, is that there will exist acoustic solutions with a single frequency ν_0. This means that the time evolution of the background flow is completely neglected. The strategy is to use Eqns. (56) and (57) to get an autonomous equation for p_s, and then apply standard tools from wave propagation in random media . We shall also consider an ideal gas from now on.

One thing that must be done in the left hand side of (124) is to relate the acoustic pressure p_s with the acoustic density fluctuations ρ_s. Their ratio is given by the speed of sound squared,

$$c^2(\boldsymbol{x}) = \gamma \frac{P_0(\boldsymbol{x})}{\rho_0(\boldsymbol{x})}$$

where P_0 and ρ_0 are hydrodynamic pressure and density associated with the background flow, and γ is the ratio of specific heats. Note that the speed of sound depends on position, since hydrodynamic pressure fluctuations will generate density fluctuations. They are of order M^2, as we now show, and must be kept. Calling

$$\delta c^2(\boldsymbol{x}) \equiv \frac{c^2(\boldsymbol{x}) - c_0^2}{c_0^2}$$

we have

$$-\frac{1}{c_0^2}\nabla \delta c^2 = \frac{1}{\gamma}\nabla\left(\frac{\rho_0}{P_0}\right) \tag{125}$$

$$= \frac{1-\gamma}{\rho_0 c_0^2}\nabla P_0 \tag{126}$$

$$= \frac{\gamma - 1}{c_0^2}(\boldsymbol{u}\cdot\nabla\boldsymbol{u}) \tag{127}$$

where we have used (52) to get the last equality, and $\partial \boldsymbol{u}/\partial t$ has been neglected with respect to $\boldsymbol{u}\cdot\nabla\boldsymbol{u}$[6].

[6] This is correct for the flows generated by the vortex structures that are considered in subsequent sections.

As already stated, all quantities of interest will be assumed to evolve harmonically in time, with a single frequency ν_0:

$$\boldsymbol{v}_s, p_s, \rho_s \sim e^{-i\nu_0 t}.$$

Under this assumptions Eqn. (57) becomes, neglecting its last term on the right hand side because it is of order M^3,

$$-i\nu_0 \boldsymbol{v}_s + (\boldsymbol{u}\cdot\nabla)\boldsymbol{v}_s + (\boldsymbol{v}_s\cdot\nabla)\boldsymbol{u} = -\frac{1}{\rho}\nabla p_s. \tag{128}$$

This equation is now solved, order by order in M to get $\boldsymbol{v}_s$ in terms of p_s. The result is, to order M^2,

$$\boldsymbol{v}_s = -\frac{i}{\nu_0\rho_0}\nabla p_s - \frac{1}{\nu_0^2\rho_0}\left((\boldsymbol{u}\cdot\nabla)\nabla p_s + (\nabla p_s\cdot\nabla)\boldsymbol{u}\right). \tag{129}$$

Replacing this result into (124) and using (127) we get the following equation for acoustic pressure as a function of position $p_s(\boldsymbol{x})$, with a harmonic time dependence implied:

$$(\nabla^2 + k_0^2 - V(\boldsymbol{x}))p_s = 0, \tag{130}$$

where $k_0 = \nu_0/c_0$ and

$$V(\boldsymbol{x})p_s = \left(V^{(1)} + V^{(2)}\right)p_s$$

with

$$\begin{aligned} V^{(1)}p_s &= \frac{2i}{\nu_0}\partial_i\partial_j(u_i\partial_j p_s) \\ V^{(2)}p_s &= -\frac{1}{c_0^2}\partial_i\partial_j(p_s u_i u_j) \\ &\quad + k_0^2\delta c^2 p_s \\ &\quad + \frac{2}{\nu_0^2}\partial_i\partial_j(u_i(u_l\partial_l\partial_j p_s + (\partial_l p_s)(\partial_l u_j))) \end{aligned} \tag{131}$$

This equation is the main object of study in this section. It resembles the equation for a quantum particle in a potential V. The physics is, however, different being, of course, the propagation of an acoustic wave in a background flow. The consequence in the mathematics is that V is a differential operator acting on p_s rather than a function. Once V is allowed to be random, Eqn. (130) becomes the type of equation that has been exhaustively studied in the literature of wave propagation in random media. However, most such studies have been confined to media in which the statistical properties are prescribed, most often imposing that the probability distribution be a Gaussian, without a physical model for the random medium itself. Later in this Section, we will consider a background flow made up of many vortex filaments and explore the consequences of this model on wave propagation.

A first step is to find out whether there is coherent wave propagation. To this end one writes first, the equation for the impulse response function (alias bare Green's function) for free wave propagation[7]:

$$(\nabla^2 + k_0^2)G^{(0)} = -\delta(\boldsymbol{x}), \tag{132}$$

whose solution is, in Fourier space,

$$\mathcal{G}(\boldsymbol{k}) = \frac{1}{k^2 - (k_0 + i\eta)^2}. \tag{133}$$

where the limit $\eta \to 0^+$ is implied. This means that wave propagation of the form

$$G^{(0)} \sim e^{i(\boldsymbol{k}\cdot\boldsymbol{x} - \nu_0 t)}$$

is possible as long as $|\boldsymbol{k}| = \pm k_0$, corresponding to the poles of (133). The impulse response of the medium filled with many scatterers is described by

$$(\nabla^2 + k_0^2 - V(\boldsymbol{x}))G = -\delta(\boldsymbol{x}), \tag{134}$$

whose solution is much harder to find since it depends on the random operator $V(\boldsymbol{x})$. Ideally, given a probability distribution for V, one would hope to find a corresponding probability distribution for G.

In practice one settles for the first few moments of G. Say, its average $\langle G\rangle$ and its two point correlation function $\langle GG\rangle$, computed over all possible realizations of the random operator V. In general, there is no reason to expect that coherent wave propagation, whose existence would be signaled by poles of $\langle G\rangle$ in Fourier space, exists at all. However, physical experience shows that coherent wave propagation does exist in nature (for instance, through a window glass) and the mathematics should be able to describe this physical reality, at least when the interaction between wave and scatterer is weak. This weak interaction case invites a solution by perturbation around the free space case, whose solution is given by $G^{(0)}$. In order to do this, Eqn. (134) is turned into an integral equation

$$G(\boldsymbol{x}) = G^{(0)}(\boldsymbol{x}) - \int d\boldsymbol{x}' G^{(0)}(\boldsymbol{x} - \boldsymbol{x}')V(\boldsymbol{x}')[G(\boldsymbol{x}')] \tag{135}$$

that we shall write in a more compact, if a little ambiguous, form

$$G = G^{(0)} - G^{(0)} * (VG) \tag{136}$$

where the star means a convolution in the space variables. The dimensionality of space has not been specified, and the formalism in this Subsection is valid both in two and three dimensions. Taking the average of (136) over all possible realizations of V gives

$$\langle G\rangle = G^{(0)} - G^{(0)} * \langle VG\rangle \tag{137}$$

[7] Note the change in sign with respect to the convention used in previous Sections.

from which, of course, there is no hope of disentangling $\langle G\rangle$ in any simple way. However, if, in Fourier space, it were possible to find a quantity[8] $\mathcal{M}(k)$ such that the average of the impulse response obeys the equation ("Dyson equation")

$$\langle G\rangle = \mathcal{G} + \mathcal{G}\mathcal{M}\langle G\rangle, \tag{138}$$

then it would be possible to simply solve this algebraic equation, using (133), to get

$$\langle G\rangle(\boldsymbol{k}) = \frac{1}{k^2 - \mathcal{M}(k) - (k_0 + i\eta)^2} \tag{139}$$

which would imply that coherent wave propagation is possible, with wavevector $\boldsymbol{k}$ of magnitude given by the solution(s) of

$$\begin{aligned} k^2 &= \mathcal{M}(k) + k_0^2 \\ &\approx \mathcal{M}(k_0) + k_0^2 \end{aligned} \tag{140}$$

where the approximate expression is consistent with the assumption of weak interaction, $\mathcal{M}$ is a small correction to k_0^2.

The problem is then how to find $\mathcal{M}$. In our case this is not hard in perturbation theory, since we are working only to order M^2. So, take (136) and iterate:

$$G = G^{(0)} - G^{(0)} * (VG^{(0)}) + G^{(0)} * (V(G^{(0)} * (VG^{(0)}))) + \cdots \tag{141}$$

Going to Fourier space and taking averages this becomes

$$\langle G\rangle = \mathcal{G} - \mathcal{G}\langle V\rangle\mathcal{G} + \mathcal{G}\langle V\rangle\mathcal{G}\langle V\rangle\mathcal{G} + \cdots \tag{142}$$

which can be inverted order by order to get $\mathcal{G}$ in terms of $\langle G\rangle$. Replacing this into (142) we get, to order M^2,

$$\langle G\rangle = \mathcal{G} + (\langle V\rangle - \langle V\rangle\mathcal{G}\langle V\rangle + \langle V\mathcal{G}V\rangle)\langle G\rangle \tag{143}$$

from which the mass operator , to order M^2, can be read off (see the Appendix for details):

$$\mathcal{M} = \langle V\rangle - \langle V\rangle\mathcal{G}\langle V\rangle + \langle V\mathcal{G}V\rangle \tag{144}$$

The next step is to actually compute this mass operator for the operator[9] V given by (131). An important assumption that we now make is that the background flow, over which acoustic waves propagate, whose properties and possible coherence we are studying, has zero average. That is, there is no net

---

[8] $\mathcal{M}(k)$ is called a mass operator after its origins in quantum field theory. We shall asume that the statistical properties of the random medium are rotation invariant so $\mathcal{M}$ is a function of the magnitude of $\boldsymbol{k}$ only.

[9] The operator V is a differential operator, so its Fourier transform is a multiple of the identity, an algebraic function.

flow associated with the disordered background: $\langle V\rangle = 0$. Thus, using (131), Eqn. (144) becomes

$$\mathcal{M} = \langle V^{(2)}\rangle + \langle V^{(1)}\mathcal{G}V^{(1)}\rangle. \tag{145}$$

So, going back to (140), we have (with B_1 and B_2 defined in detail in the next Section)

$$k^2 = k_0^2 + B_1 + B_2 \tag{146}$$

or,

$$k \approx k_0 + \frac{B_1 + B_2}{2k_0}. \tag{147}$$

Consequently, if a mass operator can be found, there will be a coherent wave that propagates with an effective (phase) velocity c^* related to c_0 through

$$c^* = c_0\left(1 - \frac{B_1}{2k_0^2} - \frac{\mathrm{Re}(B_2)}{2k_0^2}\right). \tag{148}$$

Since k has an imaginary part, the coherent wave attenuates with an attenuation length Λ given by

$$\Lambda^{-1} = \frac{1}{2k_0}\mathrm{Im}(B_2) \tag{149}$$

5.2 Appendix

In this Appendix we provide details of the computation of the mass operator described in the previous Section..

Derivation of Eqn. (144). A more precise way of writing Eqn. (132) is

$$(\nabla^2 + k_0^2)G^{(0)}(\boldsymbol{x}, \boldsymbol{x}') = -\delta(\boldsymbol{x} - \boldsymbol{x}'). \tag{150}$$

And it is easy to show that, for any function $F(\boldsymbol{x}, \boldsymbol{x}')$ invariant under translations, i.e. satisfying

$$F(\boldsymbol{x}, \boldsymbol{x}') = F(\boldsymbol{x} - \boldsymbol{x}') \tag{151}$$

its Fourier transform, defined by

$$\tilde{F}(\boldsymbol{k}, \boldsymbol{k}') \equiv \int d\boldsymbol{x}d\boldsymbol{x}' e^{-i\boldsymbol{k}\cdot\boldsymbol{x}} F(\boldsymbol{x}, \boldsymbol{x}') e^{i\boldsymbol{k}'\cdot\boldsymbol{x}'} \tag{152}$$

satisfies

$$\tilde{F}(\boldsymbol{k}, \boldsymbol{k}') = \tilde{F}(\boldsymbol{k})\delta(\boldsymbol{k} - \boldsymbol{k}'). \tag{153}$$

This is obviously the case for the bare Green's function $G^{(0)}$, and we will *assume* it is also the case for averaged quentities, such as the averaged Green's funciton for the scatterer-filled medium $\langle G\rangle$, as well as for the mass operator. Taking the average of Eqn. (135) means writing

$$\langle G(\boldsymbol{x}-\boldsymbol{x}'')\rangle = G^{(0)}(\boldsymbol{x}-\boldsymbol{x}'') - \int d\boldsymbol{x}' G^{(0)}(\boldsymbol{x}-\boldsymbol{x}')\langle V(\boldsymbol{x}')[G(\boldsymbol{x}'-\boldsymbol{x}'')]\rangle \quad (154)$$

which, as already noted, is solved by iteration, upon which the averages become (see (141))

$$\langle G\rangle = G^{(0)} - G^{(0)} * (\langle V\rangle G^{(0)}) + G^{(0)} * (\langle V(G^{(0)} * (VG^{(0)}))\rangle) + \cdots \quad (155)$$

Inverting this order by order yields, to second order,

$$\begin{aligned}\langle G\rangle = G^{(0)} &- G^{(0)} * (\langle V\rangle\langle G\rangle) - G^{(0)} * (\langle V\rangle(G^{(0)} * (\langle V\rangle\langle G\rangle)))\\ &+ G^{(0)} * (\langle V(G^{(0)} * \langle V\rangle\langle G\rangle)\rangle) + \cdots \end{aligned} \quad (156)$$

This can be written as

$$\langle G(\boldsymbol{x}-\boldsymbol{x}')\rangle = G^{(0)}(\boldsymbol{x}-\boldsymbol{x}') + \int d\boldsymbol{x}''d\boldsymbol{x}''' G^{(0)}(\boldsymbol{x}-\boldsymbol{x}'')K(\boldsymbol{x}'',\boldsymbol{x}''')\langle G(\boldsymbol{x}'''-\boldsymbol{x}')\rangle \quad (157)$$

with

$$\begin{aligned}K(\boldsymbol{x}'',\boldsymbol{x}''') \equiv -\langle V(\boldsymbol{x}'')\rangle\delta(\boldsymbol{x}''-\boldsymbol{x}''') &- \langle V(\boldsymbol{x}'')\rangle G^{(0)}(\boldsymbol{x}''-\boldsymbol{x}''')\langle V(\boldsymbol{x}''')\rangle\\ &+ \langle V(\boldsymbol{x}'')G^{(0)}(\boldsymbol{x}''-\boldsymbol{x}''')V(\boldsymbol{x}''')\rangle \end{aligned} \quad (158)$$

Taking the Fourier transform of (157), as in (153), gives

$$\langle\tilde{G}(\boldsymbol{k})\rangle\delta(\boldsymbol{k}-\boldsymbol{k}') = \mathcal{G}(\boldsymbol{k})\delta(\boldsymbol{k}-\boldsymbol{k}') + \mathcal{G}(\boldsymbol{k})\tilde{K}(\boldsymbol{k},\boldsymbol{k}')\langle\tilde{G}(\boldsymbol{k}')\rangle \quad (159)$$

and the assumption that averages are invariant under translation yields (144) with

$$\tilde{K}(\boldsymbol{k},\boldsymbol{k}') = \mathcal{M}(\boldsymbol{k})\delta(\boldsymbol{k}-\boldsymbol{k}') \quad (160)$$

with the mass operator $\mathcal{M}$ given by

$$\mathcal{M}(k_0) = B_1 + B_2 \quad (161)$$

with

$$B_1 = -\frac{1}{V}\int d\boldsymbol{x} e^{-i\boldsymbol{k}_0\cdot\boldsymbol{x}}\langle V^{(2)}(\boldsymbol{x})\rangle\left[e^{i\boldsymbol{k}_0\cdot\boldsymbol{x}}\right] \quad (162)$$

and

$$B_2 = \frac{1}{V}\int\int d\boldsymbol{x}d\boldsymbol{x}' e^{-i\boldsymbol{k}_0\cdot\boldsymbol{x}}\langle V^{(1)}(\boldsymbol{x})G^{(0)}(\boldsymbol{x}-\boldsymbol{x}')V^{(1)}(\boldsymbol{x}')\rangle\left[e^{i\boldsymbol{k}_0\cdot\boldsymbol{x}'}\right]. \quad (163)$$

The result is

$$B_1 = \frac{(\gamma-4)k_0^2}{c_0^2}\frac{1}{V}\int_V d\boldsymbol{x}\langle(\hat{k}_0\cdot\boldsymbol{u}(\boldsymbol{x}))^2\rangle \quad (164)$$

and

$$B_2 = \frac{4k_0^2}{c_0^2}\frac{1}{V}\int\frac{d\boldsymbol{q}}{(2\pi)^d}(\hat{k}_0\cdot\boldsymbol{q})^2\mathcal{G}(\boldsymbol{q})\langle(\hat{k}_0\cdot\tilde{\boldsymbol{u}}(\boldsymbol{q}-\boldsymbol{k}_0))(\hat{k}_0\cdot\tilde{\boldsymbol{u}}(\boldsymbol{k}_0-\boldsymbol{q}))\rangle \quad (165)$$

where d is the dimensionality of space.

Calculation of B_1. We first compute B_1, which is given by

$$B_1 = B_1^{(i)} + B_1^{(ii)} + B_1^{(iii)} \tag{166}$$

where

$$B_1^{(i)} = \frac{1}{Vc_0^2} \int dx e^{-ik_0\cdot x} \langle \partial_i \partial_j \left[u_i u_j e^{ik_0\cdot x} \right] \rangle \tag{167}$$

$$B_1^{(ii)} = -\frac{1}{V} \int dx e^{-ik_0\cdot x} \langle k_0^2 \delta c^2(x) e^{ik_0\cdot x} \rangle \tag{168}$$

$$B_1^{(iii)} = -\frac{2}{V\nu_0^2} \int dx e^{-ik_0\cdot x} \langle \partial_i \partial_j \left[u_i (u_l \partial_l \partial_j e^{ik_0\cdot x} + \left(\partial_l e^{ik_0\cdot x} \right) \partial_l u_j) \right] \rangle . \tag{169}$$

Take $B_1^{(i)}$ first. Integrating by parts twice in (167) it is easy to get

$$b_1^{(i)} = -\frac{k_0^2}{c_0^2} \frac{1}{V} \int dx \langle (\hat{k}_0 \cdot u)^2 \rangle . \tag{170}$$

Note that this is always negative.

We next show that $B_1^{(ii)}$ and $B_1^{(iii)}$ are proportional to $B_1^{(i)}$. From Eqn. (127) we have

$$\begin{aligned} \int dx e^{-iq\cdot x} \nabla \delta c^2 &= i \int dx q e^{-iq\cdot x} \delta c^2 \\ &= \frac{1-\gamma}{c_0^2} \int dx e^{-iq\cdot x} (u \cdot \nabla) u . \end{aligned} \tag{171}$$

Using the fact that, to the order of approximmation we are working $\nabla \cdot u = 0$, we have

$$i \int dx q^2 e^{-iq\cdot x} \delta c^2 = \frac{1-\gamma}{c_0^2} \int dx e^{-iq\cdot x} (q \cdot u)^2 \tag{172}$$

so that

$$i \int dx e^{-iq\cdot x} \delta c^2 = \frac{1-\gamma}{c_0^2} \int dx e^{-iq\cdot x} (\hat{q} \cdot u)^2 , \tag{173}$$

an equality that is true for all q. Choosing $|q| = 0$ and $\hat{q} = \hat{k}_0$ we finally get

$$\begin{aligned} B_1^{(ii)} &= -\frac{(1-\gamma) k_0^2}{c_0^2} \frac{1}{V} \int dx \langle (\hat{k}_0 \cdot u)^2 \rangle \\ &= (1-\gamma) B_1^{(i)} . \end{aligned} \tag{174}$$

Now take $B_1^{(iii)}$. From Eqn. (169) we see that the second term on the right hand side vanishes if we consider flows whose averages are invariant under space reflections. This gives,

$$\begin{aligned} B_1^{(iii)} &= -\frac{2k_0^2}{c_0^2} \frac{1}{V} \int dx \langle (\hat{k}_0 \cdot u)^2 \rangle \\ &= 2 B_1^{(i)} . \end{aligned} \tag{175}$$

Putting together (170), (174) and (175) with (166), we get the announced result (164). A noteworthy feature is that it is negative for an ideal gas.

Calculation of B_2. From (163) we have

$$B_2 = -\frac{4}{\nu_0^2 V} \int d\boldsymbol{x} e^{-i\boldsymbol{k}_0\cdot\boldsymbol{x}} \langle \partial_i \partial_j \left[u_i \partial_j I(\boldsymbol{x}) \right] \rangle \tag{176}$$

where the function I is a convolution of two functions: $\partial_k \partial_l G^{(0)}$, and $u_k \exp(i\boldsymbol{k}_0 \cdot \boldsymbol{x})$, defined by

$$I(\boldsymbol{x}) \equiv \int d\boldsymbol{x}' G^{(0)}(\boldsymbol{x} - \boldsymbol{x}') \partial'_k \partial'_l (u_k(\boldsymbol{x}') \partial'_l e^{i\boldsymbol{k}_0\cdot\boldsymbol{x}'}). \tag{177}$$

Since it is a convolution, it can be evaluated using the convolution theorem, namely, it is the inverse Fourier transform of the product of the Fourier transform of the factors. This gives

$$\begin{aligned} I(\boldsymbol{x}) &= k_{0l} \int \frac{d\boldsymbol{q}}{(2\pi)^d} e^{i\boldsymbol{q}\cdot\boldsymbol{x}} q_k q_l \mathcal{G}(\boldsymbol{q}) \int d\boldsymbol{x}' e^{-i\boldsymbol{q}\cdot\boldsymbol{x}'} u_k(\boldsymbol{x}') e^{i\boldsymbol{k}_0\cdot\boldsymbol{x}'} \\ &= \int \frac{d\boldsymbol{q}}{(2\pi)^d} e^{i\boldsymbol{q}\cdot\boldsymbol{x}} (\boldsymbol{k}_0 \cdot \boldsymbol{q})(\boldsymbol{q} \cdot u(\boldsymbol{q} - \boldsymbol{k}_0)) \mathcal{G}(\boldsymbol{q}) \end{aligned} \tag{178}$$

and, consequently,

$$\begin{aligned} B_2 &= \frac{-4i}{\nu_0^2 V} \langle \int d\boldsymbol{x} k_{0i} k_{0j} u_i(\boldsymbol{x}) e^{-i\boldsymbol{k}_0\cdot\boldsymbol{x}} \partial_j I(\boldsymbol{x}) \rangle \\ &= \frac{4k_0^2}{c_0^2 V} \int \frac{d\boldsymbol{q}}{(2\pi)^d} (\hat{k}_0 \cdot \boldsymbol{q})^2 \mathcal{G}(\boldsymbol{q}) \langle \left(\hat{k}_0 \cdot \tilde{\boldsymbol{u}}(\boldsymbol{q} - \boldsymbol{k}_0) \right) \left(\hat{k}_0 \cdot \tilde{\boldsymbol{u}}(\boldsymbol{k}_0 - \boldsymbol{q}) \right) \rangle \end{aligned} \tag{179}$$

where we have used (178) to get the second equality. This is (165), as promised.

5.3 Example: a "Gas" of Vortices

In order to push things further we take as background flow the flow generated by many vortices, each generating a velocity $\boldsymbol{v}^{(a)}(\boldsymbol{x})$:

$$\boldsymbol{u}(\boldsymbol{x}) = \sum_{a=1}^{N} \boldsymbol{v}^{(a)}(\boldsymbol{x}), \tag{180}$$

and in order to simplify things as much as possible we take identical vortices, randomly located at $\boldsymbol{x}_a$ and randomly oriented along $\hat{a}$:

$$\boldsymbol{v}_a(\boldsymbol{x}) = \boldsymbol{v}(\boldsymbol{x} - \boldsymbol{x}_a, \hat{a}). \tag{181}$$

Both random variables will be assumed to be uniformly distributed over their respective ranges. Also, cross correlations between different vortices will be assumed to vanish, a good approximation if the density $n \equiv N/V$ of vortices is low. Substitution of (180) into (164), (165) and using (148), (149) gives

$$\Lambda^{-1} = \frac{2k_0 n}{c_0^2} \langle \mathrm{Im}\mathcal{I} \rangle \tag{182}$$

and

$$c^* = c_1^* + c_2^* \tag{183}$$

where

$$\frac{c_1^*}{c_0} = \frac{n(\gamma - 4)}{2c_0^2} \int d\boldsymbol{x} \langle (\hat{k}_0 \cdot \boldsymbol{v}(\boldsymbol{x}))^2 \rangle$$

and

$$\frac{c_2^*}{c_0} = \frac{2n}{c_0^2} \langle \mathrm{Re}\mathcal{I} \rangle$$

with

$$\mathcal{I} \equiv \int \frac{d\boldsymbol{q}}{(2\pi)^d} \frac{\left| \hat{k}_0 \cdot \tilde{\boldsymbol{v}}(\boldsymbol{k}_0 - \boldsymbol{q}) \right|^2 (\hat{k}_0 \cdot \boldsymbol{q})^2}{q^2 - (k_0 + i\eta)^2} \tag{184}$$

and we have used the property $v(-\boldsymbol{q}) = v^*(\boldsymbol{q})$ for the Fourier transform of real velocities. This integral es evaluated using the formula, valid in the sense of distributions,

$$\frac{1}{x + i\eta} = \mathcal{P}\frac{1}{x} - i\pi\delta(x) \tag{185}$$

where $\mathcal{P}$ means principal value.

5.4 Two Dimensions

A simple example in two dimensions is a "gas" of vortex dipoles:

$$\boldsymbol{\omega}(\boldsymbol{x}) = \frac{\Gamma}{\pi\epsilon} \left(e^{-|\boldsymbol{x}+\boldsymbol{l}|^2/\epsilon^2} - e^{-|\boldsymbol{x}-\boldsymbol{l}|^2/\epsilon^2} \right) \hat{z}. \tag{186}$$

This is a set of two counter rotating vortices a distance $2l$ apart, with vorticity Gaussianly distributed within a core of radius ϵ, each with a circulation Γ. The Mach number is defined at the vortex core,

$$M = \frac{\Gamma}{2\pi\epsilon c_0}.$$

This example has the algebraic advantage that the Fourier transform of (186), and consequently that of the velocity, can be analytically computed:

$$\tilde{\boldsymbol{\omega}}(\boldsymbol{q}) = -2i\Gamma e^{-q^2\epsilon^2/4} \sin(\boldsymbol{q} \cdot \boldsymbol{l}) \hat{z} \tag{187}$$

$$\tilde{\boldsymbol{v}}(\boldsymbol{q}) = -2\Gamma e^{-q^2\epsilon^2/4} \frac{\sin(\boldsymbol{q} \cdot \boldsymbol{l})}{q} \hat{q}_\perp \tag{188}$$

where $\hat{q}_\perp$ is a unit vector perpendicular to $\boldsymbol{q}$: $\hat{q}_\perp \cdot \boldsymbol{q} = 0$. Substitution into (184) gives an expression that can be computed numerically. In this calculation, the one thing one must be careful about is the numerical treatment of the principal part. Asymptotic behaviours for long and short wavelengths can be analytically calculated to provide a check in the numerics. The resulting attenuation length and effective phase velocity are given as a function of wavelength (scaled by the dipole size l) in Figure (3).

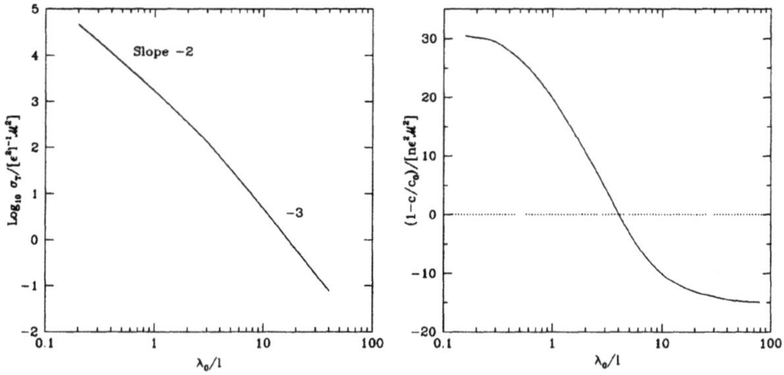

Fig. 3. Inverse attenuation length (left) and effective phase velocity (right) as a function of wavelength for an acoustic wave travelling in a medium made up of many vortex dipoles randomly distributed in an ideal gas in two dimensions. Note the λ^3 behaviour for large wavelengths of the attenuation length, as expected for Rayleigh scattering. The effective phase velocity becomes larger that the speed of sound in the unperturbed medium for wavelengths larger than a length on the order of a dipole size.

The attenuation length turns out to be inversely proportional to the total scattering cross section for an individual scatterer, it is not surprising to find that, at long wavelengths, it decays as λ^{-3}. This is just Rayleigh scattering from dipoles in two dimensions. There is a crossover to a different power law when the wavelength becomes comparable to the dipole size, $\lambda \sim l$. A more surprising feature appears in the effective phase velocity: it is smaller than c_0 for short wavelengths but it becomes larger than c_0 at long wavelengths, the crossover appearing again at wavelengths comparable to a few times the dipole size l. This may appear surprising at first sight, in view of theorems[10] that state that the effective phase velocity in a medium that is described by a Helmholtz equation with random index of refraction with zero mean is

[10] See, for example, S. M. Rytov, Y. A. Kravtzov and V. I. Tatarskii, *Principles of Statistical Radio Physics, Part 4. Wave Propagation through Random Media,* (Springer, Berlin, 1989).

lower than the phase velocity without the randomness. But, the system we are studying is not of this type.

5.5 Three Dimensions

An example in three dimensions that is closely analogous to the dipoles in two dimensions is a "gas" of vortex rings of radius L and thickness ϵ with vorticity Gaussianly distributed. In cylindrical coordinates (ρ,ϕ,z):

$$\boldsymbol{\omega}(\rho, \phi, z) = \frac{\Gamma}{\pi\epsilon^2} e^{-(z^2+\rho-L)^2/\epsilon^2} \hat{\phi}. \tag{189}$$

The Fourier transform of this expression is

$$\tilde{\boldsymbol{\omega}}(\boldsymbol{q}) = -i\tilde{\omega}(q_\rho, q_z)\hat{\phi} \tag{190}$$

with (when $L \gg \epsilon$)

$$\omega(q_\rho, q_z) = \Gamma e^{-q_z^2\epsilon^2/4} \int_0^{2\pi} d\phi \cos\phi \; e^{-q_\rho^2\epsilon^2 \cos^2\phi/4} \left[L \sin(q_\rho L \cos\phi) + \frac{q_\rho \epsilon^2 \cos\phi}{2} \cos(q_\rho L \cos\phi) \right] . \tag{191}$$

The attenuation length as a function of wavelength is plotted in Figure (4) for various ratios of ring radius to thickness. We see that it is largely independent of that ratio. For short wavelengths, it scales as λ^2 just as in two dimensions, while for long wavelengths it scales as λ^4 as expected on the basis of Rayleigh scattering for dipoles in three dimensions. The effective phase velocity, plotted in (4), on the other hand, does depend on the radius-to-thickness ratio and, as in two dimensions, it becomes "supersonic" for long wavelengths with a crossover at a wavelength a few times the vortex radius.

5.6 Bibliographical Notes

There is considerable evidence that vorticity concentrates along thin filaments in a number of turbulent flows: E. Siggia, "Numerical study of small-scale intermittency in three-dimensional turbulence", *J. Fluid Mech.* **107**, 375-406 (1981); Z. She, E. Jackson and S. Orszag, "Intermittent vortex structures in homogeneous isotropic turbulence", *Nature* **344**, 226-228 (1990); A. Vincent and M. Meneguzzi, "The spatial structure and statistical properties of homogeneous turbulence", *J. Fluid Mech.* **225**, 1-20 (1991). S. Kida and K. Ohkitani, "Spatiotemporal intermittency and instability of a forced turbulence", *Phys. Fluids A* **4**, 1018-1027 (1992); M. Tanaka and S. Kida, "Characterization of vortex tubes and sheets", *Phys. Fluids A* **5**, 2079-2082 (1993). J. Jiménez, A. A. Wray, P. G. Saffman and R. S. Rogallo, "The structure of intense vorticity in isotropic turbulence", *J. Fluid Mech.* **255**, 65-90 (1993).

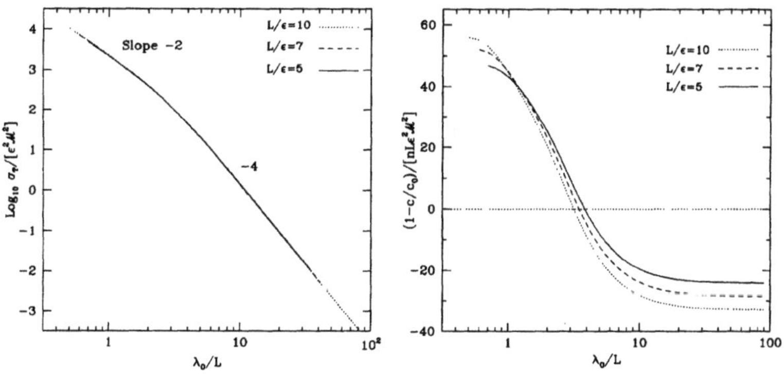

Fig. 4. Inverse attenuation length (left) and effective phase velocity (right) as a function of wavelength for an acoustic wave travelling in a medium made up of many vortex rings randomly distributed in an ideal gas in three dimensions, for three values of the inner-to-outer ring radius ratio. Note the λ^4 behaviour for large wavelengths of the attenuation length, as expected for Rayleigh scattering. The effective phase velocity becomes larger that the speed of sound in the unperturbed medium for wavelengths larger than a length on the order of the outer radius.

P. Abry, S. Fauve, P. Flandrin and C. Laroche, "Analysis of pressure fluctuations in swirling turbulent flows", *J. Physique II France* **4**, 725-733 (1994); S. Douady, Y. Couder and M. E. Brachet, "Direct observation of the intermittency of intense vorticity filaments in turbulence", *Phys. Rev. Lett.* **67**, 983-986 (1991); O. Cadot, S. Douady and Y. Couder, "Characterization of the low-pressure filaments in a three-dimensional turbulent shear flow", *Phys. Fluids* **7**, 630-646 (1995); F. Belin, J. Maurer, P. Tabeling and H. Willaime, "Observation of intense filaments in fully developed turbulence", *J. Phys. II France* **6**, 573-584 (1996).

As mentioned in the Introduction to this Section, the subject of wave propagation in complex, or more precisely, random, media, has a long and distinguished history of scholarship and the literature is vast. The standard reference is A. Ishimaru, "Wave Propagation and Scattering in Random Media", IEEE/OUP, New York (1997). A more specialized monograph, close to the subject of this Section, is V. E. Ostachev, "Acoustics in Moving Inhomogeneous Media", E&FN SPON, London (1997). Early papers that can still be read with much profit include L. L. Foldy, "The multiple scattering of waves", *Phys. Rev.* **67**, 107-119 (1945); M. Lax, "Multiple scattering of waves", *Rev. Mod. Phys.* **23** 287-310 (1951); M. Lax, "Multiple scattering of waves II. The effective field in dense systems", *Phys. Rev.* **85**, 621-629 (1952).

Recent years have seen the realization that the study of wave propagation in random media provides a solid theoretical underpinning for the century-old field of radiation transfer. See, for example, L. A. Apresyan and Yu. A. Kravtsov, "Radiation Transfer: Statistical and Wave Aspects", Gordon and

Breach (1996). Brownian motion attests to the fact that a particle moving deterministically — with travelled distance proportional to time elapsed — in between randomly placed obstacles changes this motion, in the long run and on average, to diffusive, when travelled distance is proportional to the square root of elapsed time. Something similar happens to waves under certian conditions: they may propagate in betwen obstacles randomly placed, and, on average and in the long run, diffuse. It was shown for de Broglie waves (aka electrons) in one dimension that even a small amount of disorder would prevent diffusion, the waves staying within a localized region of space: P. W. Anderson, "Abscence of diffusion in certain random lattices", *Phys. Rev.* **109**, 1492-1505 (1958). This work spawned a large body of research in "Anderson localization", whereby a similar effect is hoped for in two or three dimensions and for other types of waves such as electromagnetic or acoustic. For an inroad into this fascinating field see P. Sheng (Ed.), "Scattering and Localization of Classical Waves in Random Media", World Science, Singapore (1990) and P. Sheng, "Introduction to Wave Scattering, Localization and Mesoscopic Phenomena", Academic (1995).

6 Loose Ends

It was mentioned in Section 3 that the optical theorem is not fulfilled in a first Born approximation. Here we show that, to second order it is, using the results derived in Section 5 to order M^2. Also, the results derived in Section 5 have a limited range of validity: terms of order M^3 and higher must be negligible. At first sight one would say it is enough to have low background flow velocities. Looking at Eqn. (59), however, we see that terms that couple flow with acoustic wave involve gradients of the acoustic wave perturbation. Say, the term of order M is actually of order $M(k/k_0)$ so that for small enough wavelengths these terms may become of order one. Consequently, the validity criterion must also involve the acoustic wavelength.

6.1 Optical Theorem

In this section we reconsider the scattering of an acoustic wave by a single vortex, the problem that was treated is Section 4, but this time to second order in the Born series. We saw then that the optical theorem (81) did not hold to first order or, rather, that it was an identity. Going back to (130), we now take V to be generated by a deterministic background flow (180) with $N = 1$, i.e., a single vortex. Turn (130) into an integral equation

$$p_s(\boldsymbol{x}) = p_s^{\text{inc}}(\boldsymbol{x}) - \int d\boldsymbol{x}' G^{(0)}(\boldsymbol{x} - \boldsymbol{x}') V(\boldsymbol{x}')[p_s(\boldsymbol{x}')]. \tag{192}$$

The Green's function for the Helmholtz equation was discussed in (132-133). In coordinate space it is

$$G^{(0)}(\boldsymbol{x}) = \frac{e^{ik_0|\boldsymbol{x}|}}{4\pi|\boldsymbol{x}|}.$$

With this expression, (192) gives, in the far field ($|\boldsymbol{x}| \gg |\boldsymbol{x}'|$),

$$p_s(\boldsymbol{x}) = p_s^{\mathrm{inc}}(\boldsymbol{x}) + p_s^{\mathrm{scatt}}(\boldsymbol{x})$$

with

$$p_s^{\mathrm{scatt}}(\boldsymbol{x}) \sim -\frac{e^{ik_0|\boldsymbol{x}|}}{4\pi|\boldsymbol{x}|}\int d\boldsymbol{x}' e^{-i\boldsymbol{k}\cdot\boldsymbol{x}'}V(\boldsymbol{x}')[p_s(\boldsymbol{x}')]. \tag{193}$$

Using the solution of (130) to second order in the Born approximation, this gives

$$p_s^{\mathrm{scatt}}(\boldsymbol{x}) = -\frac{e^{ik_0|\boldsymbol{x}|}}{4\pi|\boldsymbol{x}|}\int d\boldsymbol{x}' V(\boldsymbol{x}')$$
$$\times\left[p_s^{\mathrm{inc}}(\boldsymbol{x}') - \int d\boldsymbol{x}'' G^{(0)}(\boldsymbol{x}' - \boldsymbol{x}'')e^{-i\boldsymbol{k}\cdot\boldsymbol{x}'}V(\boldsymbol{x}'')[p_s^{\mathrm{inc}}(\boldsymbol{x}'')]\right]. \tag{194}$$

With the notation of Eqn. (131), this gives the following expression for the scattering amplitude, to second order in Mach number:

$$f(\theta) = -\frac{1}{4\pi}\int d\boldsymbol{x}' e^{-i\boldsymbol{k}\cdot\boldsymbol{x}'}(V^{(1)} + V^{(2)})(\boldsymbol{x}')$$
$$\times\left[e^{i\boldsymbol{k}_0\cdot\boldsymbol{x}'} - \int d\boldsymbol{x}'' G^{(0)}(\boldsymbol{x}' - \boldsymbol{x}'')V^{(1)}(\boldsymbol{x}'')[e^{i\boldsymbol{k}_0\cdot\boldsymbol{x}'}]\right]. \tag{195}$$

We have already seen in Section 3.2 that $V^{(1)}$ gives a real contribution to the forward scattering amplitude, and it is easy to check that the same is true, to order M^2, for $V^{(2)}$. We thus have

$$\mathrm{Im} f(0) = \mathrm{Im}\left[\frac{1}{4\pi}\int d\boldsymbol{x}' e^{-i\boldsymbol{k}\cdot\boldsymbol{x}'}V^{(1)}(\boldsymbol{x}')\right.$$
$$\left.\times\int d\boldsymbol{x}'' G^{(0)}(\boldsymbol{x}' - \boldsymbol{x}'')V^{(1)}(\boldsymbol{x}'')[e^{i\boldsymbol{k}_0\cdot\boldsymbol{x}''}]\right]_{\boldsymbol{k}=\boldsymbol{k}_0} \tag{196}$$

$$= \mathrm{Im}\left[\frac{k_0^2}{\pi c_0^2}\int \frac{d\boldsymbol{q}}{(2\pi)^3}\frac{(\hat{k}_0\cdot\boldsymbol{q})^2|\hat{k}_0\cdot\tilde{\boldsymbol{u}}(\boldsymbol{k}_0 - \boldsymbol{q})|^2}{q^2 - (\boldsymbol{k}_0 + i\eta)^2}\right] \tag{197}$$

$$= \frac{k_0^5}{2c_0^2}\int \frac{d\Omega_q}{(2\pi)^3}(\hat{k}_0\cdot\hat{q})^2|\hat{k}_0\cdot\tilde{\boldsymbol{u}}(\boldsymbol{k}_0 - k_0\hat{q})|^2 \tag{198}$$

$$= \frac{k_0}{4\pi}\sigma_T \tag{199}$$

where the second line is obtained through a calculation almost identical to the one in Section 5.2. The third is obtained through Eqn. (185), and the last line comparing with Eqn. (80).

6.2 Validity of the Born Approximation

As noted in Section 3, there is cause for concern regarding the convergence of the Born series, even for low Mach number, when the wavelength of the acoustic radiation is very small. Consequently, here we study the behaviour of terms of order M^2 and M^4 in that limit, and impose that the latter be much smaller than the former. We shall take vortex dipoles in two dimensions when a specific example is needed. The reasoning in three dimensions is similar.[11]

From Eqns. (144) and (145) we see that a typical term of order M^2 is

$$\mathrm{TWO} \equiv \langle V^{(1)} \mathcal{G} V^{(1)} \rangle, \tag{200}$$

and a typical term of order four is

$$\mathrm{FOUR} \equiv \langle V^{(1)} \mathcal{G} V^{(1)} \mathcal{G} V^{(1)} \mathcal{G} V^{(1)} \rangle. \tag{201}$$

Let us study FOUR first. it is defined by

$$\begin{aligned}\mathrm{FOUR} = \frac{1}{V} \int d\boldsymbol{x}_1 d\boldsymbol{x}_2 d\boldsymbol{x}_3 d\boldsymbol{x}_4 e^{-i\boldsymbol{k}_0 \cdot \boldsymbol{x}_1} \langle V^{(1)}(\boldsymbol{x}_1) G^{(0)}(\boldsymbol{x}_{12}) V^{(1)}(\boldsymbol{x}_2) \\ \times G^{(0)}(\boldsymbol{x}_{23}) V^{(1)}(\boldsymbol{x}_3) G^{(0)}(\boldsymbol{x}_{34}) V^{(1)}(\boldsymbol{x}_4) \rangle [e^{i\boldsymbol{k}_0 \cdot \boldsymbol{x}_4}]\end{aligned} \tag{202}$$

where $\boldsymbol{x}_{ij} \equiv \boldsymbol{x}_i - \boldsymbol{x}_j$. Noting that these integrals are one convolution after another, FOUR can be writen in Fourier space as

$$\mathrm{FOUR} = -\frac{16n}{c_0^4 k_0^4 (2\pi)^{3d}} \int \Pi_{j=1}^3 \frac{d\boldsymbol{q}_j}{q_j^2 - (k_0 + i\eta)^2} \mathcal{A}\mathcal{B} \tag{203}$$

where

$$\mathcal{A} \equiv (\boldsymbol{k}_0 \cdot \boldsymbol{q}_1)(\boldsymbol{q}_1 \cdot \boldsymbol{q}_2)(\boldsymbol{q}_2 \cdot \boldsymbol{q}_3)(\boldsymbol{q}_3 \cdot \boldsymbol{k}_0)$$

and

$$\mathcal{B} \equiv \langle (\boldsymbol{k}_0 \cdot \boldsymbol{v}(\boldsymbol{k}_0 - \boldsymbol{q}_1))(\boldsymbol{q}_1 \cdot \boldsymbol{v}(\boldsymbol{q}_1 - \boldsymbol{q}_2))(\boldsymbol{q}_2 \cdot \boldsymbol{v}(\boldsymbol{q}_2 - \boldsymbol{q}_3))(\boldsymbol{q}_3 \cdot \boldsymbol{v}(\boldsymbol{q}_3 - \boldsymbol{k}_0)) \rangle.$$

Consequently, for large k_0 we have

$$\mathrm{Im}(\mathrm{FOUR}) \sim \frac{2\pi^3 n k_0^{3d+2}}{c_0^4 (2\pi)^{3d}} \int d\hat{q}_1 d\hat{q}_2 d\hat{q}_3 \hat{\mathcal{A}} \hat{\mathcal{B}} \tag{204}$$

where

$$\hat{\mathcal{A}} \equiv (\hat{k}_0 \cdot \hat{q}_1)(\hat{q}_1 \cdot \hat{q}_2)(\hat{q}_2 \cdot \hat{q}_3)(\hat{q}_3 \cdot \hat{k}_0)$$

and

$$\hat{\mathcal{B}} \equiv \langle (\hat{k}_0 \cdot \boldsymbol{v}(k_0(\hat{k}_0 - \hat{q}_1)))(\hat{q}_1 \cdot \boldsymbol{v}(k_0(\hat{q}_1 - \hat{q}_2)))(\hat{q}_2 \cdot \boldsymbol{v}(k_0(\hat{q}_2 - \hat{q}_3)))(\hat{q}_3 \cdot \boldsymbol{v}(k_0(\hat{q}_3 - \hat{k}_0))) \rangle.$$

[11] See D. Boyer and F. Lund, "Propagation of acoustic waves in disordered flows composed of many vortices II. Examples", *Phys. Fluids* **11**, 3829-3845 (1999).

For large k_0, contributions to the integral in (204) come mainly from regions of integration where all unit vectors are parallel, so that $\hat{\mathcal{A}} \sim 1$. Also, when $\hat{q}_j$ is almost parallel to $\hat{k}_0$, $\hat{q}_j - \hat{k}_0$ is almost perpendicular to $\hat{k}_0$, and, consequently, most of the contributions to the integral come from the $(d-1)$ dimensional space perpendicular to $\hat{k}_0$. Consequently, changing variables to

$$\tilde{\boldsymbol{q}}_j \equiv k_0(\hat{q}_j - \hat{k}_0)$$

we have

$$\mathrm{Im}(\mathrm{FOUR}) \sim \frac{2\pi^3 n k_0^5}{c_0^4(2\pi)^{3d}} \int_{\boldsymbol{q}_j \cdot \boldsymbol{k}_0 = 0} d\boldsymbol{q}_1 d\boldsymbol{q}_2 d\boldsymbol{q}_3 \hat{\mathcal{B}}' \tag{205}$$

where

$$\hat{\mathcal{B}}' \equiv \langle (\hat{k}_0 \cdot \boldsymbol{v}(-\boldsymbol{q}_1))(\hat{k}_0 \cdot \boldsymbol{v}(\boldsymbol{q}_1 - \boldsymbol{q}_2))(\hat{k}_0 \cdot \boldsymbol{v}(\boldsymbol{q}_2 - \boldsymbol{q}_3))(\hat{k}_0 \cdot \boldsymbol{v}(\boldsymbol{q}_3)) \rangle.$$

In the case of vorttex dipoles in two dimensions this gives

$$\mathrm{Im}(\mathrm{FOUR}) \sim n k_0^5 \epsilon^4 l M^4. \tag{206}$$

Similarly, one gets

$$\mathrm{Im}(\mathrm{TWO}) \sim n k_0^3 \epsilon^2 l M^2 \tag{207}$$

so that, finally, the condition for the fourth order contribution to be much smaller that the second order contribution

$$\mathrm{Im}(\mathrm{FOUR}) \ll \mathrm{Im}(\mathrm{TWO})$$

is

$$\lambda_0 \gg \epsilon M \tag{208}$$

a condition that is not hard to respect.

Acknowledgements

Much of the work reported herein has been performed in collaboration with M. Baffico, D. Boyer, H. Contreras, C. Coste, F. Melo, C. Rojas and M. Umeki. Over the years, I have greatly profited from discussions with S. Fauve. The support of FONDECYT Grant 1990035 and FONDAP Grant 11980002 is gratefully acknowledged.

Geometrical Acoustics in Moving Media

Vladimir E. Ostashev

NOAA/Environmental Technology Laboratory, 325 Broadway, Boulder, CO 80305 and Department of Physics, New Mexico State University, Las Cruces, NM, USA

Abstract. A complete set of linearized equations of fluid dynamics is considered, which is a starting point of geometrical acoustics in inhomogeneous moving media. The linearized equations of fluid dynamics are solved by using the Debye series. As a result the eikonal equation, dispersion equation, transport equation, and the law of acoustic energy conservation are obtained. The sound propagation in three-dimensional and stratified moving media are studied.

1 Introduction

Flow in a medium changes the refractive index of the medium. Therefore, a sound wave propagating in a medium with flow exhibits refraction. Refraction causes the path of a sound wave to deviate from line-of-sight; the time interval of sound propagation from source to receiver changes in comparison to that in a motionless medium; the amplitude of a sound wave also changes; etc. These and other phenomena caused by sound refraction can be treated by using geometrical acoustics in moving media.

The main approaches and results of geometrical acoustics in inhomogeneous moving media are systematically presented in Chapter 3 from [4], which also provides a detailed historical background of these approaches and results. The present chapter gives a concise treatment of the main results known in geometrical acoustics in inhomogeneous moving media, and the equations that are most widely used in this field. Derivations of these equations are not presented here since they are given in detail in [4]. However, the main steps in these derivations are explained.

The chapter is organized as follows. In Section 2, a complete set of linearized equations of fluid dynamics is considered, which is a starting point of geometrical acoustics in inhomogeneous moving media. In Section 3, the linearized equations of fluid dynamics are solved by using the Debye series. As a result the eikonal equation, dispersion equation, transport equation, and the law of acoustic energy conservation are obtained. Sections 4 and 5 are devoted to studies of sound propagation in three-dimensional and stratified (layered) moving media, respectively. Phase and group velocities of a sound wave, ray path, the time of sound propagation along the path, refraction laws for a sound ray and normal to the wave front are considered in these sections.

2 Linearized Equations of Fluid Dynamics

A complete set of fluid dynamic equations is given by:

$$\frac{d\mathbf{v}}{dt} + \frac{\nabla P}{\varrho} = 0\,, \tag{1}$$

$$\frac{d\varrho}{dt} + \varrho \nabla \cdot \mathbf{v} = 0\,, \tag{2}$$

$$P = P(\varrho, S)\,, \tag{3}$$

$$\frac{dS}{dt} = 0\,. \tag{4}$$

Here, $\mathbf{v}$ is the velocity, P is the pressure, ϱ is the density, and S is the entropy in a medium. All these quantities are functions of both the Cartesian coordinates $\mathbf{R} = (x, y, z)$ and the time t. Furthermore, in (1) – (4), $\nabla = (\partial/\partial x, \partial/\partial y, \partial/\partial z)$, and $d/dt = \partial/\partial t + \mathbf{v} \cdot \nabla$. Equations (1) – (4) describe the ambient state of the medium and a propagating sound wave. To obtain equations only for a sound wave, we write: $P = P_\mathrm{a} + p$, $\varrho = \varrho_\mathrm{a} + \eta$, $\mathbf{v} = \mathbf{v}_\mathrm{a} + \mathbf{w}$, and $S = S_\mathrm{a} + s$. Here, P_a, ϱ_a, $\mathbf{v}_\mathrm{a}$, and S_a are the ambient values of the pressure, density, medium velocity, and entropy, and p, η, $\mathbf{w}$, and s are their fluctuations due to the sound wave. In linear acoustics, it is assumed that a sound wave disturbs the ambient state of a medium only slightly. In this approximation, (1) – (4) can be linearized with respect to p, η, $\mathbf{w}$, and s:

$$\frac{d\mathbf{w}}{dt} + (\mathbf{w} \cdot \nabla)\mathbf{v}_\mathrm{a} + \frac{\nabla p}{\varrho_\mathrm{a}} - \frac{\eta \nabla P_\mathrm{a}}{\varrho_\mathrm{a}^2} = 0\,, \tag{5}$$

$$\frac{d\eta}{dt} + (\mathbf{w} \cdot \nabla)\varrho_\mathrm{a} + \varrho_\mathrm{a} \nabla \cdot \mathbf{w} + \eta \nabla \cdot \mathbf{v}_\mathrm{a} = 0\,, \tag{6}$$

$$\frac{ds}{dt} + (\mathbf{w} \cdot \nabla) S_\mathrm{a} = 0\,, \tag{7}$$

$$p = c^2 \eta + hs. \tag{8}$$

Here, $h = (\partial P_\mathrm{a}/\partial S_\mathrm{a})_\varrho$ is a parameter, $c = \sqrt{(\partial P_\mathrm{a}/\partial \varrho_\mathrm{a})_S}$ is the adiabatic sound speed, and $d/dt = \partial/\partial t + \mathbf{v}_\mathrm{a} \cdot \nabla$.

Equations (5) – (8) are a complete set of linearized equations of fluid dynamics. This set allows us to describe the propagation of sound waves in moving media if the ambient quantities P_a, ϱ_a, $\mathbf{v}_\mathrm{a}$, and S_a are known. However, (5) – (8) are rather involved and cannot be solved analytically in a general case.

3 Solution of Linearized Equations of Fluid Dynamics

In this section, (5) – (8) are solved in the approximation of geometrical acoustics. In what follows, we assume that the ambient quantities P_a, ϱ_a, $\mathbf{v}_a$, and S_a do not depend on time t, and that a propagating sound wave is monochromatic.

3.1 Debye Series

Let p, η, $\mathbf{w}$, and s be expressed in the form:

$$\begin{aligned} p(\mathbf{R},t) &= \exp(-i\omega t + ik_0\Psi(\mathbf{R}))p_0(\mathbf{R}), \\ \eta(\mathbf{R},t) &= \exp(-i\omega t + ik_0\Psi(\mathbf{R}))\eta_0(\mathbf{R}), \\ \mathbf{w}(\mathbf{R},t) &= \exp(-i\omega t + ik_0\Psi(\mathbf{R}))\mathbf{w}_0(\mathbf{R}), \\ s(\mathbf{R},t) &= \exp(-i\omega t + ik_0\Psi(\mathbf{R}))s_0(\mathbf{R}) . \end{aligned} \tag{9}$$

Here, ω is the frequency of a sound wave; $\Psi(\mathbf{R})$ is its eikonal; p_0, η_0, $\mathbf{w}_0$, and s_0 are the amplitudes of acoustic fluctuations that depend only on $\mathbf{R}$; and k_0 is the mean (or reference) value of the wave number. Note that there is some freedom in choosing the value of k_0.

Substituting (9) into the linearized equations of fluid dynamics, (5) – (8), we obtain a set of four equations for five unknown quantities:

$$\Psi(\mathbf{R}), \qquad p_0(\mathbf{R}), \qquad \eta_0(\mathbf{R}), \qquad \mathbf{w}_0(\mathbf{R}), \qquad s_0(\mathbf{R}) . \tag{10}$$

Then, the amplitudes p_0, η_0, $\mathbf{w}_0$, and s_0 are expressed as series in the small parameter $1/(k_0L)$ (the Debye series):

$$\begin{aligned} p_0 &= \sum_{n=1}^{\infty} \frac{p_n}{(ik_0L)^{n-1}} , \qquad \mathbf{w}_0 = \sum_{n=1}^{\infty} \frac{\mathbf{w}_n}{(ik_0L)^{n-1}}, \\ \eta_0 &= \sum_{n=1}^{\infty} \frac{\eta_n}{(ik_0L)^{n-1}} , \qquad s_0 = \sum_{n=1}^{\infty} \frac{s_n}{(ik_0L)^{n-1}} . \end{aligned} \tag{11}$$

Here, L is the spatial scale of variations of the ambient quantities P_a, ϱ_a, $\mathbf{v}_a$, and S_a, which is assumed to be much greater than the sound wavelength $\lambda = 2\pi/k_0$.

The Debye series, (11), are substituted into the set of equations for quantities given by (10). In the equations obtained, we equate coefficients of k_0^n, where $n = 1, 2, 3, \ldots$. The resulting equations allow us to derive equations for the eikonal $\Psi(\mathbf{R})$ and the quantity $p_1(\mathbf{R})$, which is the first term of the Debye series for p_0. In the approximation of geometrical acoustics, p_1 is considered the amplitude of a sound wave.

3.2 Eikonal Equation

For simplicity hereinafter, the letter "a" in subscripts of the ambient quantities P_a, ϱ_a, $\mathbf{v}_a$, and S_a is omitted. With this notation, the equation for $\Psi(\mathbf{R})$ is given by:

$$\frac{c_0}{c(\mathbf{R})} - \frac{\mathbf{v}(\mathbf{R}) \cdot \nabla\Psi(\mathbf{R})}{c(\mathbf{R})} = |\nabla\Psi(\mathbf{R})| \, . \tag{12}$$

Here, $c_0 = \omega/k_0$ is the mean (reference) value of the adiabatic sound speed. Equation (12) is called the eikonal equation. It was derived by Blokhintzev [2].

3.3 Dispersion Equation

By definition, the wave vector $\mathbf{k}$ of a sound wave is given by $\mathbf{k} = k_0\nabla\Psi$. Using this formula, the eikonal equation (12) can be written in the form:

$$\omega = k(\mathbf{R})c(\mathbf{R}) + \mathbf{k}(\mathbf{R}) \cdot \mathbf{v}(\mathbf{R}) \, . \tag{13}$$

This equations is called the dispersion equation.

3.4 Transport Equation

The equation for the amplitude p_1 of a sound wave reads:

$$2\mathbf{u} \cdot \nabla p_1 + \left[\nabla \cdot \mathbf{u} - \mathbf{u} \cdot \nabla \ln(\varrho c^2(\omega - \mathbf{k} \cdot \mathbf{v}))\right] p_1 = 0 \, . \tag{14}$$

Here, $\mathbf{u} = c\mathbf{n} + \mathbf{v}$, where $\mathbf{n} = \mathbf{k}/k$ is the unit vector in the direction of the wave vector $\mathbf{k}$. The vector $\mathbf{n}$ is normal to the wave front of a sound wave since $\mathbf{k}/k = \nabla\Psi/|\nabla\Psi|$. (The wave front is a surface where the eikonal $\Psi(\mathbf{R})$ is constant.) Equation (14) is called the transport equation.

Equations (12) and (14) allow us, in principle, to calculate the eikonal Ψ and the amplitude p_1. In the approximation of geometrical acoustics, the sound pressure p is expressed in terms of Ψ and p_1:

$$p(\mathbf{R}, t) = \exp(-i\omega t + ik_0\Psi(\mathbf{R}))p_1(\mathbf{R}) \, . \tag{15}$$

3.5 Conservation of Acoustic Energy

Using (15) for the sound pressure and the transport equation (14), the law of acoustic energy conservation can be derived:

$$\frac{\partial E}{\partial t} + \nabla \cdot \mathbf{I} = 0 \, , \tag{16}$$

where

$$E = \frac{p^2}{\varrho c^2} \left(1 + \frac{\mathbf{n} \cdot \mathbf{v}}{c}\right) . \tag{17}$$

In these equations, E is the instantaneous value of the sound energy density, and $\mathbf{I} = E\mathbf{u}$ is its flux. It follows from this formula that $\mathbf{u}$ is the velocity of sound energy propagation (the group velocity). Blokhintzev [2] was the first to derive the law of acoustic energy conservation in the approximation of geometrical acoustics. Equation (16) enables one to calculate E, $\mathbf{I}$, and p in an inhomogeneous moving medium. Note that, in a general case, acoustic energy is not conserved. For example, a sound wave can be amplified or attenuated by a flow.

4 Geometrical Acoustics in a Three-Dimensional Inhomogeneous Medium

Using the equations obtained above, in this section we consider the group and phase velocities, ray path, eikonal, and amplitude of a sound wave propagating in a three-dimensional inhomogeneous moving medium.

4.1 Group and Phase Velocities

The group $\mathbf{u}$ and phase $\mathbf{U}$ velocities of a wave propagating in an inhomogeneous medium can be obtained with the use of the following relationships:

$$\mathbf{u} = \left(\frac{\partial\omega}{\partial k_x}, \frac{\partial\omega}{\partial k_y}, \frac{\partial\omega}{\partial k_z}\right) , \qquad \mathbf{U} = \frac{\mathbf{k}\omega}{k^2} , \tag{18}$$

where $\mathbf{k} = (k_x, k_y, k_z)$. Substituting the dispersion equation (13) into (18), one obtains formulas for the group and phase velocities of a sound wave propagating in an inhomogeneous moving medium:

$$\mathbf{u} = c\mathbf{n} + \mathbf{v} , \qquad \mathbf{U} = (c + \mathbf{n} \cdot \mathbf{v})\, \mathbf{n} . \tag{19}$$

As should be the case, the value of $\mathbf{u}$ given by this equation coincides with that from the transport equation (14). Formulas for $\mathbf{U}$ and $\mathbf{u}$ were obtained by Rayleigh [5] and Barton [1], respectively.

The relative position of the group and phase velocities of a sound wave in a moving medium is shown in Fig. 1. Note that the phase velocity is perpendicular to the wave front. It follows from figure and (19) that, in a general case, the directions of the phase and group velocities do not coincide, and that $U \leq u$.

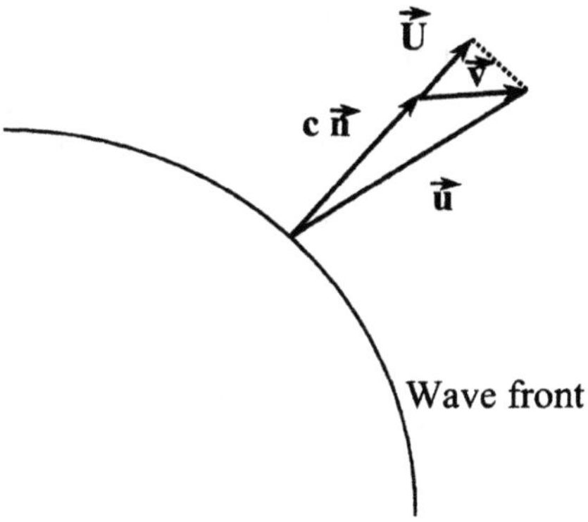

Fig. 1. Geometry of the group and phase velocities of a sound wave in a moving medium.

4.2 Ray Path

Let us consider the eikonal equation (12). This is the Hamilton-Jacobi equation. Therefore, the solution of (12) can be reduced to the integration of the characteristic system of the differential equations:

$$\frac{d\mathbf{R}}{d\tau} = \frac{\mathbf{B}}{B} + \frac{\mathbf{v}}{c} , \tag{20}$$

$$\frac{d\mathbf{B}}{d\tau} = -\frac{c_0 \nabla c}{c^2} - \frac{\nabla(\mathbf{B} \cdot \mathbf{v})}{c} , \tag{21}$$

$$\frac{d\Psi}{d\tau} = B + \frac{\mathbf{B} \cdot \mathbf{v}}{c} . \tag{22}$$

In these equations, τ is the independent variable, and the vector $\mathbf{B} = \nabla\Psi$. The vector $\mathbf{R}(\tau)$ determines the ray path of a sound wave in an inhomogeneous moving medium. The ray path $\mathbf{R}(\tau)$ can be obtained as a solution of (20) and (21). The unit vector $\mathbf{s} = \mathbf{u}/u$ is tangential to the ray path.

In many cases, it is convenient to deal with the ray length l rather than with the independent variable τ. The relationship between l and τ can be obtained with:

$$d\tau = c\,dl/u . \tag{23}$$

4.3 Eikonal and Time of Sound Propagation

The eikonal Ψ is calculated by integrating both sides of (22). Changing from $d\tau$ to dl in the equation obtained, we have:

$$\Psi(l) = \Psi_0 + \int_{l_0}^{l} \frac{c_0\, dl'}{u} . \tag{24}$$

Here, the integration is taken along the ray path, and Ψ_0 is the value of the eikonal at the initial point of the ray path where the ray length is equal to l_0.

Since dl is an increment of the ray path, the time interval of sound propagation along this increment is given by $dt = dl/u$. Integrating both sides of this formula along the ray path and using (24), one obtains the time of sound propagation along the ray path:

$$t = \frac{\Psi(l) - \Psi_0}{c_0} = \int_{l_0}^{l} \frac{dl'}{u} \, . \qquad (25)$$

4.4 Amplitude of the Sound Pressure

Using the law of acoustic energy conservation, (16), the following formula for the sound pressure amplitude, p_1, is obtained:

$$p_1^2 \frac{(1 + \mathbf{n} \cdot \mathbf{v}/c)\sqrt{1 + 2\mathbf{n} \cdot \mathbf{v}/c + v^2/c^2}\, d\sigma}{2\varrho c} = \text{const} \, . \qquad (26)$$

Here, $d\sigma$ is the cross-sectional area of an infinitesimal narrow ray tube. $d\sigma$ can be calculated if the ray path of a sound wave is known. Then, p_1 is easily obtained with (26).

5 Geometrical Acoustics in a Stratified Medium

In both the atmosphere and ocean, the horizontal scale of variations in c, ϱ, and $\mathbf{v}$ is usually much greater than the vertical scale of these variations. Therefore, for many problems of sound propagation in the atmosphere and ocean, one can assume that $c = c(z)$, $\varrho = \varrho(z)$, and $\mathbf{v} = \mathbf{v}(z)$. Such a medium is called a stratified (layered) moving medium. In this section, we consider geometrical acoustics in a stratified moving medium. For simplicity, we assume that the vertical component of the vector $\mathbf{v}$ is zero.

5.1 Refraction Law for the Unit Vector n

In a stratified moving medium, the unit vector $\mathbf{n}(z)$ normal to the wave front of a sound wave depends only on z. Let us express this vector in the following form: $\mathbf{n}(z) = (\mathbf{e}\cos\theta(z), \sin\theta(z))$. Here, $\mathbf{e}\cos\theta$ and $\sin\theta$ are the horizontal and vertical components of the vector $\mathbf{n}$; $\mathbf{e}$ is the unit vector in the direction of the projection of $\mathbf{n}$ on the horizontal plane, and θ is the grazing angle between $\mathbf{n}$ and the horizontal plane (see Fig. 2). Using the dispersion equation (13), it can be shown that the vector $\mathbf{e}$ remains constant along a ray path, and the angle θ satisfies the equation:

$$\frac{c(z)}{\cos\theta(z)} + \mathbf{e} \cdot \mathbf{v}(z) = \text{const} \, . \qquad (27)$$

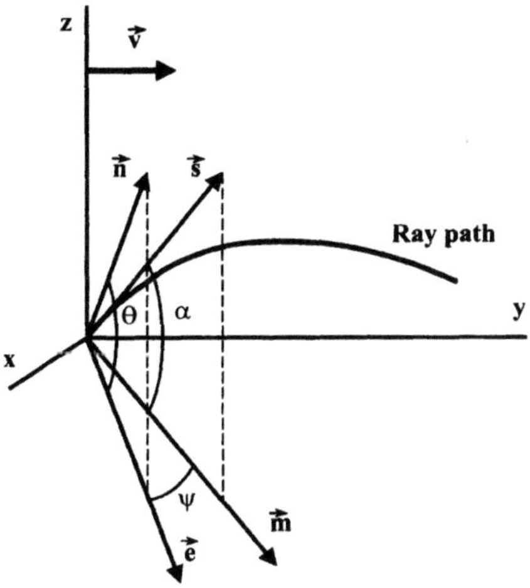

Fig. 2. Geometry of the unit vector **n** normal to the wave front and the unit vector **s** tangential to the ray path in a stratified moving medium.

This equation is the refraction law for the normal **n** to the wave front in a stratified moving medium. It allows one to determine the grazing angle θ of the vector **n** along the sound propagation path. If $\mathbf{v} = 0$, (27) coincides with Snell's law. Rayleigh [5] was the first to derive (27).

5.2 Refraction Law for the Unit Vector s

The unit vector $\mathbf{s}(z)$ tangential to the ray path can be written as $\mathbf{s}(z) = (\mathbf{m}(z)\cos\alpha(z), \sin\alpha(z))$. Here, $\mathbf{m}\cos\alpha$ and $\sin\alpha$ are the horizontal and vertical components of the vector **s**. Furthermore, **m** is the unit vector in the direction of the projection of **s** on the horizontal plane, and α is the grazing angle of this vector (see Fig. 2). Note that the vector **m** does not coincide with the vector **e**; the angle between these vectors is denoted by $\psi(z)$. Since **e** remains constant along the sound propagation path, it is convenient to characterize the azimuthal direction of the vector **s** by the angle ψ.

It follows from (19) that $\mathbf{s} = \mathbf{u}/u = (c\mathbf{n} + \mathbf{v})/|c\mathbf{n} + \mathbf{v}|$. Using this formula and (27), one can obtain formulas for the grazing angle α and the azimuthal angle ψ of the unit vector **s**:

$$\sin\alpha(z) = \frac{c(z)\sqrt{(C - \mathbf{e}\cdot\mathbf{v}(z))^2 - c^2(z)}}{\sqrt{[C - \mathbf{e}\cdot\mathbf{v}(z)]\,[(c^2(z) + v^2(z))C + (c^2(z) - v^2(z))\,\mathbf{e}\cdot\mathbf{v}(z)]}} \tag{28}$$

and

$$\sin\psi(z) = \frac{|\mathbf{v}(z) - (\mathbf{e}\cdot\mathbf{v}(z))\mathbf{e}|}{\sqrt{c^2(z)\cos^2\theta(z) + 2c(z)\cos\theta(z)\,\mathbf{e}\cdot\mathbf{v}(z) + v^2(z)}}\,. \tag{29}$$

In (28), C is a constant. The value of this constant can be determined by solving (28) for C at a given level $z = z_0$. If at this level $\mathbf{v}(z_0) = 0$, the formula for C simplifies: $C = c(z_0)/\cos\alpha(z_0)$.

Equations (28) and (29) are the refraction law for a sound ray in a stratified moving medium. These equations allow one to calculate the grazing angle $\alpha(z)$ and the azimuthal angle $\psi(z)$ if the vertical profiles $c(z)$ and $\mathbf{v}(z)$ are known. Equations (28) and (29) were obtained in [3]. For a motionless medium, (28) coincides with Snell's law; furthermore, it follows from (29) that $\psi = 0$.

5.3 The Ray Path

In a stratified moving medium, it is convenient to express the ray path in the form: $\mathbf{r} = \mathbf{r}(z)$, where $\mathbf{r} = (x, y)$ are the horizontal coordinates. Using (20) and (21), one can obtain the equation for the ray path:

$$\mathbf{r} = \mathbf{r}_0 + \int_{z_0}^{z} \frac{c\mathbf{e} + (C - \mathbf{e}\cdot\mathbf{v})\mathbf{v}/c}{\sqrt{(C - \mathbf{e}\cdot\mathbf{v})^2 - c^2}}\, dz' \,. \tag{30}$$

Here, $\mathbf{r}_0$ and z_0 are initial horizontal and vertical coordinates of the ray path, and the constant C and the unit vector $\mathbf{e}$ are essentially the same as those in the refraction laws given by (27) – (29).

5.4 Eikonal

In (24) for the eikonal Ψ, we change from the ray length l to the height z. As a result, we obtain the formula for the eikonal of a sound wave propagating in a stratified moving medium:

$$\Psi(z) = \Psi_0 + \int_{z_0}^{z} \frac{c_0(C - \mathbf{e}\cdot\mathbf{v})}{c\sqrt{(C - \mathbf{e}\cdot\mathbf{v})^2 - c^2}}\, dz' \,, \tag{31}$$

where Ψ_0 is the value of Ψ at the level $z = z_0$.

It follows from (30) and (31) that the ray path and eikonal can be obtained by evaluating the integrals on the right-hand side of these equations.

References

1. E.H. Barton: Phil. Mag. **1**, 159 (1901)
2. D.I. Blokhintzev: *Acoustics of an inhomogeneous moving medium* [in Russian], (Nauka, Moscow 1946), [English translation, Physics Dept. Brown Univ., Providence, RI 1956]
3. V.E Ostashev: Sov. Phys. Acoust. **31** (2), 130 (1985)
4. V.E. Ostashev: *Acoustics in Moving Inhomogeneous Media* (E & FN SPON, London 1997)
5. J.W.S. Rayleigh: *The theory of sound* (Dover, New York 1945)

Sound Propagation and Scattering in Random Moving Media

Vladimir E. Ostashev

NOAA/Environmental Technology Laboratory, 325 Broadway, Boulder, CO 80305 and Department of Physics, New Mexico State University, Las Cruces, NM, U.S.A.

Abstract. This paper gives a handbook-type overview of the theories of sound propagation and scattering in random moving media and the major phenomena known in this field.

1 Introduction

Many types of media can be considered as randomly inhomogeneous. Examples include the turbulent atmosphere, the ocean with internal gravity waves, turbulent flows, and liquid marine sediments. In these media, the adiabatic sound speed $c(\mathbf{R})$, the medium velocity $\mathbf{v}(\mathbf{R})$, and the density $\varrho(\mathbf{R})$ are random functions of the Cartesian coordinates $\mathbf{R} = (x, y, z)$, which are also called random fields.

A sound wave propagating in such a medium can be significantly affected by random inhomogeneities in $c(\mathbf{R})$, $\mathbf{v}(\mathbf{R})$, and $\varrho(\mathbf{R})$. Due to these random inhomogeneities, the amplitude and phase of the sound wave exhibit fluctuations; the wave loses its coherence and is scattered; the direction of sound propagation changes; etc. These phenomena can be described by using theories of sound propagation and scattering in random moving media such as those presented in Ostashev [10] and some recent papers. These theories provide mathematical methods for calculating different statistical characteristics of a sound wave propagating in a medium with random inhomogeneities in $c(\mathbf{R})$, $\mathbf{v}(\mathbf{R})$, and $\varrho(\mathbf{R})$. Note that the classical theories of waves in random media as a rule only deal with the effects of random inhomogeneities in $c(\mathbf{R})$ on wave propagation (e.g., see [6,16,18]).

The present chapter gives a handbook-type overview of the theories of sound propagation and scattering in random moving media and the major phenomena known in this field. These theories have been primarily developed with an application to sound propagation in a turbulent atmosphere. However, the theories are general and also have been used for studies of sound propagation in other random moving media (e.g., the turbulent ocean and turbulent flows).

In Section 2, starting equations for describing the propagation of a sound wave in a random moving medium are presented. Section 3 deals with a statistical description of random inhomogeneities in a medium and presents spectra of turbulence that are used in the literature. Sections 4 – 7 consider the statistical moments of a sound field for different geometries of sound propagation in a random moving medium: scattering of sound (Section 4); line-of-sight sound propagation (Section 5); interference of direct and ground-reflected waves in a turbulent atmosphere (Section 6); and sound propagation near impedance ground in a turbulent, refractive atmosphere (Section 7).

2 Equations for a Sound Wave in a Random Moving Medium

In this sections, equations describing the propagation of a sound wave in a random moving medium are presented. These equations are used in the literature for both analytical and numerical studies of sound propagation and scattering.

2.1 Helmholtz-Type Equation

An equation for the pressure $p(\mathbf{R})$ of a monochromatic sound wave with frequency ω is given by (see (6.1) from [10]):

$$\left[\Delta + k_0^2(1+\varepsilon) - \left(\nabla \ln \frac{\varrho}{\varrho_0}\right) \cdot \nabla - \frac{2i}{\omega}\frac{\partial v_i}{\partial x_j}\frac{\partial^2}{\partial x_i \partial x_j} + \frac{2ik_0}{c_0}\mathbf{v}\cdot\nabla\right] p(\mathbf{R}) = 0. \tag{1}$$

Here, $\Delta = \partial^2/\partial x^2 + \partial^2/\partial y^2 + \partial^2/\partial z^2$; $\nabla = (\partial/\partial x, \partial/\partial y, \partial/\partial z)$; $\varepsilon = c_0^2/c^2 - 1$ is the deviation from 1 of the square of the refractive index in a motionless medium; k_0, c_0 and ϱ_0 are the reference values of the wave number, adiabatic sound speed, and density (the first two of these quantities are related by $\omega = k_0 c_0$); x_1, x_2, x_3 stand for x, y, z; $v_1 = v_x$, $v_2 = v_y$, $v_3 = v_z$ are the components of the medium velocity vector $\mathbf{v}$; and repeated subscripts are summed from 1 to 3. Hereinafter, the dependence of sound pressure on time factor $\exp(-i\omega t)$ is omitted.

Equation (1) is a Helmholtz-type equation. There is no restriction on the value of the wave number k_0 in this equation. If $\mathbf{v} = 0$, (1) is an exact consequence of linearized equations of fluid dynamics. In a moving medium, this equation accounts for terms of order v/c_0. The range of applicability of (1) is considered in detail in [10].

Let $c = c_0 + \tilde{c}$ and $\varrho = \varrho_0 + \tilde{\varrho}$, where $\tilde{c}$ and $\tilde{\varrho}$ are fluctuations in the adiabatic sound speed and density due to random inhomogeneities in a medium. To an accuracy of $\tilde{c}/c_0$ and $\tilde{\varrho}/\varrho_0$, we have

$$\varepsilon = -\frac{2\tilde{c}}{c_0} = -\beta_c \frac{\tilde{T}}{T_0}, \qquad \ln\frac{\varrho}{\varrho_0} = \frac{\tilde{\varrho}}{\varrho_0} = \beta_\varrho \frac{\tilde{T}}{T_0}. \tag{2}$$

Here, T_0 and $\tilde{T}$ are the reference value and fluctuations of the temperature in a medium: $T = T_0 + \tilde{T}$. Furthermore, in (2), $\beta_c = (2T_0/c_0)(\partial c_0/\partial T_0)$ and $\beta_\varrho =$

$(T_0/\varrho_0)(\partial\varrho_0/\partial T_0)$. For sound propagation in the atmosphere, $\beta_c = -\beta_\varrho = 1$. Equation (2) expresses $\tilde{c}$ and $\tilde{\varrho}$ in terms of fluctuations in temperature $\tilde{T}$ that are usually measured experimentally. In the atmosphere and ocean, $\tilde{c}$ and $\tilde{\varrho}$ may also depend on fluctuations in humidity and salinity. This dependence is considered in details in Section 6.1 from [10].

Substituting (2) into (1) yields

$$\left[\Delta + k_0^2\left(1 - \frac{\beta_c\tilde{T}}{T_0}\right) - \frac{\beta_\varrho\nabla\tilde{T}}{T_0}\cdot\nabla - \frac{2i}{\omega}\frac{\partial v_i}{\partial x_j}\frac{\partial^2}{\partial x_i\partial x_j} + \frac{2ik_0}{c_0}\mathbf{v}\cdot\nabla\right]p(\mathbf{R}) = 0. \tag{3}$$

This equation describes the propagation of a sound field in a medium with fluctuations in temperature $\tilde{T}(\mathbf{r})$ and velocity $\mathbf{v}(\mathbf{r})$.

2.2 Parabolic Equation

Equations (1) and (3) can be simplified if the sound wavelength $\lambda = 2\pi/k_0$ is smaller than the scale of random inhomogeneities in a medium. Without loss of generality assume that the predominant direction of sound propagation coincides with the x-axis. In this case, it is convenient to express the Cartesian coordinates as $\mathbf{R} = (x, \mathbf{r})$, where $\mathbf{r} = (y, z)$ are the transverse coordinates, and to express the sound pressure $p(x, \mathbf{r})$ as

$$p(x, \mathbf{r}) = e^{ik_0x}A(x, \mathbf{r}) , \tag{4}$$

where A is called the complex amplitude of a sound wave. $A(x, \mathbf{r})$ satisfies a parabolic-type equation (see (7.28) from [10]):

$$2ik_0\frac{\partial A}{\partial x} + \Delta_\perp A + k_0^2\varepsilon_{\text{mov}}A = 0 . \tag{5}$$

Here, $\Delta_\perp = \partial^2/\partial y^2 + \partial^2/\partial z^2$, and ε_{mov} is twice the fluctuations in acoustic refractive index in a moving medium, given by

$$\varepsilon_{\text{mov}} = \varepsilon - 2\frac{v_x}{c_0} = -\beta_c\frac{\tilde{T}}{T_0} - 2\frac{v_x}{c_0} , \tag{6}$$

where v_x is the velocity component along the x-axis. Equation (5) should be supplemented with the initial condition

$$A(x = 0, \mathbf{r}) = p_0(\mathbf{r}) . \tag{7}$$

Here, $p_0(\mathbf{r})$ is the sound field due to a source, which is calculated in the plane $x = 0$.

If $\mathbf{v} = 0$, (5) coincides with the parabolic equation widely used in the theories of waves in random media [6,16]. In a random moving medium, (5) takes into account terms of order v/c_0. The ranges of applicability of this equation are studied in [10].

2.3 Parabolic Equation in a Refractive Medium

Sound propagation is often affected by both random and regular inhomogeneities in a medium. In this case, the adiabatic sound speed can be expressed as a sum $c(\mathbf{R})+\tilde{c}(\mathbf{R})$, and the medium velocity as a sum $\mathbf{V}(\mathbf{R})+\mathbf{v}(\mathbf{R})$. Here, $c(\mathbf{R})$ and $\mathbf{V}(\mathbf{R})$ describe the mean values or profiles of the adiabatic sound speed and medium velocity.

The complex amplitude $A(x,\mathbf{r})$ of a sound wave propagating in a refractive, random medium satisfies the following parabolic equation [10,20]:

$$2ik_0\frac{\partial A}{\partial x} + \hat{L}_{\mathbf{r}}\, A + k_0^2\varepsilon_{\mathrm{mov}}A = 0 \ , \tag{8}$$

where the operator $\hat{L}_{\mathbf{r}}$ operates on the transverse coordinates $\mathbf{r}$ and is given by

$$\hat{L}_{\mathbf{r}} = \Delta_\perp + k^2(\mathbf{R}) - k_0^2 + \frac{2ik_0\mathbf{V}_\perp(\mathbf{R})\cdot\nabla_\perp}{c_0} \ . \tag{9}$$

Here, $\nabla_\perp = (\partial/\partial y, \partial/\partial z)$, $k(\mathbf{R}) = k_0c_0/\left[c\left(\mathbf{R}\right) + V_x\left(\mathbf{R}\right)\right]$ is the sound wave number in a refractive medium, and V_x and $\mathbf{V}_\perp$ are the components of the mean velocity in the direction of sound propagation and perpendicular to it. The complex amplitude A is still related to the sound pressure p by (4), and the initial condition to (8) is given by (7).

3 Statistical Description of Random Inhomogeneities

To calculate the statistical moments of a sound field p propagating in a random moving medium, one must know statistical characteristics of the random fields $\tilde{T}(\mathbf{R})$ and $\mathbf{v}(\mathbf{R})$ appearing in (3) and (6). In this section, the correlation functions and spectral densities of the random fields $\tilde{T}$ and $\mathbf{v}$ are considered. Hereinafter, it is assumed that the mean values of these random fields are zero, and they do not correlate.

3.1 Correlation Functions and Spectral Densities

The correlation function of temperature fluctuations is given by

$$B_T(\mathbf{R}) = \langle\tilde{T}(\mathbf{R}'+\mathbf{R})\tilde{T}(\mathbf{R}')\rangle = \int d^3K\, e^{i\mathbf{K}\cdot\mathbf{R}}\Phi_T(\mathbf{K}) \ . \tag{10}$$

Here, the brackets $\langle\ \rangle$ denote ensemble average, $\mathbf{K}$ is the turbulence wave number, and $\Phi_T(\mathbf{K})$ is the three-dimensional spectral density or the spectrum of temperature fluctuations. Equation (10) is valid for homogeneous, anisotropic turbulence. For homogeneous and isotropic turbulence, B_T and Φ_T depend only on the moduli of their arguments so that (10) simplifies:

$$B_T(R) = \frac{4\pi}{R}\int_0^\infty dK\, K\sin\left(KR\right)\Phi_T(K) \ . \tag{11}$$

For homogeneous, anisotropic turbulence, the correlation tensor of velocity fluctuations is given by

$$B_{ij}(\mathbf{R}) = \langle v_i(\mathbf{R}' + \mathbf{R}) v_j(\mathbf{R}') \rangle = \int d^3K \, e^{i\mathbf{K}\cdot\mathbf{R}} \Phi_{ij}(\mathbf{K}) \, , \tag{12}$$

where $i = 1,2,3$, $j = 1,2,3$, and $\Phi_{ij}(\mathbf{K})$ is the three-dimensional spectral tensor of velocity fluctuations. For homogeneous and isotropic turbulence, $B_{ij}(\mathbf{R})$ and $\Phi_{ij}(\mathbf{K})$ still depend on the vectors $\mathbf{R}$ and $\mathbf{K}$, respectively (e.g. see [4,18]). However, in this case, Φ_{ij} can be expressed in terms of the specific turbulence kinetic energy spectrum $E(K)$:

$$\Phi_{ij}(\mathbf{K}) = \frac{(K^2\delta_{ij} - K_i K_j) E(K)}{4\pi K^4} \, . \tag{13}$$

Here, $\delta_{i,j}$ is the Kronecker symbol.

3.2 Energy, Inertial, and Dissipation Subranges

Turbulence can be subdivided into three subranges: energy, inertial, and dissipation.

The energy subrange contains the largest inhomogeneities or eddies in a turbulent medium. These inhomogeneities are created by different forces and processes in the medium. The energy subrange is defined by the inequality $K < 1/L_0$, where L_0 is the outer scale of turbulence.

In the inertial subrange of turbulence, $1/L_0 < K < 1/l_0$. Here, l_0 is the inner scale of turbulence. The inertial subrange is characterized by a cascade mechanism of energy transfer from large eddies in a medium to smaller ones.

Turbulence motion decays in the dissipation subrange, where $K > 1/l_0$. This subrange contains the smallest inhomogeneities in a medium, which usually do not affect sound propagation.

The Kolmogorov, Gaussian, and von Kármán spectra have been most widely used in atmospheric acoustics to describe spectra of homogeneous and isotropic turbulence. These spectra are considered below.

3.3 Kolmogorov Spectrum

The Kolmogorov spectrum is a realistic spectrum of locally homogeneous and isotropic turbulence in the inertial subrange. The spectrum of temperature fluctuations is given by

$$\Phi_T^K(K) = \frac{\mathcal{A} C_T^2}{K^{11/3}} \, . \tag{14}$$

Here, the superscript K stands for the Kolmogorov spectrum, C_T^2 is the structure parameter of temperature fluctuations, and $\mathcal{A} = 5/(18\pi\Gamma(1/3)) \approx 0.0330$ is a numerical coefficient, where Γ is the gamma function.

In the inertial subrange, the energy spectrum of turbulence is given by

$$E^K(K) = \frac{22\pi A}{3} \frac{C_v^2}{K^{5/3}} , \tag{15}$$

where C_v^2 is the structure parameter of velocity fluctuations. Note that C_T^2 and C_v^2 characterize the intensity of temperature and velocity fluctuations.

3.4 Gaussian Spectrum

The Gaussian spectrum is a model spectrum used to approximately model a spectrum of turbulence in the energy subrange.

The spectrum of temperature fluctuations is given by

$$\Phi_T^G(K) = \frac{\sigma_T^2 l_T^3}{8\pi^{3/2}} \, e^{-K^2 l_T^2/4} . \tag{16}$$

Here, the superscript G stands for the Gaussian spectrum, σ_T^2 is the variance of temperature fluctuations, and l_T is of the order of the outer scale of temperature fluctuations.

For the energy spectrum of turbulence, we have

$$E^G(K) = \frac{\sigma_v^2 l_v^5 K^4}{8\pi^{1/2}} \, e^{-K^2 l_v^2/4} . \tag{17}$$

Here, σ_v^2 is the variance of velocity fluctuations, and l_v is of the order of the outer scale of velocity fluctuations.

3.5 von Kármán Spectrum

The von Kármán spectrum can be considered as a combination of the Kolmogorov and Gaussian spectra. In the inertial subrange, it coincides with the Kolmogorov spectrum. In the energy subrange, the dependence of the von Kármán spectrum on K is somewhat similar to that for the Gaussian spectrum.

The spectrum of temperature fluctuations is given by

$$\Phi_T^{vK}(K) = \frac{A C_T^2 L_T^{11/3}}{(1 + K^2 L_T^2)^{11/6}} , \tag{18}$$

where the superscript vK stands for the von Kármán spectrum, and L_T is of the order of the outer scale of temperature fluctuations. Strictly speaking, the von Kármán spectrum should also contain the factor $\exp(-(K/5.92\, l_0)^2)$, which describes inhomogeneities in the dissipation subrange [16]. However, this factor may be omitted in acoustics because a sound wave usually is not affected by these inhomogeneities.

For the von Kármán spectrum, the variance of the temperature fluctuations calculated with the use of (11) and (18) is given by (see (6.49) from [10])

$$\sigma_T^2 = \frac{\Gamma^2(1/3)}{2^{4/3}\sqrt{3}\pi} C_T^2 L_T^{2/3} \,. \tag{19}$$

Solving this equation for C_T^2 and substituting the result into (18), $\Phi_T^{vK}(K)$ can be expressed in terms of σ_T^2 and L_T:

$$\Phi_T^{vK}(K) = \frac{2^{1/3}5}{3^{3/2}\Gamma^3(1/3)} \frac{\sigma_T^2 L_T^3}{(1+K^2L_T^2)^{11/6}} \,. \tag{20}$$

Equations (18) and (20) are equivalent forms of the temperature spectrum $\Phi_T^{vK}(K)$.

The energy spectrum of turbulence is given by

$$E^{vK}(K) = \frac{22\pi\mathcal{A}}{3} \frac{C_v^2 L_v^{17/3} K^4}{(1+K^2L_v^2)^{17/6}} \,. \tag{21}$$

Here, L_v is of the order of the outer scale of velocity fluctuations. Using (21), one can calculate the variance of velocity fluctuations (see (6.49) from [10]):

$$\sigma_v^2 = \frac{\Gamma^2(1/3)}{2^{4/3}\sqrt{3}\pi} C_v^2 L_v^{2/3} \,. \tag{22}$$

Solving this equation for C_v^2 and substituting the result into (21), we have

$$E^{vK}(K) = \frac{\pi 2^{4/3} 55}{9\sqrt{3}\Gamma^3(1/3)} \frac{\sigma_v^2 L_v^5 K^4}{(1+K^2L_v^2)^{17/6}} \,. \tag{23}$$

This formula expresses $E^{vK}(K)$ in terms of σ_v^2 and L_v.

Inhomogeneous, Isotropic Atmospheric Turbulence. It is well known that atmospheric turbulence is anisotropic in the energy subrange, and its statistical parameters depend on height z above the ground. Some progress has been made recently in modeling three-dimensional spectral densities of such inhomogeneous, anisotropic turbulence.

A three-dimensional model of inhomogeneous, isotropic turbulence in the daytime atmospheric boundary layer was developed in papers [13,21,22]. In this model, spectra of temperature $\tilde{T}$ and velocity $\mathbf{v}$ fluctuations are chosen as von Kármán spectra given by (20) and (23), whose parameters depend on height z.

Using turbulence similarity scaling theories, the following expressions for $\sigma_T^2(z)$ and $L_T^2(z)$ were obtained:

$$\sigma_T^2(z) = \frac{4.0\, T_*^2}{\left[1+10(-z/L_{mo})\right]^{2/3}} \,, \tag{24}$$

and

$$L_T(z) = 2.0\, z\, \frac{1 + 7.0(-z/L_{mo})}{1 + 10(-z/L_{mo})} \, . \tag{25}$$

Here, T_* is a temperature scale, and $L_{mo} = u_*^3/(0.4\beta Q_s)$ is the Monin-Obukhov length. In the latter formula, u_* is the friction velocity, Q_s is the kinematic surface heat flux, and $\beta = g/T_s$ is the Boussinesq buoyancy parameter, where g is gravitational acceleration and T_s is surface temperature. Equations (24) and (25) can be used for both shear- and buoyancy-produced turbulence.

Making use of turbulence similarity theories, the following formulas for σ_v^2 and L_v were obtained for shear-produced turbulence,

$$\sigma_v^2 = 3.0 u_*^2, \qquad L_v = 1.8 z \, , \tag{26}$$

and for buoyancy-produced turbulence,

$$\sigma_v^2 = 0.35 w_*^2, \qquad L_v = 0.23 z_i \, . \tag{27}$$

Here, z_i is the boundary layer inversion height, and $w_* = (Q_s \beta z_i)^{1/3}$. For both shear- and buoyancy-produced turbulence, the corresponding energy spectra $E(K)$ should be added.

The three-dimensional spectra of temperature and velocity fluctuations given by (20) and (23) – (27) account for realistic vertical inhomogeneity of shear- and buoyancy-produced turbulence, coincide with Kolmogorov spectra in the inertial subrange, and are finite in the energy subrange. Also, the spectra are given by relatively simple formulas that is convenient for analytical calculations of the statistical moments of a sound field.

Inhomogeneous, Anisotropic Atmospheric Turbulence. Using the rapid distortion theory, Mann [8] obtained formulas for 3D velocity spectral tensor $\Phi_{i,j}^M(\mathbf{K})$ for homogeneous, anisotropic, shear-produced turbulence in a free space. The tensor $\Phi_{i,j}^M(\mathbf{K})$ has three free parameters: σ_v^2, L_v, and the parameter Γ (not to be confused with the gamma-function above and the coherence function of a sound field below), which do not depend on z. Mann's spectrum is rather involved and is not presented here for the sake of brevity. Furthermore, the spectral tensor $\Phi_{i,j}^M(\mathbf{K})$ is homogeneous (i.e., its parameters do not depend on z), which is unrealistic. Accounting for flow blocking by the ground, Mann [8] also obtained a 3D spectral tensor $\Phi_{i,j}^M(z;\mathbf{K})$ for inhomogeneous, anisotropic turbulence. But this spectrum is even more involved than $\Phi_{i,j}^M(\mathbf{K})$ and is difficult to use in calculations of sound propagation.

To overcome this problem, Wilson [23] suggested to generalize Mann's homogeneous spectrum $\Phi_{i,j}^M(\mathbf{K})$ by choosing the free parameters as follows: $\sigma_v^2 = 1.52 u_*^2$, $L_v = 0.587 z$, and $\Gamma = 3.58$. Due to realistic dependence of L_v on z, the generalized Mann's spectrum is a good model for a spectrum of inhomogeneous, anisotropic, shear-produced turbulence.

4 Scattering of Sound

In this section, sound scattering by temperature and velocity fluctuations is considered.

The geometry of the problem is shown in Fig. 1. Random inhomogeneities $\tilde{T}(\mathbf{R})$ and $\mathbf{v}(\mathbf{R})$ are located in the scattering volume V. The sound wave incident on the scattering volume propagates in the direction of the unit vector $\mathbf{n}_0$. A sound wave scattered by random inhomogeneities $\tilde{T}(\mathbf{R})$ and $\mathbf{v}(\mathbf{R})$ propagates in the direction of the unit vector $\mathbf{n}$.

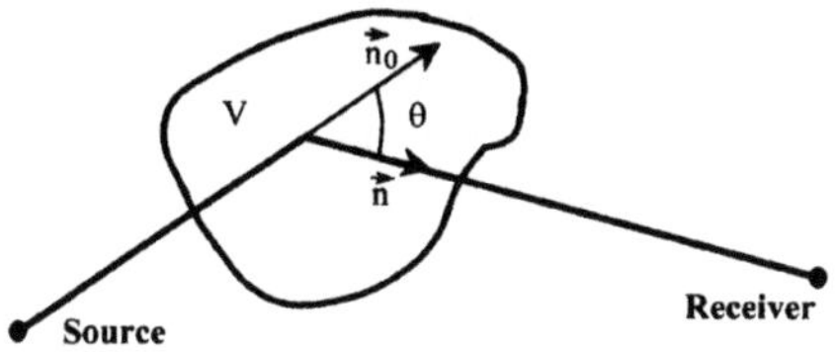

Fig. 1. Geometry of sound scattering. $\mathbf{n}_0$ and $\mathbf{n}$ are the unit vectors in the directions of propagation of the incident and scattered waves. θ is the scattering angle.

4.1 Sound Scattering Cross-Section

The sound pressure $p_s(\mathbf{R})$ of the scattered wave can be calculated as the Born approximation of the solution of (3). The intensity I_s of the scattered wave is related to p_s by

$$I_s = \frac{k_0 \, p_s p_s^*}{2\omega \varrho_0} . \tag{28}$$

The sound scattering cross section per unit volume into a unit solid angle in the direction of the vector $\mathbf{n}$ is defined as

$$\sigma(\mathbf{n} - \mathbf{n}_0) = \frac{\langle I_s \rangle R^2}{I_0 V} . \tag{29}$$

Here, I_0 is the intensity of the wave incident on the scattering volume, R is the distance from the scattering volume to the receiver, and V is the value of the scattering volume.

With the use of (28) and (29), one can derive the formula for the sound scattering cross section for the case of isotropic turbulence (see (7.16) from [10]):

$$\sigma(\theta) = \sigma_T(\theta) + \sigma_v(\theta) = \frac{\pi k_0^4 \beta^2(\theta) \Phi_T(2k_0 \sin(\theta/2))}{2T_0^2} + \frac{k_0^2 \cos^2\theta \cot^2(\theta/2) E(2k_0 \sin(\theta/2))}{8c_0^2} . \tag{30}$$

Here, θ is the scattering angle between the unit vectors $\mathbf{n}_0$ and $\mathbf{n}$ (see Fig. 1), and $\beta(\theta) = \beta_c + 2\beta_\varrho \sin^2(\theta/2)$. For sound scattering in the atmosphere, $\beta(\theta) = \cos\theta$. Furthermore, in (30), $\sigma_T(\theta)$ and $\sigma_v(\theta)$ are the scattering cross sections due to temperature and velocity fluctuations.

Note that sound scattering in turbulent media has also been considered elsewhere [5,9,7,18].

4.2 von Kármán Spectrum

Using (30) for $\sigma(\theta)$ and formulas for the Kolmogorov, Gaussian, and von Kármán spectra of temperature and velocity fluctuations presented above, it is straightforward to calculate the sound scattering cross section for these spectra. For example, for the von Kármán spectrum, the temperature and velocity contributions to $\sigma(\theta)$ are given by

$$\sigma_T^{vK}(\theta) = \frac{5k_0^{1/3}\beta^2(\theta)}{2^{17/3}9\Gamma(1/3)\left[\sin^2(\theta/2) + 1/(2k_0L_T)^2\right]^{11/6}} \frac{C_T^2}{T_0^2}, \tag{31}$$

and

$$\sigma_v^{vK}(\theta) = \frac{55k_0^{1/3}\cos^2\theta\sin^2\theta}{2^{17/3}54\Gamma(1/3)\left[\sin^2(\theta/2) + 1/(2k_0L_T)^2\right]^{17/6}} \frac{C_v^2}{c_0^2}. \tag{32}$$

Using these equations, $\sigma_T^{vK}(\theta)$ and $\sigma_v^{vK}(\theta)$, normalized to

$$5k_0^{1/3}/(2^{17/3}9\Gamma(1/3)),$$

are plotted in Fig. 2 as functions of the scattering angle θ for the following values of parameters: $C_T^2/T_0^2 = 1$, $C_v^2/c_0^2 = 2$, $k_0L_T = k_0L_v = 1/2$, and $\beta(\theta) = \cos\theta$. It is clearly seen from Fig. 2 that $\sigma_T^{vK}(\theta)$ and $\sigma_v^{vK}(\theta)$ have qualitatively different dependences on the scattering angle θ.

5 Line-of-Sight Sound Propagation in a Random Moving Medium

If λ is smaller than the outer scale of turbulence and random inhomogeneities in a medium are relatively weak, the direction of sound propagation differs only slightly from that in a non-turbulent medium. This case is often referred to as line-of-sight sound propagation in a random medium. Line-of-sight sound propagation can be treated with the use of the parabolic equation (5). This equation allows one to describe fluctuations in the amplitude and phase of a sound wave, the coherence of this wave, intensity fluctuations, and other phenomena.

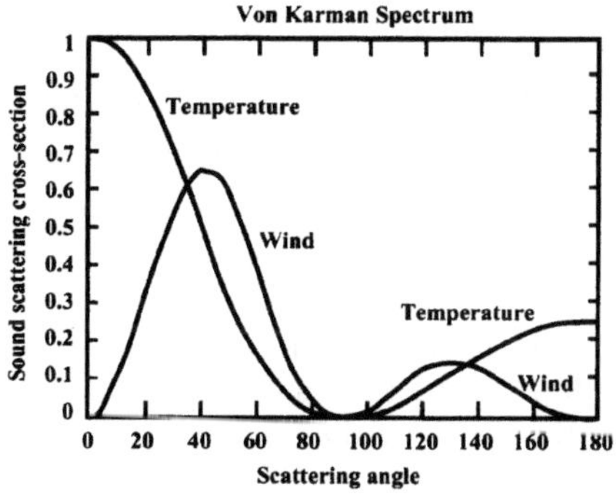

Fig. 2. The normalized temperature (σ_T^{vK}) and wind velocity (σ_v^{vK}) contributions to the scattering cross section versus the scattering angle θ for the von Kármán spectrum of turbulence.

5.1 Statistical Moments of a Sound Field

For line-of-sight sound propagation, it is convenient to express the sound pressure in the form:

$$p(x, \mathbf{r}) = p_0(x, \mathbf{r}) \exp(i\phi(x, \mathbf{r}) + \chi(x, \mathbf{r})) \, . \tag{33}$$

Here, the x-axis coincides with the predominant direction of sound propagation, $p_0(x, \mathbf{r})$ is the sound field propagating in a medium without random inhomogeneities, and $\chi(x, \mathbf{r})$ and $\phi(x, \mathbf{r})$ are the log-amplitude and phase fluctuations due to random inhomogeneities.

For line-of-sight propagation, the statistical moments of a sound field that are most often of interest are the following:

$$\langle \chi^2(x) \rangle, \quad \langle \phi^2(x) \rangle, \quad B_\chi(x, \mathbf{r}), \quad B_\phi(x, \mathbf{r}), \quad \langle p(x, \mathbf{r}) \rangle, \quad \Gamma(x, \mathbf{r_1}, \mathbf{r_2}) \, . \tag{34}$$

Here, $\langle \chi^2(x) \rangle$ and $\langle \phi^2(x) \rangle$ are the variances of log-amplitude and phase fluctuations;

$$B_\chi(x, \mathbf{r_1} - \mathbf{r_2}) = \langle \chi(x, \mathbf{r_1}) \chi(x, \mathbf{r_2}) \rangle$$

and

$$B_\phi(x, \mathbf{r_1} - \mathbf{r_2}) = \langle \phi(x, \mathbf{r_1}) \phi(x, \mathbf{r_2}) \rangle$$

are the transverse correlation functions of log-amplitude and phase fluctuations; $\langle p(x, \mathbf{r}) \rangle$ is the mean sound field; and $\Gamma(x, \mathbf{r_1}, \mathbf{r_2}) = \langle p(x, \mathbf{r_1}) p^*(x, \mathbf{r_2}) \rangle$ is the transverse coherence function. For $\mathbf{r_1} = \mathbf{r_2}$, the transverse coherence function becomes the mean intensity of a sound field $I = \Gamma(x, \mathbf{r_1}, \mathbf{r_1}) = \langle p(x, \mathbf{r_1}) p^*(x, \mathbf{r_1}) \rangle$.

5.2 Effective Correlation Function and Spectral Density

Using (6), we obtain a formula for the correlation function of the random field ε_{mov}, which appears in the parabolic equation (5):

$$B_{\text{eff}}(\mathbf{R}) = \langle \varepsilon_{\text{mov}}(\mathbf{R}' + \mathbf{R}) \varepsilon_{\text{mov}}(\mathbf{R}') \rangle = \frac{\beta_c^2 B_T(\mathbf{R})}{T_0^2} + \frac{4 B_{11}(\mathbf{R})}{c_0^2} \,. \tag{35}$$

Here, B_{11} is the component of the velocity tensor B_{ij} for $i = j = 1$. The correlation function B_{eff} can be expressed in terms of the three-dimensional spectral density $\Phi_{\text{eff}}(\mathbf{K})$ of the random field ε_{mov}:

$$B_{\text{eff}}(\mathbf{R}) = \int d^3K \, e^{i\mathbf{K}\cdot\mathbf{R}} \Phi_{\text{eff}}(\mathbf{K}) \,. \tag{36}$$

$B_{\text{eff}}(\mathbf{R})$ and $\Phi_{\text{eff}}(\mathbf{K})$ are called the effective functions.

5.3 Markov Approximation

In the Markov approximation, $B_{\text{eff}}(\mathbf{R})$ is written as

$$B_{\text{eff}}^M(x, \mathbf{r}) = \delta(x) b_{\text{eff}}(\mathbf{r}) \,, \tag{37}$$

where $\delta(x)$ is the delta-function, and b_{eff} is called the effective transverse correlation function, given by

$$b_{\text{eff}}(\mathbf{r}) = 2\pi \int d^2K_\perp \, e^{i\mathbf{K}_\perp\cdot\mathbf{r}} \Phi_{\text{eff}}(0, \mathbf{K}_\perp) \,. \tag{38}$$

For isotropic turbulence, in (38) Φ_{eff} can be expressed in terms of Φ_T and E:

$$\Phi_{\text{eff}}(0, K_\perp) = \frac{\beta_c^2 \Phi_T(K_\perp)}{T_0^2} + \frac{E(K_\perp)}{\pi K_\perp^2 c_0^2} \,. \tag{39}$$

The Markov approximation is usually valid provided that the parabolic equation (5) can be used. The range of applicability of the Markov approximation is presented in detail elsewhere [16].

5.4 Rytov Method

In the Rytov method, the complex amplitude A is expressed in terms of χ and ϕ with the use of (4) and (33). Then, A is substituted into the parabolic equation (5), and nonlinear terms in χ and ϕ are omitted. The resulting equation can be solved for χ and ϕ. As a result, one obtains analytical formulas for χ and ϕ for plane and spherical waves in terms of the random field ε_{mov}. The Rytov method is valid if $\langle \chi^2 \rangle$ is less than 1. (The range of applicability of the Rytov method is given in detail elsewhere [16].)

For a plane sound wave, formulas for χ and ϕ read

$$\chi(x,\mathbf{r}) = \frac{k_0^2}{4\pi}\int_0^x dx_1 \int d^2r_1 \cos\left(\frac{k_0(\mathbf{r}-\mathbf{r}_1)^2}{2(x-x_1)}\right)\frac{\varepsilon_{\text{mov}}(x_1,\mathbf{r}_1)}{x-x_1}, \tag{40}$$

and

$$\phi(x,\mathbf{r}) = \frac{k_0^2}{4\pi}\int_0^x dx_1 \int d^2r_1 \sin\left(\frac{k_0(\mathbf{r}-\mathbf{r}_1)^2}{2(x-x_1)}\right)\frac{\varepsilon_{\text{mov}}(x_1,\mathbf{r}_1)}{x-x_1}. \tag{41}$$

For a spherical sound wave, χ and ϕ are given by equations similar to (40) and (41).

5.5 Formulas for Statistical Moments of a Sound Field

Starting from (40), (41) and similar equations for a spherical sound wave and using the Markov approximation (37), formulas for the variances and correlation functions of the log-amplitude and phase fluctuations of plane and spherical sound waves have been derived [10]. Here, these formulas are presented for the case of a spherical wave propagating through homogeneous and isotropic turbulence. Analogous formulas for a plane wave can be found in Section 7.2.5 from [10].

The variances of log-amplitude and phase fluctuations are given by

$$\langle\chi^2(x)\rangle = \frac{\pi^2k_0^2x}{2}\int_0^1 d\eta\int_0^\infty dK\,K\left[1-\cos\left(\frac{K^2\eta(1-\eta)}{K_F^2}\right)\right]\Phi_{\text{eff}}(0,K) \tag{42}$$

and

$$\langle\phi^2(x)\rangle = \frac{\pi^2k_0^2x}{2}\int_0^1 d\eta\int_0^\infty dK\,K\left[1+\cos\left(\frac{K^2\eta(1-\eta)}{K_F^2}\right)\right]\Phi_{\text{eff}}(0,K). \tag{43}$$

Here, $K_F = k_0/x$.

Formulas for the correlation functions of the log-amplitude and phase fluctuations read

$$B_\chi(x,r) = \frac{\pi^2k_0^2x}{2}\int_0^1 d\eta\int_0^\infty dK\,K\left[1-\cos\left(\frac{K^2\eta(1-\eta)}{K_F^2}\right)\right]J_0(Kr\eta)\Phi_{\text{eff}}(0,K) \tag{44}$$

and

$$B_\phi(x,r) = \frac{\pi^2k_0^2x}{2}\int_0^1 d\eta\int_0^\infty dK\,K\left[1+\cos\left(\frac{K^2\eta(1-\eta)}{K_F^2}\right)\right]J_0(Kr\eta)\Phi_{\text{eff}}(0,K). \tag{45}$$

If $r = 0$, (44) and (45) coincide with (42) and (43), respectively, as they should since $B_\chi(x,0) = \langle\chi^2(x)\rangle$ and $B_\phi(x,0) = \langle\phi^2(x)\rangle$.

The mean sound field and the coherence function of a sound field can be obtained starting from the parabolic equation (5) and using the Markov approximation (37). For the mean sound field, we have

$$\langle p(x,\mathbf{r})\rangle = \exp(-\gamma x) p_0(x,\mathbf{r}) \,, \tag{46}$$

where γ is the extinction coefficient:

$$\gamma = \frac{\pi^2 k_0^2}{2} \int_0^\infty dK \, K \Phi_{\text{eff}}(0,K) \,. \tag{47}$$

According to (46), the mean sound field attenuates exponentially as a function of the distance x of sound propagation in a random medium. Equations (46) and (47) hold both for plane and spherical sound waves. For a spherical wave, $p_0(\mathbf{R}) = \exp(ik_0|\mathbf{R}-\mathbf{R}_0|)/|\mathbf{R}-\mathbf{R}_0|$, where $\mathbf{R}_0$ are the coordinates of a point source. For a plane wave, $p_0 = \exp(ik_0 x)$.

When considering the coherence function $\Gamma(x,\mathbf{r}_1,\mathbf{r}_2) = \langle p(x,\mathbf{r}_1)p^*(x,\mathbf{r}_2)\rangle$, it is convenient to introduce new transverse coordinates $\mathbf{r} = \mathbf{r}_1 - \mathbf{r}_2$ and $\mathbf{r}_+ = (\mathbf{r}_1 + \mathbf{r}_2)/2$, and to denote the coherence function as $\Gamma(x,\mathbf{r},\mathbf{r}_+)$. Then, the formula for the coherence function of a spherical sound wave reads

$$\Gamma(x,\mathbf{r},\mathbf{r}_+) = \tag{48}$$
$$\frac{1}{(4\pi x)^2} \exp\left\{ \frac{ik_0 \mathbf{r}\cdot\mathbf{r}_+}{x} - \pi^2 k_0^2 x \int_0^1 d\eta \int_0^\infty dK \, K[1 - J_0(Kr\eta)] \Phi_{\text{eff}}(0,K) \right\}.$$

For a plane sound wave, a formula for the coherence function is given by (7.63) from [10].

Equations (42) - (48) express the statistical moments of a sound field in terms of the effective spectrum $\Phi_{\text{eff}}(0,K)$. According to (39), $\Phi_{\text{eff}}(0,K)$ is a linear combination of $\Phi_T(K)$ and $E(K)$, which are given by (14) - (23) for the Kolmogorov, Gaussian, and von Kármán spectra. Therefore, for these spectra, calculation of the statistical moments of a sound field given by (34) is reduced to the evaluation of two-dimensional integrals on the right-hand sides in (42) - (45), (47) and (48). Most of these integrals have been evaluated analytically and the resulting formulas have been summarized in reference [10]. In Section 5.6, we present three of these formulas for the coherence functions of a spherical sound wave for the Kolmogorov, Gaussian, and von Kármán spectra.

Formulas for the statistical moments (34) of plane and spherical sound waves propagating through inhomogeneous and anisotropic turbulence have been also derived. These formulas are summarized in reference [12].

5.6 Coherence Function

For the Kolmogorov spectrum, the coherence function of a spherical sound wave is given by

$$\Gamma^K(x, \mathbf{r}, \mathbf{r}_+) = \frac{1}{(4\pi x)^2} \exp\left\{ \frac{ik_0 \mathbf{r} \cdot \mathbf{r}_+}{x} - \frac{3}{8}\mathcal{B}\left(\frac{\beta_c^2 C_T^2}{T_0^2} + \frac{22}{3}\frac{C_v^2}{c_0^2}\right) k_0^2 r^{5/3} x \right\}, \tag{49}$$

where $\mathcal{B} = \sqrt{3}\Gamma^2(1/3)/(5 \cdot 2^{7/3}\Gamma(2/3)) \approx 0.364$ is a numerical coefficient.

For the Gaussian spectrum, the formula for the coherence function reads

$$\Gamma^G(x, \mathbf{r}, \mathbf{r}_+) = \frac{1}{(4\pi x)^2} \exp\left\{ \frac{ik_0 \mathbf{r} \cdot \mathbf{r}_+}{x} - 2\gamma_T^G x \left[1 - \frac{\sqrt{\pi} l_T}{2r} \mathrm{erf}\left(\frac{r}{l_T}\right)\right] - 2\gamma_v^G x \left[1 - \frac{\sqrt{\pi} l_v}{4r} \mathrm{erf}\left(\frac{r}{l_v}\right) - \frac{1}{2} e^{-r^2/l_v^2}\right] \right\} \tag{50}$$

Here, $\mathrm{erf}(r/l_v)$ is the error-function, and γ_T^G and γ_v^G are the temperature and velocity contributions to the extinction coefficient $\gamma^G = \gamma_T^G + \gamma_T^G$ of the mean sound field, for the Gaussian spectrum:

$$\gamma_T^G = \frac{\sqrt{\pi}\beta_c^2 k_0^2 l_T \sigma_T^2}{8T_0^2}, \qquad \gamma_v^G = \frac{\sqrt{\pi} k_0^2 l_v \sigma_v^2}{2c_0^2}. \tag{51}$$

For the von Kármán spectrum, the coherence function is given by

$$\Gamma^{vK}(x, \mathbf{r}, \mathbf{r}_+) = \frac{1}{(4\pi x)^2} \exp\left\{ \frac{ik_0 \mathbf{r} \cdot \mathbf{r}_+}{x} - \frac{2x\gamma_T^{vK} L_T}{r} \int_0^{r/L_T} d\eta \left[1 - \frac{2^{1/6}\eta^{5/6}}{\Gamma(5/6)} K_{5/6}(\eta)\right] - \frac{2x\gamma_v^{vK} L_v}{r} \int_0^{r/L_v} d\eta \left[1 - \frac{2^{1/6}\eta^{5/6}}{\Gamma(5/6)} \left(K_{5/6}(\eta) - \frac{\eta}{2} K_{1/6}(\eta)\right)\right] \right\}. \tag{52}$$

Here, $K_{5/6}$ and $K_{1/6}$ are the modified Bessel functions, and

$$\gamma_T^{vK} = \frac{3}{10}\pi^2 \mathcal{A} k_0^2 L_T^{5/3} \beta_c^2 C_T^2 / T_0^2, \qquad \gamma_v^{vK} = \frac{3}{10}\pi^2 \mathcal{A} k_0^2 L_v^{5/3} \frac{4C_v^2}{c_0^2} \tag{53}$$

are the temperature and velocity contributions to the extinction coefficient $\gamma^{vK} = \gamma_T^{vK} + \gamma_v^{vK}$, for the von Kármán spectrum.

The normalized coherence functions Γ^K, Γ^G, and Γ^{vK} versus r/L_v are plotted in Fig. 3 (reproduced from [15]) for $\gamma_v^{vK} x = 0.1$ and $\gamma_v^{vK} x = 100$. When plotting these functions, it is assumed that the temperature fluctuations are zero; the variances of velocity fluctuations for the von Kármán and Gaussian spectra are the same; and $l_v = (2\Gamma(5/6)/\Gamma(13))L_v \approx 0.843 L_v$. (The latter two assumptions are explained in Section 6.2.6 from [10].) It follows from Fig. 3 that the coherence functions for the Kolmogorov, Gaussian, and von Kármán spectra decrease as r increases.

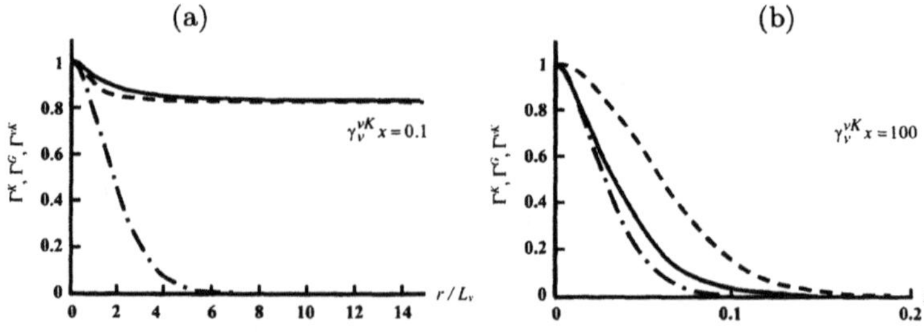

Fig. 3. The normalized coherence functions of a spherical sound wave for the von Kármán (Γ^{vK}, solid line), Kolmogorov (Γ^{K}, dash-dotted line) and Gaussian (Γ^{G}, dashed line) spectra. (a) $\gamma_v^{\mathrm{vK}} x = 0.1$, and (b) $\gamma_v^{\mathrm{vK}} x = 100$.

It is mentioned in Section 3.5 that the von Kármán spectrum can be considered as a combination of the Kolmogorov and Gaussian spectra. Therefore, one would expect that, in certain limiting cases, the coherence function for the von Kármán spectrum should coincide with that for the Kolmogorov or Gaussian spectrum. Figure 3 illustrates this fact. For relatively small distances of sound propagation in a turbulent medium ($\gamma_v^{vK} x = 0.1$), the coherence function for the von Kármán spectrum almost coincides with that for the Gaussian spectrum; however, it significantly differs from the coherence function for the Kolmogorov spectrum. On the other hand, for the relatively large distances of sound propagation ($\gamma_v^{vK} x = 100$), the coherence function for the von Kármán spectrum is close to the coherence function for the Kolmogorov spectrum and differs from that for the Gaussian spectrum.

6 Interference of the Direct and Ground-Reflected Waves

For many geometries of sound propagation in the atmosphere, the sound field p is a sum of the direct wave from source to receiver and that reflected from the ground (see Fig. 4, which is reproduced from [11]). In the figure, the source and receiver are located at heights h_s and h_r above the ground, and the horizontal distance between them is L. For $L \gg h_s, h_r$, the interference between the direct and ground reflected waves could lead to very deep minima in the amplitude of the resulted sound field p if there were no turbulence. However, temperature and velocity fluctuations are always present in the atmosphere. This leads to a significant (up to tens of dB) increase of the mean squared sound pressure $\langle |p|^2 \rangle$ at interference minima. The latest theory of this phenomenon has been developed in references [11,17]. Here, we present some results obtained in these references.

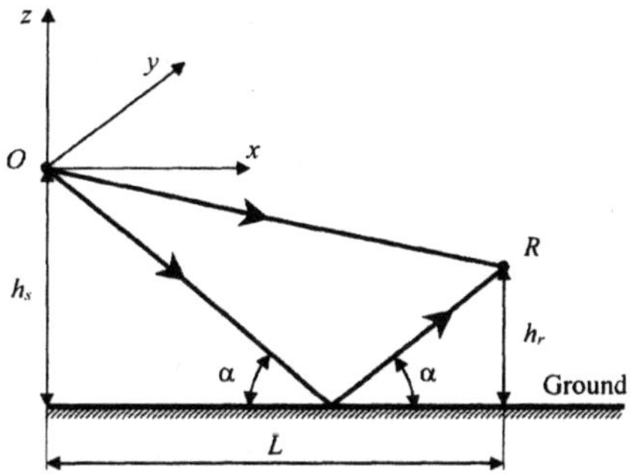

Fig. 4. Geometry of the interference of the direct and ground-reflected waves. Source and receiver are located at points O and R respectively.

6.1 Mean Squared Sound Pressure

Using the Rytov method and the Markov approximation, the following formula for the mean squared sound pressure was obtained

$$\langle |p|^2 \rangle = \frac{1}{R_d^2} + \frac{|Q|^2}{R_r^2} + \frac{2|Q|C}{R_d R_r} \cos\left[(R_r - R_d)k_0 + \Omega\right]. \tag{54}$$

Here, R_d and R_r are the path lengths of the direct and ground-reflected waves, $Q = |Q|e^{i\Omega}$ is the spherical-wave reflection coefficient, and C is called the coherence factor which describes the coherence between the direct and reflected waves:

$$C = \exp\left\{-\pi^2 k_0^2 L \int_0^1 d\eta \int_0^\infty \left[\frac{\beta_c^2 \Phi_T(K)}{T_0^2} + \frac{E(K)}{\pi K^2 c_0^2}\right] [1 - J_0(hK\eta)] K dK\right\}. \tag{55}$$

Here, J_0 is the Bessel function of zero order, and $h = h_s h_r/(h_s + h_r)$. It follows from (55) that $0 < C \leq 1$.

The last term on the right-hand side of (54) describes interference minima and maxima. If temperature and velocity fluctuations are zero, the coherence factor $C = 1$. In this case, the interference minima are very deep for near grazing sound propagation ($L \gg h_s, h_r$) since $Q \approx -1$. In a turbulent atmosphere, C is less than 1, so that the interference minima are suppressed. For relatively large distances L between source and receiver, C becomes much smaller than 1. In this case, the interference minima and maxima almost disappear and the last term on the right-hand side of (54) can be omitted.

The coherence factor C is closely related to the coherence function of a spherical sound wave $\Gamma(x, \mathbf{r}, \mathbf{r}_+)$ given by (48). Indeed, comparison between (55) and (48) reveals that, for $L = x$ and $h = r$, the coherence factor C is equal to the coherence function $\Gamma(x, \mathbf{r}, \mathbf{r}_+)$, normalized by $(4\pi x)^{-2} \exp(ik_0 \mathbf{r} \cdot \mathbf{r}_+/x)$. Therefore, the coherence factor C can be easily calculated for the

Kolmogorov, Gaussian, and von Kármán spectra with the use of (49) - (53) for the coherence functions $\Gamma(x, \mathbf{r}, \mathbf{r}_+)$ for these spectra.

6.2 Comparison with Experimental Data

Figure 5 from reference [11] shows the scheme of an experiment that was carried out in a large anechoic chamber in Ecole Centrale de Lyon, France. The heated grid was located horizontally. A spectrum of temperature fluctuations above the grid was approximated well by the von Kármán spectrum. The source and receiver (a microphone) were located above the grid at the distances h_s and h_r from an acoustically hard wall, which was also located above the grid. As shown in Fig. 5, the sound field p at the microphone was a sum of the direct wave and that reflected from the wall. Therefore, the laboratory experiment in question modeled the geometry of sound propagation in the atmosphere shown in Fig. 4.

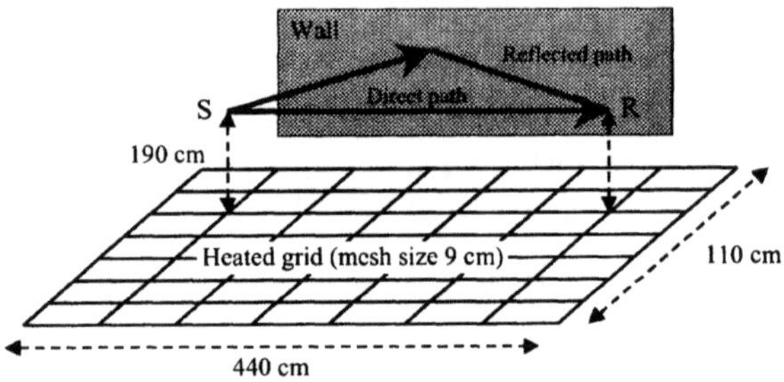

Fig. 5. Geometry of laboratory experiment that models the interference of the direct and ground-reflected waves. S and R are source and receiver (microphone), respectively.

Using sound pressure recorded by the microphone, the relative sound pressure level was calculated:

$$W = 10 \log \left(\frac{\langle |p|^2 \rangle}{p_0^2} \right) . \tag{56}$$

Here, $p_0 = 1/R_0$ is the amplitude of the sound pressure measured at a distance $R_0 = 2$ cm from the source. A more detailed description of the experiment can be found in two other papers [1,19].

The relative sound pressure level W measured in the experiment versus the distance L between source and microphone is shown in Fig. 6 (reproduced from [11]) by the "+" symbol. For these experimental data, $h_s = 10$ cm, $h_r = 7$ cm, and the sound frequency $f = 40$ kHz.

The relative sound pressure level W was also calculated with the use of (52) and (54) - (56) for the von Kármán spectrum of temperature fluctuations,

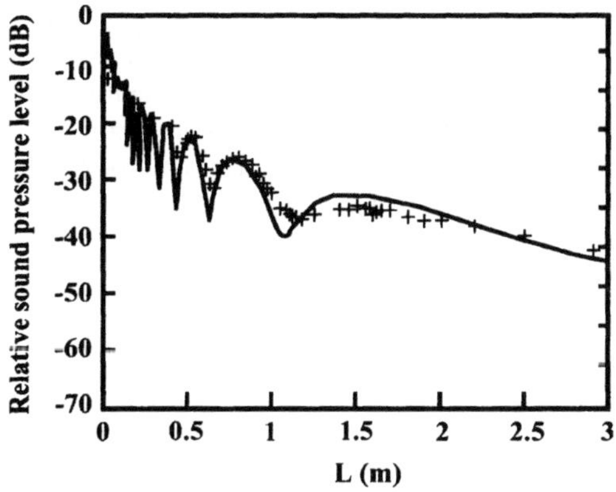

Fig. 6. Relative sound pressure level W versus the distance L. Symbols "+" correspond to experimental data obtained in the laboratory experiment. Solid line corresponds to theoretical predictions.

given by (18). The result is plotted in Fig. 6 by a solid line. The agreement between theoretical predictions and experimental data is good. It follows from Fig. 6 that W has several maxima and minima due to interference of the direct wave and that reflected from the wall. When the grid was not heated, the sound pressure level at interference minima could be 25 dB less than that shown in Fig. 6 [19].

7 Sound Propagation Near Impedance Ground in a Refractive, Turbulent Atmosphere

Sound propagation in the atmosphere is often affected by three factors: impedance ground, sound refraction due to the mean profiles of the adiabatic sound speed $c(\mathbf{R})$ and wind velocity vector $\mathbf{V}(\mathbf{R})$, and fluctuations in temperature $\tilde{T}(\mathbf{R})$ and wind velocity $v_x(\mathbf{R})$. To simultaneously account for the effect of these factors on sound propagation, numerical simulations usually have been employed (e.g., [2,3]). In this approach, a realization of the random fields $\tilde{T}(\mathbf{R})$ and $v_x(\mathbf{R})$ is synthesized numerically. These random fields are superimposed on the mean profiles of $c(\mathbf{R})$ and $\mathbf{V}(\mathbf{R})$. Then, a realization of the sound field p due to a monochromatic sound source above the impedance ground in such a refractive, turbulent atmosphere is calculated numerically by using the parabolic equation. This procedure is repeated for many realizations of $\tilde{T}(\mathbf{R})$ and $v_x(\mathbf{R})$, usually 40 or more. Finally, the statistical moments of the sound field are calculated from realizations of p.

A different approach for calculating the statistical moments of a sound field p propagating above the impedance ground in a refractive, turbulent atmosphere has been proposed recently [14,20]. In this approach first, closed analytical equations for the statistical moments of p are derived and only then are these equations solved numerically or analytically. Using this approach and the Markov approximation (37) and starting from the parabolic equation (8), closed equations for the statistical moments $\Gamma_{n,m}$ of p of arbitrary order $n+m$ were derived. Here, $\Gamma_{n,m}$ is defined as

$$\Gamma_{n,m}(x, \mathbf{r}_1, ..., \mathbf{r}_n; \mathbf{r}'_1, ..., \mathbf{r}'_m) = \langle p(x, \mathbf{r}_1)...p(x, \mathbf{r}_n) p^*(x, \mathbf{r}'_1)...p^*(x, \mathbf{r}'_m) \rangle . \quad (57)$$

7.1 Mean Sound Field

For $n = 1$ and $m = 0$ in (57), the statistical moment $\Gamma_{1,0}$ coincides with the mean sound field $\langle p \rangle$. The closed equation for the mean sound field is given by

$$\frac{\partial \langle p \rangle}{\partial x} - \frac{i}{2k_0} \left[\Delta_\perp + k^2(x, \mathbf{r}) + k_0^2 + \frac{2ik_0 \mathbf{V}_\perp(x, \mathbf{r}) \cdot \nabla_\perp}{c_0} \right] \langle p \rangle + \frac{k_0^2}{8} b_{\mathrm{eff}}(0) \langle p \rangle = 0 . \quad (58)$$

This equation should be supplemented by the initial and boundary conditions. The initial condition is given by: $\langle p(0, \mathbf{r}) \rangle = p_0(\mathbf{r})$, and the boundary condition reads:

$$\left. \frac{\langle \partial p(x, y, z) \rangle}{\partial z} \right|_{z=0} = -ik_0 \beta \langle p(x, y, z) \rangle |_{z=0} . \quad (59)$$

Here, β is the normalized admittance of the ground.

In references [14,20], (58) was solved analytically, and the following formula for the mean sound field was obtained:

$$\langle p(x, \mathbf{r}) \rangle = e^{-k_0^2 b_{\mathrm{eff}}(0) x/8} p_0(x, \mathbf{r}) . \quad (60)$$

Here, $p_0(x, \mathbf{r})$ is the sound field propagating near the impedance ground in a refractive, non-turbulent atmosphere. $p_0(x, \mathbf{r})$ satisfies the parabolic equation (8) and the same boundary and initial conditions as for $\langle p(x, \mathbf{r}) \rangle$. The first exponential term on the right-hand side of (60) describes the attenuation of the mean sound field due to turbulence. Thus, in (60), the effects of turbulence on the mean sound field are separated from those of impedance ground and refraction.

7.2 Coherence Function

For $n = 1$ and $m = 1$, the statistical moment $\Gamma_{1,1}$ coincides with the transverse coherence function $\Gamma(x, \mathbf{r}_1, \mathbf{r}_2) = \langle p(x, \mathbf{r}_1) p^*(x, \mathbf{r}_2) \rangle$. A closed equation for $\Gamma(x, \mathbf{r}_1, \mathbf{r}_2)$ reads

$$\frac{\partial \Gamma(x, \mathbf{r}_1, \mathbf{r}_2)}{\partial x} - \frac{i}{2k_0}\Big[\nabla^2_{\perp,1} - \nabla^2_{\perp,2} + k^2(x, \mathbf{r}_1) - k^2(x, \mathbf{r}_2)$$
$$+ \frac{2ik_0}{c_0}\left(\mathbf{V}_\perp(x, \mathbf{r}_1)\cdot\nabla_{\perp,1} + \mathbf{V}_\perp(x, \mathbf{r}_2)\cdot\nabla_{\perp,2}\right)\Big]\,\Gamma(x, \mathbf{r}_1, \mathbf{r}_2)$$
$$+\frac{k_0^2}{4}\left[b_{\rm eff}(0) - b_{\rm eff}(\mathbf{r}_1 - \mathbf{r}_2)\right]\Gamma(x, \mathbf{r}_1, \mathbf{r}_2) = 0. \tag{61}$$

Here, $\nabla_{\perp,1} = (\partial/\partial y_1, \partial/\partial z_1)$, and $\nabla_{\perp,2} = (\partial/\partial y_2, \partial/\partial z_2)$. The initial condition to (61) is given by: $\Gamma(0, \mathbf{r}_1, \mathbf{r}_2) = p_0(\mathbf{r}_1)p_0^*(\mathbf{r}_2)$. The boundary conditions to this equation read:

$$\left.\frac{\partial \Gamma(x, \mathbf{r}_1, \mathbf{r}_2)}{\partial z_1}\right|_{z_1=0} = -ik_0\beta\Gamma|_{z_1=0}\,, \qquad \left.\frac{\partial \Gamma(x, \mathbf{r}_1, \mathbf{r}_2)}{\partial z_2}\right|_{z_2=0} = ik_0\beta^*\Gamma|_{z_2=0}\,. \tag{62}$$

In reference [20], the two-dimensional version of (61) was solved numerically. The solid line in Fig. 7 (reproduced from that reference) shows the mean sound pressure level W versus the distance between source and receiver. When plotting the figure it is assumed that source and receiver are located at heights 5 m and 2.6 m above the ground, the impedance of the ground is $\beta = (20.8 + i\,19.2)^{-1}$, and the sound frequency $f = 40$ Hz. Furthermore, the adiabatic sound speed is constant $c(z) = c_0$, sound propagates in the upwind direction, and the mean wind velocity varies exponentially with height: $V_x(z) = (u_*/0.4)\ln(z/z_0)$, where $u_* = 0.6$ m/s, and $z_0 = 0.01$ m. Finally, it is assumed that $\tilde{T} = 0$, and a spectrum of velocity fluctuations is given by the von Kármán spectrum (23), where the values of σ_v^2 and L_v correspond to shear-produced turbulence (see, (26)).

The dash-dotted line in Fig. 7 represents the sound pressure level in a non-turbulent atmosphere. This line is significantly below the solid one in the refractive shadow zone, which starts at about 600 – 700 m from the source. This difference is due to sound scattering into the refractive shadow zone, which can increase the sound pressure level by tens of dB. The dashed line in Fig. 7 corresponds to the average of 40 runs of the standard parabolic equation, which uses realizations of the random fields $\tilde{T}$ and v_x. Dotted lines enclose the 90% confidence interval for the 40% runs. The solid and dashed lines predict similar increases in the sound pressure levels in the shadow zone. However, the solid line obtained from (61) has a much smoother dependence on the distance of sound propagation. Oscillations in the dashed line are due to the limited number of realizations of the random fields $\tilde{T}$ and v_x.

The approach for calculating the statistical moments $\Gamma_{n,m}$ proposed in papers [14,20] can be advantageous in comparison with the standard approach based on numerical solution of the parabolic equation. First, only one equation for $\Gamma_{n,m}$ should be solved, while the standard approach requires synthesizing of the random fields $\tilde{T}(\mathbf{R})$ and $v_x(\mathbf{R})$ and at least 40 solutions

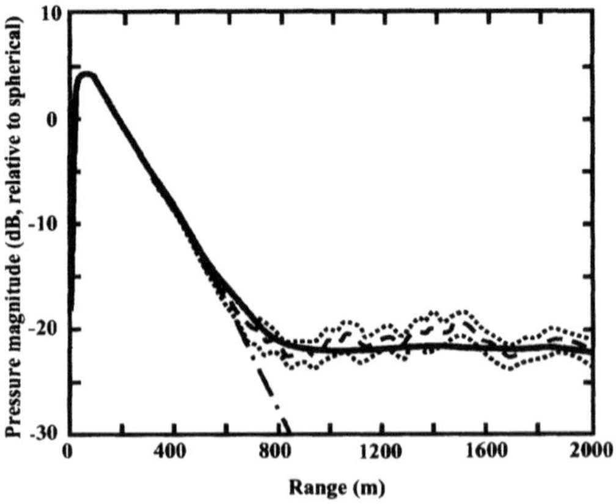

Fig. 7. Sound pressure levels versus the distance between source and receiver for upwind sound propagation near the impedance ground. Solid line represents predictions based on (61). Dashed line shows the average of 40 runs of the standard parabolic equation, which uses realizations of the random fields $\tilde{T}$ and v_x. Dotted lines enclose the 90% confidence interval for the 40 runs. Dash-dotted line represents the sound pressure level in a non-turbulent atmosphere.

of the parabolic equation. Secondly, the approach proposed in [14,20] yields a smooth dependence of $\Gamma_{n,m}$ on parameters of the problem, while the standard approach yields oscillations in these statistical moments due to the limited number of realizations of the random fields $\tilde{T}(\mathbf{R})$ and $v_x(\mathbf{R})$ (e.g., see Fig. 7). Third, in some cases, the approach proposed in [14,20] allows us to obtain exact or asymptotic formulas for the statistical moments of p.

Equations (58) and (61) are valid for sound propagation through homogeneous and isotropic turbulence. Closed equations for $\Gamma_{n,m}$ for inhomogeneous and anisotropic turbulence are given in papers [14,20].

Acknowledgments

This material is partly based upon work that was supported by the U.S. Army Research Office under contract number DAAG55-98-1-0463.

References

1. Ph. Blanc-Benon, D. Juvé: 'Experimental study of sound propagation above rigid and non rigid boundaries in the presence of turbulence'. *Proc. 9th Intern. Symp. on Long Range Sound Propagation, The Hague, Netherlands* NCPA Report Number DM0501-01, pp. 1–14 (2000).

2. P. Chevret, Ph. Blanc-Benon, D. Juvé: J. Acoust. Soc. Am. **100**, 3587 (1996)
3. X. Di, K.E. Gilbert: 'Wave propagation in a 3-D turbulent atmosphere: horizontal coherence'. In: *Proc. 8th International Symposium on Long Range Sound Propagation, Pennsylvania State University, 1998* pp.169-180
4. J. Hinze: *Turbulence* (McGraw-Hill, New York 1975)
5. M.S. Howe: *Acoustics of fluid-structure interactions* (Cambridge University Press, New York 1998)
6. A. Ishimaru: *Wave propagation and scattering in random media* (Academic Press, New York 1978)
7. M.J. Lighthill: Proc. Camb. Phil Soc. **49**, 531 (1953)
8. J. Mann: J. Fluid Mech. **273**, 141 (1994)
9. A.S. Monin: Sov. Phys. Acoust. **7**, 370 (1961)
10. V.E. Ostashev: *Acoustics in Moving Inhomogeneous Media* (E & FN SPON, London 1997)
11. V.E. Ostashev, E. Salomons, S. Clifford, R. Lataitis, D.K. Wilson, Ph. Blanc-Benon, D. Juvé: J. Acoust. Soc. Am. **109**, 1894 (2001)
12. V.E. Ostashev, D.K. Wilson: 'Line-of-sight sound propagation through anisotropic and inhomogeneous atmospheric turbulence'. In: *Int. Workshop: Tomography and Acoustics: Recent Developments and Methods, Leipzig, Germany 2001,* pp.63-70
13. V.E. Ostashev, D.K. Wilson: Acustica-acta acustica **86**, 260 (2000)
14. V.E. Ostashev, D.K. Wilson: 'Closed equations for statistical moments of a sound field in a turbulent, refractive atmosphere near an impedance ground and their solutions'. In: *Proc. 29th Inter. Congr. Noise Control Engineering, Nice, France 2000*
15. V.E. Ostashev, B. Brähler, V. Mellert, G. H. Goedecke: J. Acoust. Soc. Am. **104**, 727 (1998)
16. S.M. Rytov, Yu.A. Kravtsov, V.I. Tatarskii: *Principles of statistical radio physics. Part 4, Wave propagation through random media* (Springer, Berlin 1989)
17. E. Salomons, V.E. Ostashev, S. Clifford, R. Lataitis: J. Acoust. Soc. Am. **109** 1881 (2001)
18. V.I. Tatarskii: *The effects of the turbulent atmosphere on wave propagation* (Israel Program for Scientific Translation, Jerusalem 1971)
19. J. Wasier: Etude expérimentale des effets d'une frontière sur la propagation des ondes acoustiques à travers une turbulence thermique. Ph.D. Thesis, ECL 99-46, Ecole Centrale de Lyon, France (1999)
20. D.K. Wilson, V.E. Ostashev: J. Acoust. Soc. Am. **109**, 1909 (2001)
21. D.K. Wilson, V.E. Ostashev: 'A reexamination of acoustic scattering in the atmosphere using improved models for the turbulence spectrum'. In: *Proceedings of the Battlespace Atmospheric and Cloud Impacts on Military Operations Conference 2000.* (Ft. Collins, CO 2000)
22. D.K. Wilson: J. Acoust. Soc. Am. **108**, 2021 (2000)
23. D.K. Wilson: J. Acoust. Soc. Am. **105**, 658 (1999)

Plasma–Hydrodynamic Analogy for Waves and Vortices in Shear Flows

Anatoly Fabrikant

KLA-Tencor, One Technology Dr. Milpitas, CA 95035, USA

Abstract. A physical analogy between resonant wave-particle interactions in plasma and in fluid mechanics is considered in details. Numerous examples show that the plasma-hydrodynamical analogy is a powerful tool for classification of known phenomena, clarifying their physical nature, discovering new unexpected effects and inventing new devices. The analogy shows up both in plane parallel and in circular shear flows.

1 Research Philosophy

This research may seem a little weird. It sounds rather far from practical applications, since it is based on extremely simplified models. It aimed mostly not at the solution of any known problems, but rather at connections and comparisons of different effects, which may come from very distant fields of physics and fluid mechanics. What is the goal of such an exercise other than just fun?

The best level of analysis in classical physics was based on qualitative models. They were rarely used in fluid dynamics. Since the nonlinear partial equations of fluid mechanics almost never allow us any analytical solutions, it is hard to believe that any simple model may really help. There are few books regarding qualitative models in fluid mechanics. But most of the research in fluid dynamics was traditionally based on a formal mathematical approach.

The simple qualitative models may have even less respect nowadays. It seems more and more easy to get just what we need for any particular flow geometry and boundary conditions. A simple question often pops up, "Why would we need any analytical solutions at all? They are usually hard to obtain and they seem hardly ever practical since they almost never get close enough to reality." The answer may not seem obvious.

We are at the beginning of a digital era. According to the famous Moore's law - computing power grows exponentially, increasing twice every 5 years. As an immediate result, we face a conflict between the human brain and computer abilities. How can our natural associative thinking survive in the digital era?

With all due respect to the increasing power of Mr. Computer, I still believe that there are and there always will be problems not suitable for him. Either because of their extreme complexity, or because the entire foundation or the precise formulation of a particular problem is not quite clear. In the foreseeable future, only the human brain will be capable of dealing with these situations. Our way of thinking, based on images, simple mechanisms, and vague analogues, sometimes appealing to our unconsciousness often works successfully when the direct logic fails.

Based on this belief, I tried to develop simple qualitative models, paying less attention to their correlation with real observations. Instead I hoped they would make clear basic mechanisms of wave-flow interactions. Thus I tried to follow a famous statement of Richard Feynman's - "The purpose of computing is insight, not numbers."

I believe that clear understanding may be achieved within the framework of refined models, each one presenting a basic physical mechanism. The set of such models effectively makes for a kind of language suitable for qualitative description in physics. I feel that I understand a phenomenon when I can say: "Oh, this is just a (insert a model name), maybe with some deviations."

Most of these models are quite universal, meaning applicable to wave phenomena in different branches of physics. While I concentrated here on aero- and hydro-acoustics, virtually the same mechanisms may often be found for sea waves, or Rossby waves in the atmosphere, or sometimes even gravitational waves in general relativity. This makes these models even more attractive. At the same time it makes sense to look through already found effects in different fields and try to find acoustic analogies for these effects. Not only does this serve an educational purpose, broadening one's horizons, but this also permits natural classification of phenomena in fluid mechanics, shows a well-known phenomenon from a different viewpoint, and sometimes helps us find some unexplored opportunities, even in a well-established and seemingly complete field of research.

To the best of my knowledge, a comparison of waves in fluid flows with plasma waves was first mentioned by Case in 1962 in regard to Lin's rule of integration paths around critical points in shear flows. Since then, this analogy was occasionally considered in several publications, where it was used to give a physical interpretation of well-known results in fluid dynamics. Virtually all of those results were about either hydrodynamic instability of shear flows or about wind surface waves in the ocean.

These notes are supposed to provide an illustration of acoustical analogies that came from electrodynamics, plasma physics and electronics. I will discuss several basic types of wave-flow interaction and how they relate to plasma phenomena. Due to limited space, no references are provided here. A reader may find an extensive bibliography as well as more comprehensive considerations in the book by A.L. Fabrikant, Yu.A. Stepanyants "Wave propagation in hydrodynamic shear flows".

2 Resonant Wave-Flow Interactions

The best standard wave system in plasma to compare with is a longitudinal wave in the 1-dimensional case, when electrostatic field is $\boldsymbol{E} = (E, 0, 0)$ and $E = E_0 \cdot exp(-i\omega t + kx)$, and the wave propagates in a system of free electrons with the distribution function $f(x, v)$. Kinetic equation for the distribution function is:

$$\frac{\partial f}{\partial t} + \nu \frac{\partial f}{\partial x} + \frac{eE}{m} \frac{\partial f}{\partial v} = 0 \tag{1}$$

This equation is almost obvious: the number of particles with the speed v changes by two reasons: these particles may come from a neighboring location, or they may appear after slower particles are accelerated by electric field. The second reason increases the number of such particles only if the number of slower particles exceeds the number of faster ones. Then, there are more particles that are accelerated to a given speed, increasing the distribution function for this speed, than those that are accelerated to a higher speed, thus decreasing the number of particles with a given speed.

Proceeding from the linearized version of the kinetic equation we determine the plasma conductivity:

$$\sigma = \frac{E}{m} \int \frac{v(df_0/dv)}{i(\omega - kv)} dv \tag{2}$$

This real part of conductivity characterizes the dissipation rate of an electromagnetic wave. If it is negative, then electron plasma will amplify the wave. Without collisions or any other dissipative additions this value seems pure imaginary. The only way we can get a real part may come from a pole, where $\omega = kv$. This corresponds to a resonance between particles velocity and the wave phase speed.

How can we know for sure what to do with the singularity in the integrand? It was Lev Landau who found the answer in 1947 in his usual elegant manner. And the effect that he discovered this way was named Landau damping of plasma waves. The idea seems quite obvious now. Let us assume we are considering an initial problem. We should use the Laplace transform for this linear problem. That means we should use exponentially increasing waves, i.e. integrate in the upper half of a complex ω plane, so that the integration path would pass around any singularities from above. Then in order to get analytical continuation of the solution to the real ω , we may lower the path down to the real axis, just passing around the pole (Fig.1).

Integration along the real axis brings nothing to the real part of conductivity. Only the semi-circle loop around the pole gives us some contribution. The rule of integration around the singular point, called Landau rule, may be written as:

$$\frac{1}{\omega - kv} = P \frac{1}{\omega - kv} + i\pi\delta(\omega - kv) \tag{3}$$

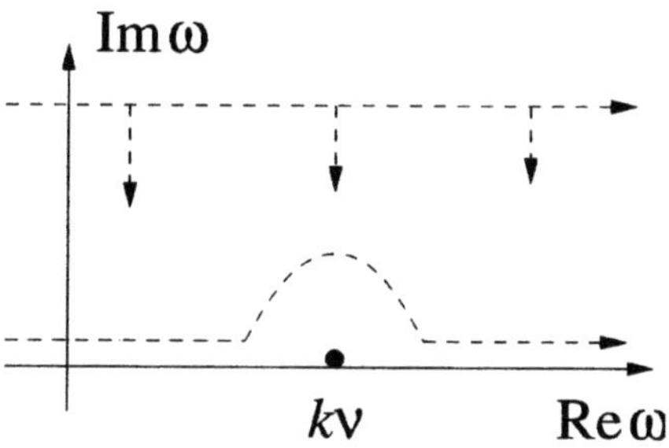

Fig. 1. Path in the complex ω plane

As a result we have the expression for the real part of conductivity, determined by the distribution function derivative at the resonant point, where particle speed coincides with the wave phase speed.

If there are more slower particles than faster particles around the resonance (this is the case in equilibrium plasma), then conductivity is positive and an electromagnetic wave will dissipate. This is called Landau damping. The wave energy goes to the resonant particles, causing the wave dissipation.

If there are more faster particles than slower ones in a vicinity of the resonance, the distribution function derivative is positive and so is the conductivity. Then we have a wave amplification that may happen in a non-equilibrium plasma with particle beams.

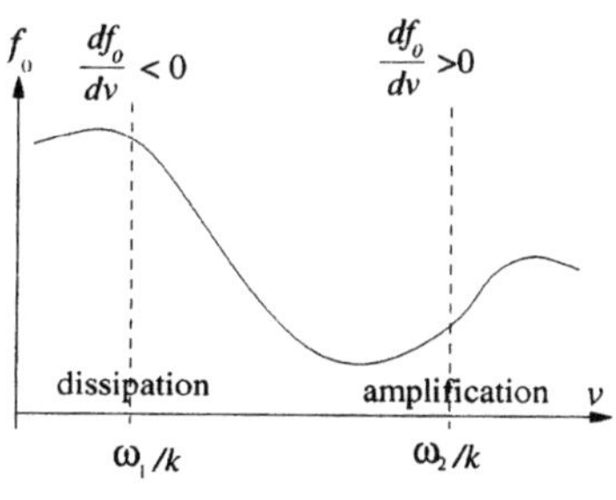

Fig. 2. Amplification or dissipation according to the sign of the distribution function derivative

I emphasize that Landau's rule was found in the collisionless approximation, neglecting dissipation effect. If we had some dissipation, like particle collisions, that would remove the singularity from the real axis in the integral and we would have the same Landau's rule without references to the initial problem. But the important point is that small dissipative effects do not influence the value of conductivity, and the Landau damping of plasma waves exists even in collisionless plasma. That was quite a shock for physicists at the time when Landau discovered it.

Similar effects can be found in fluid dynamics. A classic standard model, usually considered in this field, is a shear flow with the velocity profile $U(y)$Ȧgain, we have a variety of fluid particles, moving with different speeds. Again, we can define a velocity distribution function for the shear flow. It is almost the same definition as for electrons in plasma. In an incompressible fluid the volume moving with velocities within their interval $[U, U + dU]$ is virtually equivalent to the number of particles within a given velocity interval. This volume is defined by the formula $dy = f(U) \cdot dU$, therefore the distribution function is $f = (dU/dy)^{-1}$. Small perturbations in a 2-dimensional shear flow of ideal incompressible fluid is described by the Raleigh equation for the stream function amplitude:

$$\frac{d^2\psi}{dy^2} - k^2\psi - \frac{d^2U/dy^2}{U - \omega/k}\psi = 0 \tag{4}$$

This equation was heavily used in the theory of hydrodynamic instability. For a long time, however, it was not clear what to do with the singularity in this equation. As you can see, the singular point here is where the fluid velocity coincides with the wave phase velocity. In terms of the analytical theory of differential equations the singularity is a branching point, provided it does not coincide with an inflection point of the velocity profile. The equation has an ambiguous general solution. There is a problem of a relation between analytical solutions on both sides of the singularity.

In a strict mathematical sense, the dynamics of small perturbations within the Raleigh equation is incorrect in a singular point. A direct resolution of this problem would be a consideration of a more complicated problem at a vicinity of the singular point. This is exactly the way employed in the book by Lin, where a small viscosity was taken into account, thus removing the singularities in equations of the linear theory. In a viscous fluid the Rayleigh equation transforms into the Orr-Sommerfeld equation:

$$\frac{i\nu}{k}\left[\frac{d^4\psi}{dy^4} - 2k^2\frac{d^2\psi}{dy^2} + k^4\psi\right] + (U - \omega/k)\left(\frac{d^2\psi}{dy^2} - k^2\psi\right) - d^2U/dy^2\psi = 0 \tag{5}$$

Here ν is the kinematic viscosity. What Lin did was an analysis of the asymptotic behavior of solutions of the Orr-Sommerfeld equation in the limit $\nu \to 0$. This allows us to match solutions of the Rayleigh equation in the neighborhood of the singular point.

The Lin's approach requires quite an extensive background in the analytical theory of differential equations and also in depth knowledge of perturbation methods. His final result, however, seems rather simple and may be presented as the Lin's rule of integration. We have to choose an integration path in the complex y-plane , that goes around a singularity from below, if $dU/dy > 0$ and from above if $dU/dy < 0$. Only these solutions make a physical sense, since they are limits of viscous solutions of from the Orr-Sommerfeld equation in the limit $\nu \to 0$.

A more physical approach to the choice of the correct branch of an inviscid solution was found by Dikii and Case. They used the same method of analysis as was applied by Landau for plasma waves: the initial-value problem was solved using the Laplace transform. Complex values of ω were considered in the upper half of the complex ω plane: $Im\omega > 0$. It corresponds to growing perturbations. This way the Lin's rule of integration becomes quite obvious.

The physics of wave-flow interaction both in collisionless plasma and in an ideal fluid seems very similar. Waves exchange their energy with resonant particles, whose velocity coincides with the wave phase velocity. This is a key point of the plasma- hydrodynamic analogy. Further investigations showed that this analogy is actually more profound than it seemed before.

Of course, there are differences between phenomena in incompressible fluid and electron plasma. The most apparent distinction lies in the fact that the velocity at a given point is determined uniquely in the classical hydrodynamics, so those particles moving with different velocities must be spaced apart. Therefore, the simplest wave problems are 2-dimensional ones in hydrodynamics. In particular, different velocities are located at different levels y in a shear flow. As a result, we have resonant singularities in differential equations rather than in a simple integral like we had it in plasma. Wave problems in hydrodynamics present usually more challenge therefore.

There are more analogies between wave effects in plasma and fluid mechanics. A classical instability criterion for an inviscid shear flow is given by the Rayleigh theorem that relates the instability to the presence of an inflection point in the velocity profile. In plasma we have a similar criterion: longitudinal waves are unstable when the velocity distribution function has a minimum. Taking into account the definition of a distribution function for shear flows, we find that this function has extrema at inflection points of the velocity profile.

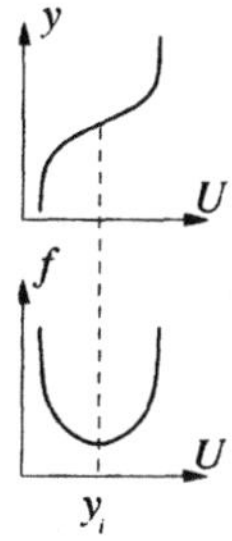

Fig. 3. Distribution function versus velocity

This gives us a well-known Rayleigh theorem, yielding a criterion of hydrodynamic instability of shear flows in ideal fluid: only velocity profiles with inflection points are unstable. Thus the Raleigh theorem is the hydrodynamic version of the plasma criterion.

Since waves, interacting with resonant particles, may dissipate or amplify, depending on the shape of distribution function near the resonance, it would be natural to expect the same phenomenon in fluid flows. Indeed, such an interaction was discovered, first for surface waves, in particular, see waves interacting with a shear wind. Although the mechanism of see waves amplification due to critical layers in the wind was found by Miles in 1956 independently, without any relation to Landau damping, later it was realized that this is basically the same effect of wave-flow resonance. Waves are amplified if the second derivative of the velocity profile is negative in the critical layer. This criterion becomes quite obvious if we follow Landau criterion. A derivative of the velocity distribution function is proportional to the second derivative of the velocity profile,

$$\frac{df}{dU} = \frac{d}{dU}\frac{1}{|dU/dy|} = -\frac{d^2U/dy^2}{|dU/dy|^2} \tag{6}$$

that yields the Miles' criterion.

The same resonant mechanism of amplification may work for sound waves, interacting with shear flows. A plane sound wave, propagating from a fluid at rest into a shear flow, is affected both wave refraction and with resonant effects in the critical layer, where the wave phase velocity coincides with the fluid velocity. Small perturbation in a shear flow most conveniently are described by the Rayleigh equation for the pressure amplitudes:

$$\frac{d^2p}{dy^2} - \frac{2dU/dy}{U-\omega/k}\cdot\frac{dp}{dy} + p\left[\frac{(\omega-kU)^2}{c_s^2} - k^2\right] = 0 \tag{7}$$

Approximate solution for this equation, found with the method of matched asymptotic expansions, gives us the energy conservation law in the following form:

$$\frac{\eta_-(1-|R|^2)}{(\omega/k)^2} - \frac{\eta_+|T|^2}{(U_0-\omega/k)^2} = \pi k^2|p|^2\left[\frac{d^2U/dy^2}{|dU/dy|^3}\right]_{y=y_c} \tag{8}$$

We can see from this equation, that the net energy gain, taking into account both reflected and transmitted waves, is again defined by the velocity profile curvature or the derivative of the velocity distribution function (6) at the critical layer, where $\omega = kU$.

The excessive energy comes to the wave from the fluid flow in the vicinity of the critical layer. Nonlinear effects, accompanying the resonant wave amplification, affect the velocity profile in the critical layer so that its curvature decreases and the wave-flow interaction decreases. These nonlinear effects work about the same way as well-known quasi-linear effects for plasma waves, interacting with resonant particles.

It is also worth here to note that resonant wave-flow interaction appeared to be a quite universal effect; it was discovered for all kinds of waves in a moving media, including internal gravity waves, Rossby waves, magneto-hydrodynamic waves in plasma, etc.

3 Negative Wave Energy and Over-Reflection

There is also another aspect that is common for waves both in plasma and in fluid flows as well as for many other types of waves in non-equilibrium systems with streams. It is waves of negative energy. Again, the concept of negative energy first emerged for electron beams. And only after decades this concept was introduced in fluid mechanics.

$$E = E_0 + \boldsymbol{PU} = \frac{\omega}{\omega - \boldsymbol{kU}} E_0 \tag{9}$$

NEW (negative energy waves) appear when the total energy of media with waves is less than the energy of media without waves. This may happen in a moving media, because wave and momentum are transformed in a moving reference system: When $\omega - kU < 0$, the energy becomes negative.

The concept of NEW is of great importance for analysis of wave generation and absorption in fluid flows. If we manage to reduce energy of NEW its amplitude will actually grow and so we generate the wave this way. If there is interaction of NEW and a wave with positive energy, both will grow since the energy will flow from positive energy waves to NEW and both amplitudes will increase. Thus we are having wave instability. These effects may actually happen in shear flows.

Simple models, suitable for analysis NEW effects are based on the algebraic method, which approximates shear flows by piece-wise profiles made of zero vorticity segments. Analytical solutions of the Rayleigh equation may be easily found for these segments, then matched at the segments' boundaries, forming the global analytical solution. The simplest profile of this type is the vortex sheet flow (Fig.4).

Reflection of a plane sound wave from the vortex sheet was analyzed first by Miles and Ribner. Matching pressure and fluid deviations for acoustic plane waves at the sheet surface yields the sound reflection coefficient:

$$R = \frac{\eta_-(\omega - kU)^2 - \eta_+\omega^2}{\eta_-(\omega - kU)^2 + \eta_+\omega^2} \tag{10}$$

where $\eta_\pm$ are y-component of the wave vectors below and above the vortex sheet, and k is the x -component of the wave vector. Taking into account that the group velocity for a transmitted wave

$$v^+_{gr-y} = \frac{\eta_+ c^2}{\omega - kU} \tag{11}$$

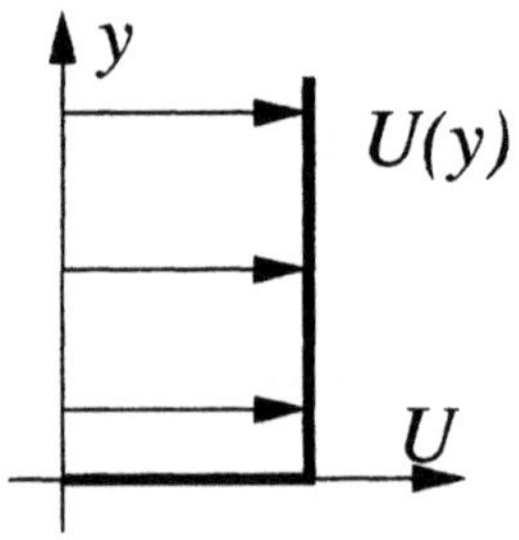

Fig. 4. Profile of the vortex sheet flow

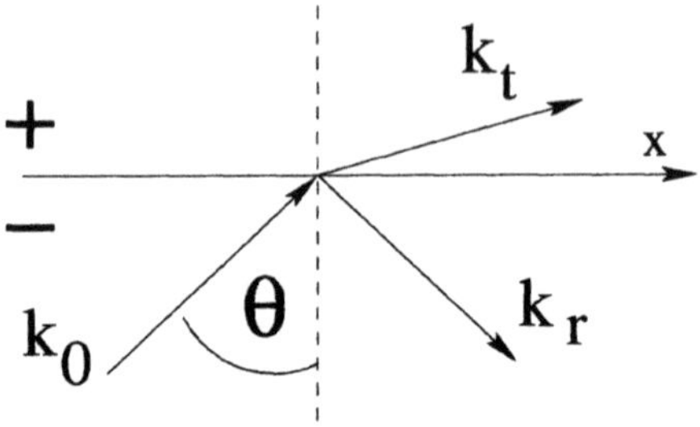

Fig. 5. Scattering by the interface

must be positive, i.e., directed outbound, the value of η_+ must be negative, when $\omega - kU < 0$. In that case the reflection coefficient absolute value $|R| > 1$, meaning that the reflected sound is amplified. The excessive energy for the reflected sound comes, obviously, from the moving fluid. In particular, the transmitted wave holds negative energy:

$$E = \frac{\omega}{\omega - kU} E_0 = \frac{p^2}{2\rho c^2} \frac{\omega}{\omega - kU} < 0 \tag{12}$$

Here p is the amplitude of pressure perturbations, c and ρ are the sound speed and the fluid density. The energy flux in the moving media is negative:

$$S_y = v_{gr\_y} E = \frac{\eta_+ \omega}{2\rho c^2 (\omega - kU)^2} |p|^2 < 0 \tag{13}$$

because the transmitted wave propagates away from the vortex sheet and takes negative energy with it. Thus the energy comes down to the vortex sheet from the moving medium, the reflected wave gains more energy than the incident wave and we have the wave amplification.

The sound over-reflection due to NEW in the moving medium is possible only for super-sonic flows, where the condition $\omega - kU < 0$ can only be met. Later the over-reflection was found also for other types of waves: water surface waves, electromagnetic waves in moving dielectric, plasma waves, etc.

It is also worth to note here, that the issue of super-sonic vortex sheet instability, although separate, had many interesting developments, including the importance of NEW as an explanation for some types of flow instability. For example, while the Kelvin- Helmholtz instability may disappear for a 2-dimensional super-sonic flow, an acoustic instability will arise instead. Indeed, the over-reflection of sound waves at a super-sonic vortex sheet increases their energy, and with a feedback, provided by a simple reflection from a rigid wall the sound oscillations may grow in a boundary layer (see. Fig.6).

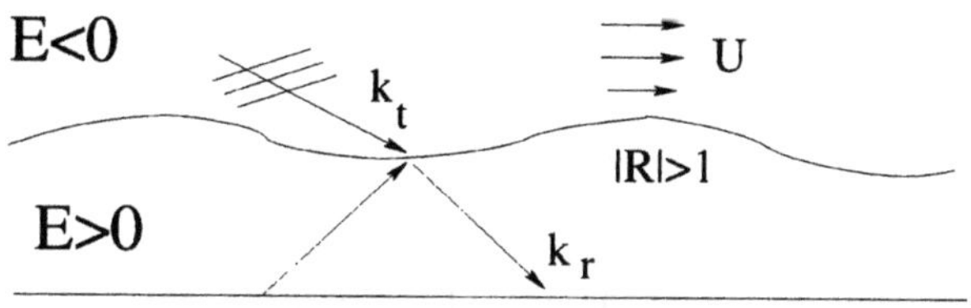

Fig. 6. Feedback mechanism

This simple configuration may also be considered as a model for so-called radiative instability. In the moving frame of reference the fluid layer next to a wall, i.e., the boundary layer, holds NEW, and the instability is caused by sound waves radiating into the outside medium. This is an example of the instability, caused by the energy withdrawal from NEW.

There is one more possibility of wave instability, related to NEW. It is when the wave dissipates its energy instead of radiating it in a moving medium. For $\omega - kU < 0$ not only the wave energy changes its sign, but also the dissipation becomes negative. The negative dissipation may cause instability in shear flows. This effect also allows a simple consideration within the algebraic method, if, for example, a vortex sheet divides ideal and viscous flows.

4 Acoustic Analogues of Microwave Devices: Monotron and Klystron

Aerodynamic generators of sound, or whistles, reveal a lot of resemblance to VHF generators in vacuum electronic. Both types of devices are based on wave interaction with streams of particles - either electron plasma or fluid. The most obvious analogy seems to be the one between the traveling wave tube (TWT) and acoustic surface wave influenced by a shear flow (Fig.7).

This analogy does not work well, though. While the TWT is based on a resonance between electromagnetic waves and Langmuir waves in an electron beam, the acoustic wave can not find a resonance with the shear flow. The reason is simple - the shear flow is unstable; its eigenmodes have complex frequencies, which can not coincide with the sound frequency. The sound wave

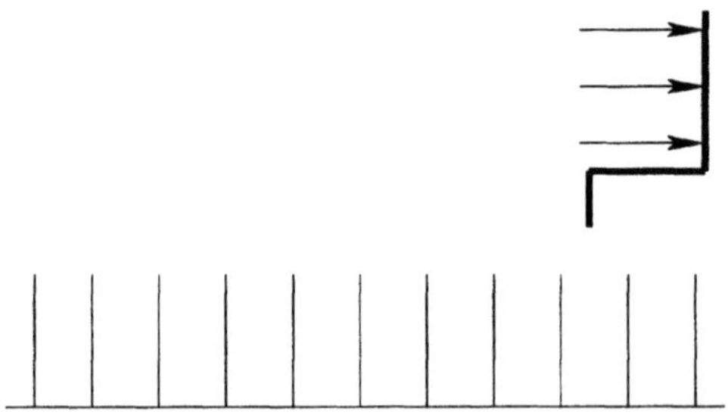

Fig. 7. Acoustic surface wave influenced by a shear flow

may interact with resonance particles in the critical layer in shear flows and gain some energy from this kind of resonant interaction. But since only a small portion of the entire flow participates in this type of wave-flow interaction, it is not as efficient as the mechanism of the TWT.

If we want the sound wave to "catch" exponentially growing flow waves, we need a wide spatial spectrum for a given sound frequency. This is achieved in an acoustic resonator, excited by a fluid flow, - a whistle (Fig.8). A direct analog of a whistle is a quasi-static microwave generator, monotron (Fig. 9), consisting of an LC circuit, penetrated by an electron beam.

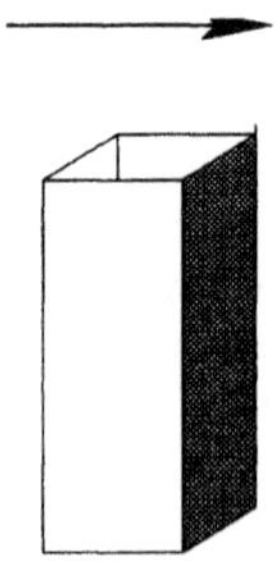

Fig. 8. Acoustic resonator excited by a fluid flow

Excitation or dissipation of electromagnetic oscillations in this circuit is determined by the impedance of the electron gap:

$$Z = \frac{8}{i\omega S} \int \frac{\sin^2(kd/2)}{k^2 \epsilon(\omega, k)} dk \tag{14}$$

where S is the area of the capacitor surfaces, d is the distance between the surfaces, and $\epsilon = 1 - \frac{\omega_p^2}{(\omega - kU)^2}$ is the permittivity of the electron plasma. The integral in (14) has poles at zeros of permittivity, $\omega = kU \pm \omega_p$, which correspond to the Langmuir plasma waves. The integration contour, passing these poles (Fig.10), may be found from the initial value problem.

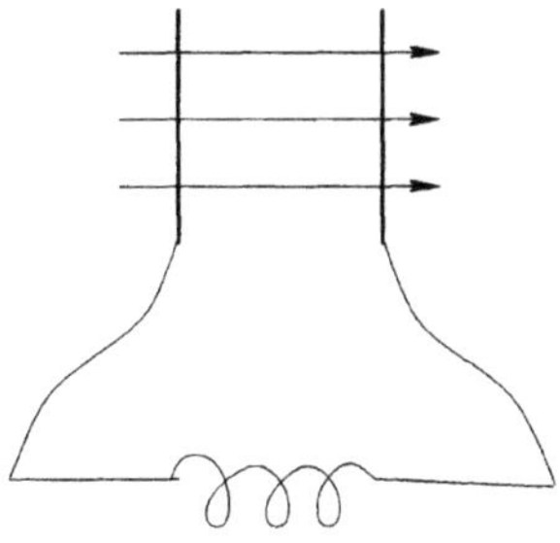

Fig. 9. Quasi-static microwave generator: monotron

The capacitor resistance:

$$R = \frac{4\pi d}{\omega S}\frac{\omega_b^2}{\omega^2}\frac{\theta \sin\theta + 2\cos\theta - 2}{\theta} \tag{15}$$

has an oscillatory dependence on the single parameter, "flight angle", $\theta = \omega d/U$.

Note that free electron waves are stable. It is the feedback, provided by the LC circuit, that causes instability. That feedback excites electron waves in the incoming electron flow. The waves propagate with the flow and induce oscillations in the downward surface of the capacitor. If the wave phase shift, defined by the parameter θ is right, the feedback is positive and the oscillations are unstable. In the opposite case, when the phase shift makes a negative feedback, the LC circuit oscillations diminish.

We can find very similar behavior for a whistle, based on an acoustic resonator. The whistle excitation is determined by a reflection coefficient, R for the main mode of an acoustic waveguide, reflected from an open end, placed next to a fluid flow. This kind of a diffraction problem may be solved analytically, under the approximation of weak interaction with the flow. The perturbation of the reflection coefficient, caused be the flow influence is

$$\Delta R = \frac{4\pi i\omega}{c_s S_0}\int \frac{R_{fl}(\omega,k)\cdot W}{\sqrt{k^2-\omega^2/c_s^2}} e^{-2\kappa(\omega,k)L} d^2\boldsymbol{k} \tag{16}$$

where S is the area of the open end, L is the distance between the open end and the flow, R_{fl} is the reflection coefficient of a plane wave from the flow, and W is the power spectrum of the y-component of oscillating velocity at the open end.

The integral in (16) have poles, as soon as the poles are present in the flow reflection coefficient, R_{fl} . These poles correspond to the eigenmodes or waves in a shear flow. This is similar to the monotron, only the flow waves are unstable and the poles lie in the complex plane, distant from the real axis. For a vortex sheet the pole $k = (1-i)\omega/U$ is far from the real axis, and for symmetric waves in a thin jet the pole will be relative close to the axis, since these waves are almost stable. In any case, we find the integration path from the initial value problem - see Fig.11. This path goes around the poles in the lower half-plane of complex k, which correspond to exponentially increasing waves.

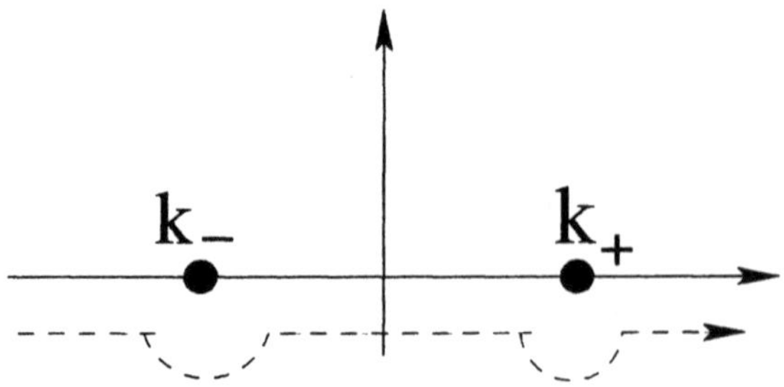

Fig. 10. Contour in the k plane

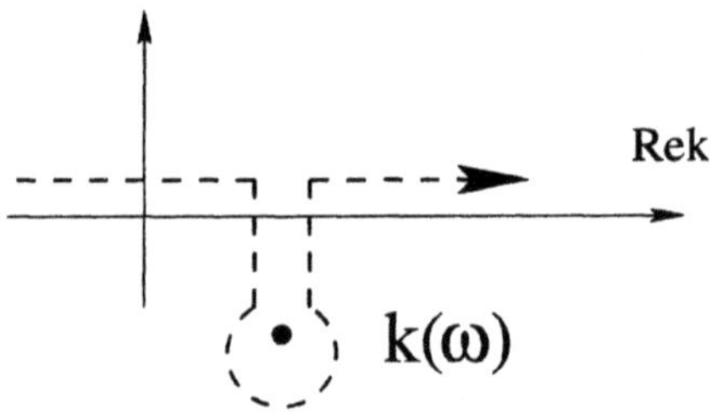

Fig. 11. Contour in the k plane

The increment for acoustic oscillations in the resonator is determined, as for the monotron, by a single parameter, $\theta = \omega d/U$, where d is the interaction distance between acoustic mode and fluid flow. Depending on the whistle geometry this distance may have different meaning, for example, the distance to the flow or the width of the open end.

Again, as for the monotron, the oscillations increment changes its sign alternatively with this parameter growing. The reason is the same: the resonator provides a feedback to the flow waves, and this feedback may be either positive or negative, depending on the phase shift, accumulated by flow waves, which propagate by the open end. Indeed, the oscillating outflow from the open end produces oscillating displacement perturbations of the flow, which are transformed into pressure perturbations while propagating with flow. These pressure perturbations hit the open end back and this closes the feedback loop (Fig.12).

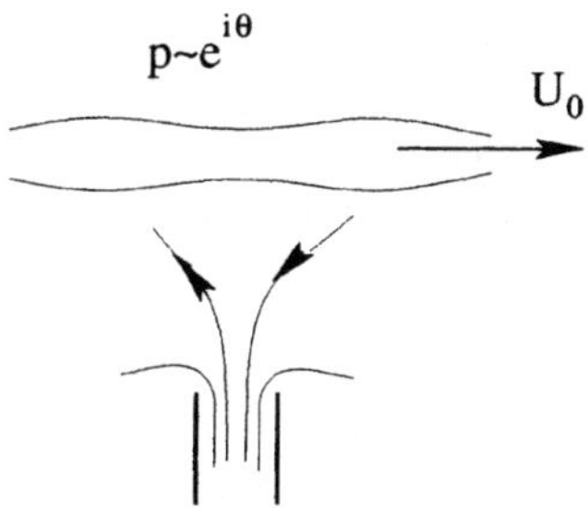

Fig. 12. Feedback loop

Proceeding from the analogy between acoustic and microwave generators, we can suggest a potentially more efficient type of whistle - acoustic klystron. It is just a resonator with two open ends in the same flow (Fig.13). Here the phase shift for flow waves is extended over a longer distance, that allows more flexibility in the feedback optimization. This idea comes from a direct analogy with the klystron, the microwave generator, comprising two capacitors, penetrated by the same electron beam (Fig.14).

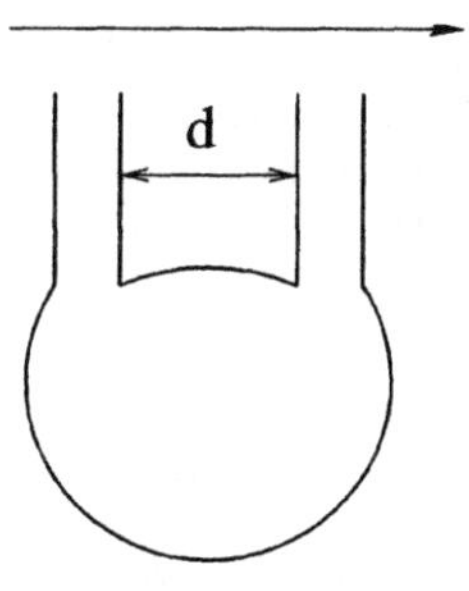

Fig. 13. Resonator with two open ends

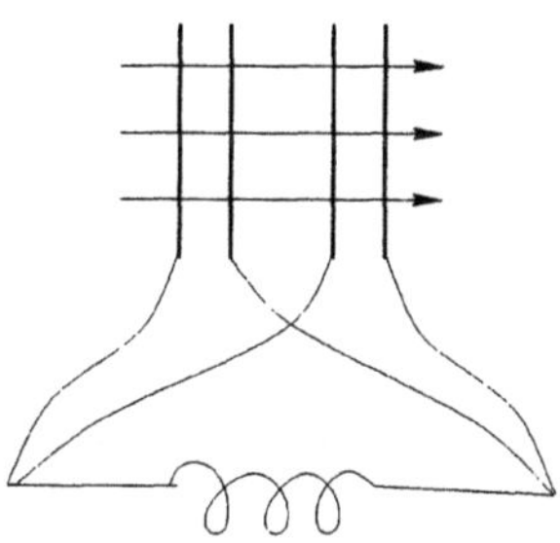

Fig. 14. Klystron

The important point, drawn from the acousto-electronic analogy, is that it is not the flow instability, that excites the resonator, but rather the flow inertia or oscillating waves, which the flow may hold. For example, in a thin jet the symmetric waves are stable, but still can excite the resonator. As for the flow instability, it does generate perturbations in the flow, but they are not synchronous with oscillations of the acoustic resonator and do not contribute in the whistle self-excitation. For practical purposes, though, an external force imposed on the whistle mouth by these oscillations, may be an important addition to the self-excitation mechanism, as well as strong nonlinear effects of vortex formation in fluid flow.

5 Wave-Vortex Interactions

The plasma-hydrodynamic analogy is not limited to plane parallel flows. Among the most important plasma resonances there is a cyclotron resonance, caused by interaction of electromagnetic waves with electrons rotating in a magnetic field. An obvious candidate for this role in aeroacoustics would be the sound interaction with rotating flows, or vortices. However, the wave-vortex interaction has its specific features, distinguishing it from plasma cyclotron effects.

A simple model for vortex oscillations is based on axial symmetric flow, a cylindrical vortex, which is defined only by the rotation frequency $\Omega(r)$. Perturbations of a cylindrical vortex in an ideal incompressible fluid are described by a cylindrical analog of the Rayleigh equation for the stream function amplitude of $e^{-i\omega t+in\theta}$ oscillations:

$$\frac{d^2\psi}{dr^2}+\frac{1}{r}\frac{d\psi}{dr}-\frac{n^2}{r^2}\psi-\frac{n}{(\omega-n\Omega)r}\frac{dQ}{dr}\psi=0 \tag{17}$$

where $Q(r)=rd\Omega/dr+2\Omega$ is the flow vorticity.

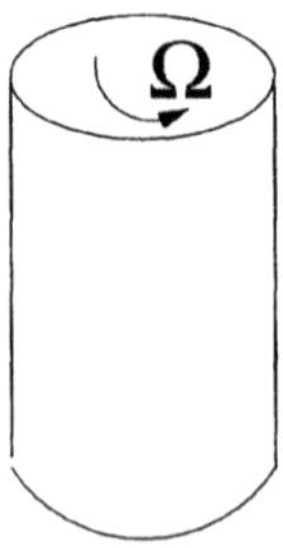

Fig. 15. Rankine vortex

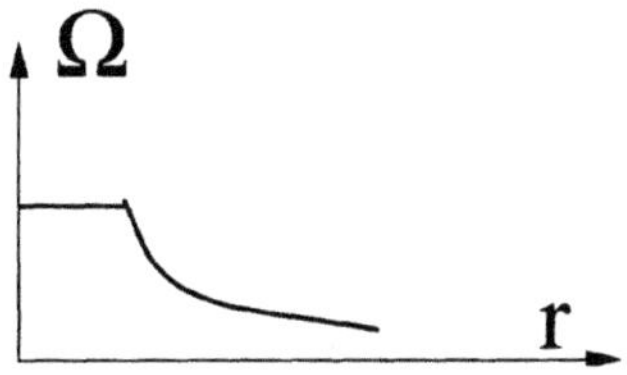

Fig. 16. Vorticityof the Rankine vortex

There are at least two basic effects, which can be analyzed within this simple model. First it is the cylindrical analog of wave amplification, based on the negative dissipation. For the Rankine vortex, where the vorticity is constant, equal to 2Ω inside the vortex core and zero outside, we can apply the algebraic method and find the reflection coefficient for a cylindrical harmonic of sound, even if the vortex core is viscous:

$$1 - |R|^2 = \frac{8\pi\nu}{(|n|-1)!(|n|-2)!a^2} \frac{\omega - n\Omega}{(\omega - n\Omega + \Omega)^2} \left[\frac{\omega a}{c_s}\right]^{2|n|} \tag{18}$$

Here a is the core radius, and ν is the core viscosity. If $\omega - n\Omega < 0$, the cylindrical wave is amplified. This is an acoustic analog of Zel'dovich effect, discovered first for electromagnetic waves, scattered at a rotating conducting cylinder. The source of amplification lies in the fact, that sound waves in the rotating cylinder may have negative energy, which dissipates in the vortex core. The same reason causes the radiative instability of the Rankine vortex, when a radiating sound takes energy away from the increasing NEW inside the vortex core.

Another basic effect, of the wave-vortex interaction is the resonance in a cylindrical critical layer r_c , where $\omega - n\Omega(r_c) = 0$. The equation (17) has a singularity in the critical layer, which must be treated the same way as for a plane shear flow. This singularity exists only for harmonics $n \geq 2$. The energy loss in the critical layer:

$$1 - |R_n|^2 \approx \left[\frac{d\Omega}{dr}\right]_{r=r_c} \tag{19}$$

is proportional to the radial gradient of the vorticity. Since for concentrated vortices the vorticity typically decreases with the radius, the resonant interaction with a vortex almost certainly amplifies the sound.

This is the acoustic analog of the cyclotron damping in magnetized plasma. The difference between hydrodynamic and plasma effects, as usual, stems from the fact, that different velocities in an incompressible fluid must have different locations. Therefore, we have singularities in differential equations and basic models are 2-dimensional.

For a plane sound wave incident at an intensive vortex, we can decompose the acoustic field in a series of cylindrical harmonics and tread each harmonic independently. For long waves only the harmonic $n = 2$ does substantial resonant interaction. Since it is amplified by a concentrated vortex, the whole absorption cross-section of the vortex may be negative. This may only happen if the rotation frequency of the vortex exceeds the sound sub-harmonic frequency: $\Omega > \omega/2$.

It is still not clear what could be the role of such kind of resonant interaction in aerodynamic generation of sound (or other wave modes) by intensive vortices or the dynamics of strong vortical turbulence. There are numerous examples of concentrated vortex flows in the atmosphere, ocean, or technology. Possible applications of the wave- vortex resonance are still waiting for a more detailed research.

6 Analogies Table

As a conclusion, the list of known analogies of the plasma-hydrodynamic type is presented below. It includes not only those, which are briefly discussed above, but plenty of others, analyzed somewhere else. Practically all of them, to the best of my knowledge, were mentioned in the book, referred to at the end of the first chapter.

FLUID DYNAMICS	PLASMA
Velocity profile	Velocity distribution
Lin's rule	Landau rule
Kelvin cat's eye flow in a critical layer	Trapped particles in a plasma wave
Vorticity waves $\Omega = \delta(y - y_0)$	Van Kampen - Case waves
Inflection point	Min distribution function
Rayleigh theorem	Plasma instability criterion
Dissipation in a critical layer	Landau damping of plasma waves
Miles mechanism for wind waves generation	"Kynetic" instability of plasma waves.
Quasi-linear effect of sea waves on the wind profile	Quasi-linear relaxation for plasma waves
Flatten vorticity in a shear flow profile	"Plato" in a velocity distribution function of plasma particles
Plasma echo	Echo effect in shear flows
Negative energy of sound in supersonic flows	Negative energy of plasma waves in electron beams.
Whistle	Monotron
Double-mouth whistle	Klystron
Radiative instability of an oscillator in a fluid flow	Anomalous Doppler effect for a moving oscillator.
Negative dissipation of sound in a viscous vortex	Zel'dovich effect in rotating dielectric cylinder.
Radiative vortex instability.	"Evaporation" of a rotating black hole
Resonant wave amplification by an intensive vortex.	Cyclotron instability in a magnetized plasma
Sound scattering by localized vortices	Scattering of electromagnetic waves by charged particles.

New Linear Mechanisms of Acoustic Wave Generation in Smooth Shear Flows (Nonmodal Study)

George D. Chagelishvili

Centre for Plasma Astrophysics, Abastumani Astrophysical Observatory, Ave. A. Kazbegi 2a, 380060 Tbilisi, Georgia ***and*** Space Research Institute, Str. Profsoyuznaya 84/32, 117810 Moscow, Russia

Abstract. This lecture presents new linear mechanisms of acoustic wave generation in smooth shear flows using a non-modal study.

1 Introduction

Flows with inhomogeneous velocity profiles, met in laboratory experiments, industrial applications, the earth's atmosphere, oceans and many astrophysical objects, form an important part of the human-natural environment. Therefore, the investigation of shear flows has received a human attention for a long time and has even become one of the principal subjects of physics for more than century. Nevertheless, shear flow phenomena are still not completely understood. In fact, not all types of shear flows successfully passed analysis of *basic linear processes.*

1.1 Basic Problems in Shear Flows

The traditional and generally acknowledged paradigm in the study of dynamics of disturbances is the eigenvalue analysis (also known as the *modal analysis/approach*). Usually, it consists of two steps:
(*) linearization about the mean (laminar) solution and afterwards
(**) search for unstable eigenmodes of the linearized problem.
The necessary condition for the flow to behave unstably (in the sense of the standard linear theory) is the existence of exponentially growing eigenmodes. For some processes (e.g., thermally driven instabilities in Rayleigh-Benard convection flow, or centrifugally driven instabilities in rotating Couette flow) modal approach well matches with laboratory studies. At the same time, for other kinds of hydrodynamic processes, especially those driven predominantly by flow velocity shear forces, *predictions of the modal approach fail to match most experiments.* For plane Couette flow, for instance, extraction of the shear flow energy by disturbances – amplification of disturbances – is observed to "switch on" for Reynolds numbers as low as $\mathcal{R}\approx350$, while the common eigenvalue analysis predicts stability for all $\mathcal{R}$. Traditionally, this

anomaly was recognized as a failure of linearization and was attributed to nonlinear effects. However, recently it was argued that the failure of modal analysis should be attributed to step (**) [1–3].

1.2 Essence of the Difficulties of the Modal Approach

The difficulties of the modal approach become evident by the juxtaposition of the following facts:

– Standard linear algebra says that, even in the case when all eigenmodes of a linear system are exponentially stable, inputs to the system may be amplified by arbitrarily large factors, if the eigenfunctions are not mutually orthogonal [1];

– Operators that arise in Poiseuille or Couette flow are not self-adjoint (i.e., its eigenfunctions are not mutually orthogonal).

Hence, the usage of the full spectral (Fourier or Laplace) expansion for this category of shear flows may be quite misleading: *the study of a single mode evolution does not give adequate description of the system behaviour.*

Nowadays, one can state, that the essence of the difficulties arising in modal analysis is rigorously revealed [1]: Operators arising in such a description of linear processes in shear flows are not self-adjoint – exponentially far from normal [1,2]. Consequently, corresponding eigenfunctions are not orthogonal to each other and *they strongly interfere.* Therefore, the knowledge drawn from the analysis of separate eigenfunctions and eigefrequencies is far from completeness. All these lead to the necessity of the exact work out of interference between different modes for the correct description of the phenomena. The latter problem appears to be insuperable as a rule.

1.3 Onto Nonmodal Approach

The difficulties of modal approach provoke change in the paradigm. The so-called *nonmodal approach* to the study of waves and instabilities in shear flows (stemming out of the 1887 paper by Lord Kelvin [4]) has become well-established and has been extensively used in the 1990's [3,5–16]. As a result, impressive progress in the understanding of shear flow phenomena has been enforced. The nonmodal analysis implicates change of independent variables from a laboratory to a moving frame and study of temporal evolution of *spatial Fourier harmonics* (SFH) of disturbances without a spectral expansion in time. The method is operative for any smooth mean velocity profile, but it is most manageable for the flows with linear, or piecewise linear velocity profiles [8,9], As an example we take the flow with the mean velocity directed along the X, and a constant shear along the Y:

$$\boldsymbol{V}_0 = \boldsymbol{V}_0(Ay, 0). \tag{1}$$

where A is the constant parameter of the shear. (For simplicity we consider only two dimensional (2D) flows inasmuch as the third dimension does not carry in a significant novelty.)

The first apparent novelty of the nonmodal approach is that *the wave number along the shear axis of each SFH varies in time.* Mathematical justification of this variation is presented later (see (6)–(12)). While, the existence of the variation follows from the fact that in the flow with constant shear rate (see (1)) disturbances cannot have a form of a simple plane wave due to effect of the shearing background on wave crests [5,6,10–23].

The time dependence of SFH's wave number resides in the following:

If at initial moment of time spatial Fourier harmonic with wave numbers k_x and $k_y(0)$

$$\psi(0) = \tilde{\psi}(k_x, k_y(0)) \exp(\mathrm{i}k_x x + \mathrm{i}k_y(0)y) \tag{2}$$

is disturbed (where ψ is a general vector composed from the SFH of disturbed physical variables), then the SFH phase evolution is defined as follows:

$$\psi(t) = \tilde{\psi}(k_x, k_y(t), t) \exp(\mathrm{i}k_x x + \mathrm{i}k_y(t)y), \tag{3}$$

$$k_y(t) = k_y(0) - k_x A t, \tag{4}$$

i.e. the SFH's wave-number along the flow shear axis (along the Y) varies in time. In other words, there is a "drift" of the SFH in the $\boldsymbol{k}$-space (in wave-number space) in the linear approximation.

Temporal variation of the wave numbers of SFH of disturbances in the shear flow is well-acknowledged in papers cultivating the nonmodal approach (see Refs. [5–7,10–24]).

The nonmodal approach greatly simplifies the mathematical description of shear flow dynamics and helps to grasp the phenomena that are overlooked in the framework of the modal approach. The nonmodal approach has already yielded some unexpected results on time evolution of both vortex [3,5,6,17–20] and acoustic [10,11] disturbances in shear flows. This approach has also been successfully applied to the study of MHD waves [7,12,13,22]. It has helped to formulate a new conjecture of transition to turbulence [15,21,24] and has led to the discovery of phenomena of linear coupling of different modes of disturbances in shear flows [13,16,22,23,25,26].

The chief peculiarity of found phenomena lies in two novel channels of energy exchange in shear flows.

- Through the first channel the mean flow exchanges energy with disturbances. This process proceeds non-exponentially and involves vortices [3,5,6,17–20] and waves [10,11]:
  (i) The energy of an infinitesimal disturbance in a shear flow can be significantly amplified even if the flow is spectrally stable.
- The second channel ensures energy exchange between different kind of disturbances:

(ii) vortices and waves (non-resonant phenomenon of abrupt emergence of waves from vortices),
(iii) different wave modes (resonant phenomenon of mutual transformations of waves).

It should be pointed out that *the both channels are at work at the linear stage of evolution of disturbances.*

2 Mathematics of the Study

The aim of this lecture is to describe new phenomena – **(i)**, **(ii)** and **(iii)** – discovered employing the nonmodal approach.

Full set of equations of our study is the following:

$$\frac{\partial \varrho}{\partial t} + \nabla(\varrho \boldsymbol{V}) = 0 , \tag{5a}$$

$$\varrho \left[\frac{\partial \boldsymbol{V}}{\partial t} + (\boldsymbol{V}\nabla)\boldsymbol{V} \right] = -\boldsymbol{\nabla} P , \tag{5b}$$

$$\frac{\partial P}{\partial t} + (\boldsymbol{V}\nabla)P - \Gamma \frac{P}{\varrho} \left[\frac{\partial \varrho}{\partial t} + (\boldsymbol{V}\nabla)\varrho \right] = 0 . \tag{5c}$$

Where ϱ, P, and $\boldsymbol{V}$ are density, pressure and velocity of the flow respectively.

Let us consider the simplest 2D compressible flow, that is: unbounded, inviscid, planar flow with uniform mean density and pressure ($\varrho_0, P_0 = \mathrm{const}$) and constant shear of velocity ($\boldsymbol{U}_0 = \boldsymbol{U}_0(Ay, 0)$ – see (1); The shear parameter A is accepted to be positive without a loss of generality). Consideration of 2D flow is justified by the fact that the third (Z) dimension does not carry in a significant novelty in the phenomena **(i)**, **(ii)** and **(iii)**.

Equation (5a)–(5c) are generally nonlinear. However, the phenomena **(i)**, **(ii)** and **(iii)**, we are planning to describe phenomena that are linear by nature. Therefore, we neglect the nonlinear terms in respect to disturbed quantities ϱ', p', $\boldsymbol{u}$ (where $\varrho' \equiv \varrho - \varrho_0$, $p' \equiv P - P_0$, $\boldsymbol{u} \equiv \boldsymbol{V} - \boldsymbol{V}_0$) in the (5a)–(5c). Hence, using ordinary linearization procedure over ϱ', p' and $\boldsymbol{u}$, in (5a)–(5c) yields:

$$\left(\frac{\partial}{\partial t} + Ay \frac{\partial}{\partial x} \right) \varrho' + \varrho_0 \left(\frac{\partial u_x}{\partial x} + \frac{\partial u_y}{\partial y} \right) = 0 , \tag{6a}$$

$$\left(\frac{\partial}{\partial t} + Ay \frac{\partial}{\partial x} \right) u_x + A u_y = -\frac{\partial p'}{\partial x} , \tag{6b}$$

$$\left(\frac{\partial}{\partial t} + Ay \frac{\partial}{\partial x} \right) u_y = -\frac{\partial p'}{\partial y} , \tag{6c}$$

$$p' = c_s^2 \varrho' , \tag{6d}$$

where $c_s \equiv \sqrt{\Gamma P_0 / \varrho_0}$ is the sound velocity.

Following the standard method of the nonmodal approach, let us introduce new variables:

$$t_1 \equiv t; \quad x_1 \equiv x - Ayt; \quad y_1 \equiv y\,, \tag{7a}$$

$$\frac{\partial}{\partial t} + Ay\frac{\partial}{\partial x} = \frac{\partial}{\partial t_1}; \quad \frac{\partial}{\partial x} = \frac{\partial}{\partial x_1}; \quad \frac{\partial}{\partial y} = \frac{\partial}{\partial y_1} - At_1\frac{\partial}{\partial x_1}\,. \tag{7b}$$

In terms of these new variables from the (6a)-(6d) may be reduced to the following form:

$$\frac{\partial \varrho}{\partial t_1} + \varrho_0\left\{\frac{\partial u_x}{\partial x_1} + \left(\frac{\partial}{\partial y_1} - At_1\frac{\partial}{\partial x_1}\right)u_y\right\} = 0\,, \tag{8a}$$

$$\frac{\partial u_x}{\partial t_1} + Au_y = -\frac{c_s^2}{\varrho_0}\frac{\partial \varrho'}{\partial x_1}\,, \tag{8b}$$

$$\frac{\partial u_y}{\partial t_1} = -\frac{c_s^2}{\varrho_0}\left(\frac{\partial}{\partial y_1} - At_1\frac{\partial}{\partial x_1}\right)\varrho'\,. \tag{8c}$$

The coefficients of the linearized set of (6a)–(6c) depend on the spatial coordinate y. As a result of mathematical transformation (7a,b) this *spatial inhomogeneity* is changed by the *inhomogeneity in time* (see (8a)–(8c)). This enables us to perform the Fourier analysis for the variables in (8a)–(8c) in respect to x_1 and y_1

$$\left\{\begin{matrix} u_x(x_1,y_1,t_1) \\ u_y(x_1,y_1,t_1) \\ \varrho'(x_1,y_1,t_1) \end{matrix}\right\} = \int_{-\infty}^{+\infty} dk_x dk_y \left\{\begin{matrix} \tilde{u}_x(k_x,k_y,t_1) \\ \tilde{u}_y(k_x,k_y,t_1) \\ \tilde{\varrho}'(k_x,k_y,t_1) \end{matrix}\right\} \exp(\mathrm{i}k_x x_1 + \mathrm{i}k_y y_1) \tag{9}$$

and trace the temporal evolution of the each SFH separately:

$$\frac{\partial d}{\partial t_1} + \mathrm{i}k_x\tilde{u}_x + \mathrm{i}(k_y - Ak_x t_1)\tilde{u}_y = 0\,, \tag{10a}$$

$$\frac{\partial \tilde{u}_x}{\partial t_1} + A\tilde{u}_y = -\mathrm{i}k_x c_s^2 d\,, \tag{10b}$$

$$\frac{\partial \tilde{u}_y}{\partial t_1} = -\mathrm{i}(k_y - Ak_x t_1)c_s^2 d\,, \tag{10c}$$

where $d \equiv \tilde{\varrho}/\varrho_0$.

As one can see from (9), dependence of each SFH on spatial coordinates in new variables x_1, y_1 is expressed as $\exp(\mathrm{i}k_x x_1 + \mathrm{i}k_y y_1)$. Returning to the old variables (t, x, y), dependence of each SFH on spatial coordinates takes the form:

$$\begin{aligned} &\exp(\mathrm{i}k_x x_1 + \mathrm{i}k_y y_1) = \exp\left[\mathrm{i}k_x(x - Ayt) + \mathrm{i}k_y y\right] = \\ &= \exp\left[\mathrm{i}k_x x + \mathrm{i}(k_y - k_x At)y\right] \end{aligned} \tag{11}$$

As one can see from this equation, the wave number along the Y (along the shear axis) varies in time: $k_y(t) = k_y - k_x At$. This time dependence is the drift of SFH in $\boldsymbol{k}$-space that is described in Sect.3 (see (2)–(4)).

Introducing non-dimensional variables and parameters:

$$D \equiv \mathrm{i}d, \quad \boldsymbol{v} \equiv \tilde{\boldsymbol{u}}/c_s \,, \tag{12a}$$

$$R \equiv \frac{A}{c_s k_x}, \quad \tau \equiv c_s k_x t, \quad \tau^* \equiv \frac{k_y}{R k_x} \,, \tag{12b}$$

$$\beta(\tau) \equiv k_y(\tau)/k_x = k_y/k_x - R\tau = (\tau^* - \tau)R \,. \tag{12c}$$

it is easy to obtain from (10a)-(10c) the following set of first order differential equations:

$$\frac{\partial D}{\partial \tau} = v_x + \beta(\tau) v_y \,, \tag{13a}$$

$$\frac{\partial v_x}{\partial \tau} = -R v_y - D \,, \tag{13b}$$

$$\frac{\partial v_y}{\partial \tau} = -\beta(\tau) D \,. \tag{13c}$$

The total energy of each SFH in $\boldsymbol{k}$-space is defined as follows:

$$E_{\boldsymbol{k}}(\tau) \equiv \frac{1}{2}(|v_x|^2 + |v_y|^2) + \frac{1}{2}|D|^2, \tag{14}$$

where the two terms correspond to kinetic and potential energies of the SFH respectively. Note that $\partial E_{\boldsymbol{k}}(\tau)/\partial \tau = -R v_x v_y$. Though, $E_{\boldsymbol{k}}(\tau)$ is a conserved quantity in the absence of the shear.

Further numerical analysis of our study is governed by (11a)–(11c). However, it is important to rewrite them in another form for a fundamental understanding the physical nature of the phenomena.

It is easy to verify that variables D, v_x and v_y obey the important *algebraic* relation:

$$v_y - \beta(\tau) v_x - RD = \mathrm{const}, \tag{15}$$

where const is some constant of integration.

Processing (11a)–(11c) one can get the following second order *inhomogeneous* differential equations:

$$\frac{\partial^2 v_x}{\partial \tau^2} + \omega^2(\tau) v_x = \beta(\tau) \times \mathrm{const} \,, \tag{16a}$$

$$\frac{\partial^2 D}{\partial \tau^2} + \frac{2R\beta(\tau)}{\omega^2(\tau)} \frac{\partial D}{\partial \tau} + \left[\omega^2(\tau) + \frac{2R^2}{\omega^2(\tau)}\right] D = -\frac{2R}{\omega^2(\tau)} \mathrm{const} \,, \tag{16b}$$

$$\frac{\partial^2 v_y}{\partial \tau^2} + \frac{2R}{\beta(\tau)} \frac{\partial v_y}{\partial \tau} + \omega^2(\tau) v_y = \mathrm{const} \,, \tag{16c}$$

and equations, expressing v_y and d through v_x and $\partial v_x/\partial \tau$:

$$v_y = \frac{1}{1+R^2}\left(\mathrm{const} + \beta(\tau) v_x - R\frac{\partial v_x}{\partial \tau}\right) \,, \tag{17a}$$

$$D = -\frac{1}{1+R^2}\left(R[\mathrm{const} + \beta(\tau) v_x] + \frac{\partial v_x}{\partial \tau}\right) \,. \tag{17b}$$

where $\omega^2(\tau) \equiv 1 + \beta^2(\tau)$. Note that the time dependence of $\beta(\tau)$ and $\omega(\tau)$ is caused by the temporal variation of wave number of SFH ($k_y(\tau) = k_y - k_x R\tau$).

2.1 Types of Disturbances

In order to classify the disturbances we choose to analyze (16a). Note that the first term in the left hand side of this equation is responsible for the flow compressibility. It is possible to classify the solution of the (16a) from mathematical and physical standpoints separately.

Mathematically, general solution of (16a) is a sum of *general* solution of the corresponding homogeneous equation (with const $= 0$) and a *particular* solution of this inhomogeneous equation. It should be emphasized that the latter is not uniquely determined. The sum of a particular solution of the inhomogeneous equation and any particular one of the corresponding homogeneous equation is a particular solution of the inhomogeneous equation too.

Physically, (16a) describes two different types of disturbances:
(a) Acoustic waves($v_x^{(W)}$) – that is described by general solution of the corresponding homogeneous equation;
(b) Aperiodic vortices ($v_x^{(V)}$) – that is originated from the equation inhomogeneity ($\beta(\tau) \times$ const) and is associated with the particular solution of the inhomogeneous equation. The aperiodic vortical solution is uniquely determined. Hence, it follows that the aperiodic vortical solution and particular solution of the inhomogeneous equation do not uniquely correspond to each other. (The first of them is only associated with the latter one.) The understanding of this fact is important for the comprehension of the phenomenon **(ii)**.

Clearly we have adhered to the physical standpoint of separation of types of disturbances. Thus:

$$v_x = v_x^{(W)} + v_x^{(V)}; \quad v_y = v_y^{(W)} + v_y^{(V)}; \quad D = D^{(W)} + D^{(V)}. \tag{18}$$

In fact the modificated initial value problem is solved at nonmodal approach. And (13a)–(13c) (or equivalently (16a)–(16c)) describes the dynamics of two types of disturbances: acoustic waves and aperiodic vortices. Clearly, the character of dynamics depends on what type of disturbance is input in (13) initially: pure acoustic waves (without admixes of aperiodic vortices) or pure aperiodic vortices (without admixes of acoustic waves). From the aforesaid follows our further steps. Firstly, in Sect. 5, the pure acoustic waves are input in (13) and the dynamics is studied using analytical as well as numerical calculations. Then, in Sect. 6, the pure aperiodic vortices are input in (13) and the dynamics is studied numerically.

2.2 Onto the Value of Shear Rate

Dynamics of shear flows crucially depends on the value of the shear rate R. Therefore we focus our attention on this parameter. The first natural question that arises in this respect is the following: what is an upper limit of the value of the shear parameter R which does not lead to the formation of the hydrodynamic shock waves. The latter may occur when the variation of the flow velocity over a dissipative length scale (characteristic width of the shock-wave front) is of the same order as the sound speed c_s. The dimensional parameter A is the rate of change of the flow velocity per unit length scale. Thus, the condition for the formation of the shock wave is

$$\frac{Al_0}{c_s} \geq 1, \tag{19}$$

where l_0 is the dissipative length scale. According to (12b) we have

$$R = Al_x/c_s = (Al_0/c_s)(l_x/l_0), \tag{20}$$

where $l_x \simeq 1/k_x$ is a characteristic length scale of disturbances along the X axis. From (20) and the inequality $l_x \gg l_0$ it follows that the critical value of the shear parameter in a sense of the shock wave formation is $R \simeq 1$. Therefore, we concentrate on the study of the dynamics of SFH of disturbances at small ($R \ll 1$) and moderate ($0.2 < R \leq 1$) values of the shear rate.

3 The Dynamics of SFH of Acoustic Waves

Let us consider the case when pure acoustic wave SFH is input in the flow initially (i.e. at $\tau = 0$). The different physical variables may be written in the following way:

$$\left\{ \begin{array}{c} v_x(\tau) \\ v_y(\tau) \\ d(\tau) \end{array} \right\} = \left\{ \begin{array}{c} a_{v_x}(\tau) \\ a_{v_y}(\tau) \\ a_d(\tau) \end{array} \right\} \exp \left\{ \begin{array}{c} i\varphi_{v_x}(\tau) \\ i(\varphi_{v_x}(\tau) + \varphi_{v_y}(\tau)) \\ i(\varphi_{v_x}(\tau) + \varphi_d(\tau)) \end{array} \right\}, \tag{21}$$

where $\varphi_{v_x}(\tau)$ it is determined by the following equation:

$$\varphi_{v_x}(\tau) = \int_0^\tau \omega(\tau) d\tau + \varphi_0. \tag{22}$$

Putting (21), (22) in (17a) and (17b) and taking into account that: const = 0 in the considered case and $D \equiv id$, it is not difficult to express all amplitudes and phases represent by $a_{v_x}(\tau)$:

$$a_{v_y}(\tau)\equiv\frac{1}{1+R^2}\sqrt{\left[\beta(\tau)a_{v_x}(\tau)-R\frac{\partial a_{v_x}(\tau)}{\partial\tau}\right]^2+R^2\omega^2(\tau)a_{v_x}^2(\tau)}\ , \quad (23a)$$

$$a_d(\tau)\equiv\frac{1}{1+R^2}\sqrt{a_{v_x}^2(\tau)\omega^2(\tau)+\left[R\beta(\tau)a_{v_x}(\tau)+\frac{\partial a_{v_x}(\tau)}{\partial\tau}\right]^2}\ , \quad (23b)$$

$$\varphi_{v_y}(\tau)\equiv\arccos\left\{\frac{\beta(\tau)a_{v_x}-R(\partial a_{v_x}(\tau)/\partial\tau)}{\sqrt{[\beta(\tau)a_{v_x}-R(\partial a_{v_x}(\tau)/\partial\tau]^2+R^2\omega^2(\tau)a_{v_x}^2}}\right\}, \quad (23c)$$

$$\varphi_d(\tau)\equiv\arcsin\left\{\frac{R\beta(\tau)a_{v_x}+(\partial a_{v_x}(\tau)/\partial\tau)}{\sqrt{\omega^2(\tau)a_{v_x}^2+[(\partial a_{v_x}(\tau)/\partial\tau)+R\beta(\tau)a_{v_x}]^2}}\right\}. \quad (23d)$$

Thus, for the solution of the problem, it is necessary to find $a_{v_x}(\tau)$. This will be done below analytically at low shear rates, and numerically at moderate shear rates.

One important peculiarity of the acoustic wave SFH dynamics is clearly seen even from these general equations: it appears that the phases of v_y and d are displaced relatively to v_x ($\varphi_{v_y}(\tau)\neq\varphi_d(\tau)\neq 0$). This fact suggests that:
– different disturbed physical quantities of acoustic wave SFH (v_x, v_y and d) propagates with different group velocity and no valid unique group velocity of the SFH may be determined;
– The basis of the theory of geometrical acoustic necessitates re-examination inasmuch as this theory is built upon the identity of phases of disturbed physical quantities.

3.1 Velocity Field of SFH of Acoustic Waves

Taking into account the identity $[\mathbf{k}\times(\mathbf{k}\times\mathbf{v})]_z/k_x^2\equiv v_y-\beta(\tau)v_x$, one can rewrite (15) for acoustic waves (const $=0$) in the following form:

$$[\mathbf{k}\times(\mathbf{k}\times\mathbf{v})]_z/k_x^2=RD=\mathrm{i}Rd. \quad (24)$$

This equation reveals an important property of acoustic waves in shear flows: there are no purely potential perturbations in shear flows – acoustic waves acquire a vortical behaviour.

This property is weakly pronounced at $R=0.1$ (see Fig. 1), while at $R=0.4$ case the vortical character of the acoustic waves becomes strongly pronounced (see Fig. 2). Generally, vorticity of the disturbance depends on the angle between the vectors $\mathbf{k}$ and $\mathbf{v}$. At $R\ll1$ right hand side of (24) is small, and, as it is seen from the Fig. 1, $\mathbf{v}$ and $\mathbf{k}$ are almost parallel to each other (vorticity of the SFH is negligible). At $0.2<R\leq1$ the direction of the velocity field of the given SFH varies considerably from point to point, the angles between $\mathbf{v}$ and $\mathbf{k}$ vary and are nonzero in the wide region of space, i.e., velocity field of the SFH gains a vortical character.

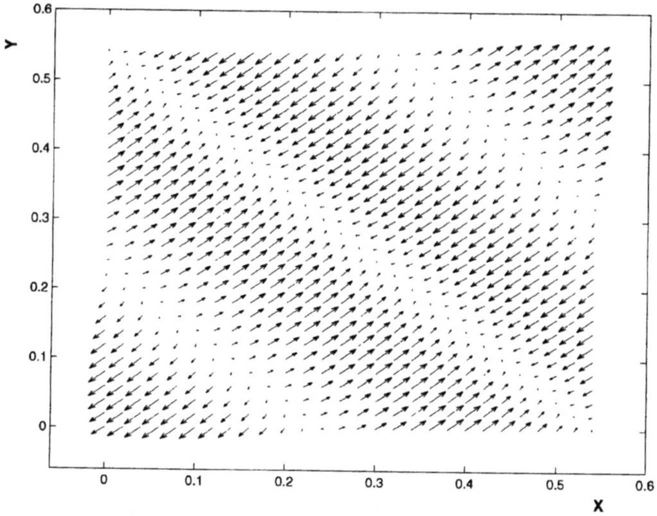

Fig. 1. Velocity field of SFH of acoustic waves at parameters: $R = 0.1$; $\beta(\tau) = 1$. Vortical character of acoustic waves is weakly pronounced.

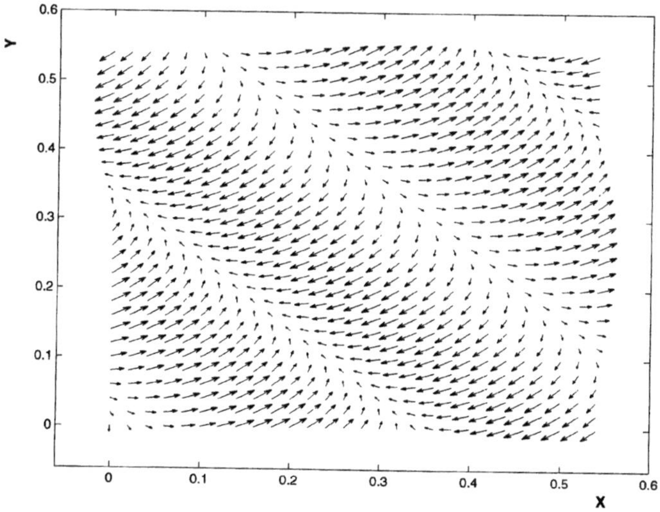

Fig. 2. Velocity field of SFH of acoustic waves at parameters: $R = 0.4$; $\beta(\tau) = 1$. Vortical character of acoustic waves is strongly pronounced.

3.2 The Dynamics of SFH of Acoustic Waves: Low Shear Rates

Let us first analyze the linear evolution of the SFH of acoustic waves in the limit $R \ll 1$. When const $= 0$ (16a) becomes a homogeneous equation that is well-known in mathematical physics. It describes the dynamics of a linear oscillator with a variable frequency $\omega(\tau)$ [27]. The equation is solved approximately when $\omega(\tau)$ depends on τ adiabatically [28]. Mathematically this condition is written simply as $|\partial\omega(\tau)/\partial\tau| \ll \omega^2(\tau)$ or, taking into account the definition of $\omega(\tau)$, as

$$R|\beta(\tau)| \ll |\omega^3(\tau)| = \left[1+\beta^2(\tau)\right]^{3/2}. \tag{25}$$

At $R \ll 1$ the condition (25) holds for all possible values of $\beta(\tau)$. It means that the condition holds during the whole linear stage of the disturbance evolution—during the "drift" of SFH in the $\boldsymbol{k}$-space. If $\beta(\tau) \gg 1$, then the adiabatic character of the frequency variation holds even for moderate values of R ($0.2 < R \leq 1$).

Since not only $\omega(\tau)$, but also $\partial\omega(\tau)/\partial\tau$ varies adiabatically, the same is true for $a_{v_x}(\tau)$ and $\partial a_{v_x}(\tau)/\partial\tau$ [28]. Therefore, there exists an *adiabatic invariant* $a^2_{v_x}(\tau)\omega(\tau) = C$, and the amplitude $a_{v_x}(\tau)$ is directly expressed by the frequency $\omega(\tau)$:

$$a_{v_x}(\tau) = \sqrt{\frac{C}{\omega(\tau)}}. \tag{26}$$

It is clear that the sign of the constant C defines the direction of the acoustic wave propagation. Below, for simplicity the positive values of C and $\omega(\tau)$ will be considered.

Taking into account (26), one can reduce (21-23) to the following expressions:

$$v_x(\tau) \equiv \sqrt{\frac{C}{\omega(\tau)}} \cdot \exp\left[\mathrm{i}\varphi_{v_x}(\tau)\right], \tag{27a}$$

$$v_y(\tau) = \frac{1}{1+R^2}\sqrt{\frac{C}{\omega(\tau)}\left\{\beta^2(\tau)\left(1-\frac{R^2}{2\omega^2}\right)^2 + R^2\omega^2(\tau)\right\}} \times \\ \times \exp[\mathrm{i}(\varphi_{v_x}(\tau)+\varphi_{v_y}(\tau))], \tag{27b}$$

$$d(\tau) = \frac{1}{1+R^2}\sqrt{\frac{C}{\omega(\tau)}\left\{R^2\beta^2(\tau)\left(1+\frac{1}{2\omega^2(\tau)}\right)^2 + \omega^2(\tau)\right\}} \times \\ \times \exp[\mathrm{i}(\varphi_{v_x}(\tau)+\varphi_d(\tau))]. \tag{27c}$$

where

$$\varphi_{v_y}(\tau) = \arccos\left\{\frac{\beta(\tau)(1 - R^2/2\omega^2(\tau))}{\sqrt{\beta^2(\tau)[1 - R^2/2\omega^2(\tau)]^2 + R^2\omega^2(\tau)}}\right\}, \qquad (28a)$$

$$\varphi_d(\tau) = \arcsin\left\{\frac{R\beta(\tau)[1 + 1/2\omega^2(\tau)]}{\sqrt{R^2\beta^2(\tau)[1 + 1/2\omega^2(\tau)]^2 + \omega^2(\tau)}}\right\} \qquad (28b)$$

are the displacements of the phases of v_y and d, respectively, relative to v_x. In the zero shear case ($R = 0$) all physical quantities oscillate with identical phases ($\varphi_{v_y}(\tau) = \varphi_d(\tau) = 0$), while for $R \neq 0$ the phase difference appears. Equations (28a) and (28b) and Figs. 3,4 show that: $\varphi_d(\tau)$ is significant only at moderate shear rate, $\varphi_{v_y}(\tau)$ – at any shear rate. In the course of time, when $\beta(\tau) \to 0$, $\varphi_{v_y}(\tau)$ tends to $\pi/2$. Henceforth, at $\beta(\tau) < 0$ a lag in phase of v_y from v_x increases and at $\tau \to \infty$ tends to $\pi - \arccos(1/\sqrt{1 + R^2})$. The difference of phases is of crucial importance in the vicinity of $\beta(\tau) = 0$, where the derivatives $\partial\varphi_{v_y}(\tau)/\partial\tau$, $\partial\varphi_d(\tau)\partial\tau$ are remarkable (see Figs. 3,4). One of the key aspects of the dynamics of acoustic waves is the process

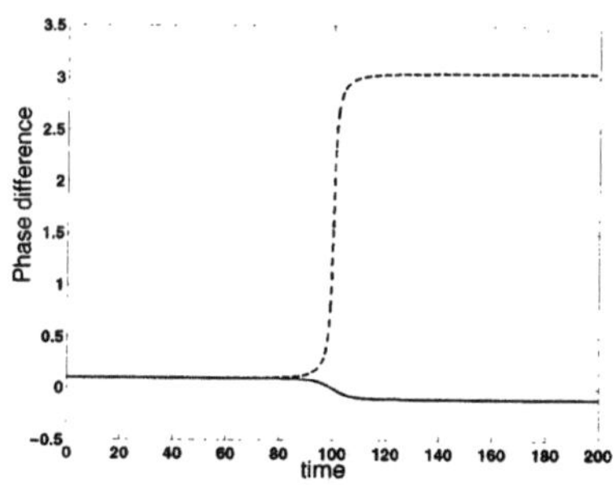

Fig. 3. Plot of phase differences at low shear rates: $R = 0.1$; $\beta_0 = 10$. Dashed line corresponds to $\varphi_{v_y}(\tau)$, solid line corresponds to $\varphi_d(\tau)$.

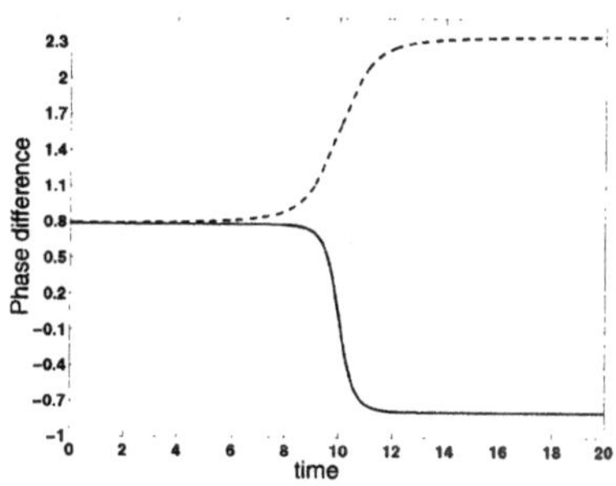

Fig. 4. Plot of phase differences at moderate shear rates: $R = 1$; $\beta_0 = 10$. Dashed line corresponds to $\varphi_{v_y}(\tau)$, solid line corresponds to $\varphi_d(\tau)$.

of energy exchange between the waves and the background sheared flow. The description of this process is quite straightforward at $R \ll 1$ – (14) and (27a)-(27c) lead to the following equations for the energy of SFH and its derivative:

$$E_{\mathbf{k}}(\tau) \approx C\omega(\tau) = C\sqrt{1 + (\beta_0 - R\tau)^2}\,, \qquad (29a)$$

$$E^{(1)} = -C\frac{R\beta(\tau)}{\omega(\tau)}\,. \qquad (29b)$$

According to (29a,b) when $k_y/k_x > 0$ ($\beta_0 > 0$) at $0 < \tau < \tau_* \equiv |\beta_0|/R$ the

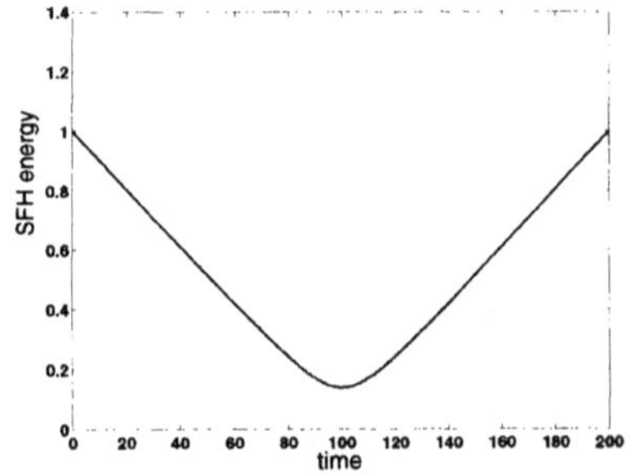

Fig. 5. Temporal evolution of normalized energy of SFH of acoustic waves $(E_{\mathbf{k}}(\tau)/E_{\mathbf{k}}(0))$ at $R = 0.1$, $\beta_0 = 10$, when adiabatic regime is realized.

SFH give energy to the mean flow. The SFH energy decreases and reaches minimum at $\tau = \tau_*$. Afterwards, when the SFH "appear" into the area of $\mathbf{k}$-space where $k_y(\tau)/k_x < 0$ (the "growth area" for the acoustic waves [11], $\tau_* < \tau < \infty$), the energy of SFH begins to increase. Evidently, if the SFH is situated in the "growth area" from the beginning ($\beta_0 < 0$) then its energy increases during the whole process of evolution. One can also see that for $R \ll 1$ the energy exchange process between SFH of acoustic wave and mean flow is symmetrical about the point $\tau = \tau_*$ (see Fig. 5).

3.3 The Dynamics of SFH of Acoustic Waves: Moderate Shear Rates

Let us analyze, now, the evolution of SFH of acoustic waves at $R \simeq 1$. In this case the adiabatic character of the frequency variation still holds if $|\beta(\tau)| \gg 1$. For the SFH with $\beta_0 \gg 1$ the process evolves in the following way: At the initial stage of the evolution it has an adiabatic character. Afterwards, during the certain interval of time the inequality (25) is violated and the frequency of the acoustic wave does not vary adiabatically. Later on, when $|\beta(\tau)|$ attains large enough values the evolution of the SFH becomes again adiabatic. The presence of the nonadiabatic stage in the process of the evolution essentially influences the temporal variability of the SFH energy. In particular, as is clear from Figs. 6,7, the slope of the energy curves after going through the nonadiabatic region depends on values of the initial phase φ_0. Figs. 6,7 represent the cases where maximum distinction between the energy variability processes was observed. The phase difference of the SFH presented on Figs. 6,7 initially is approximately $\pi/2$ initially. This ensures the same difference between the phases of the SFH, when they "enter" the nonadiabatic region. The behaviour of the energy of the SFH in both cases almost coincides on the initial/adiabatic stage of evolution. This was to be expected since in the case when the process has adiabatic character, the energy is proportional to the frequency (see Fig. 5) and does not depend on the phase of the SFH.

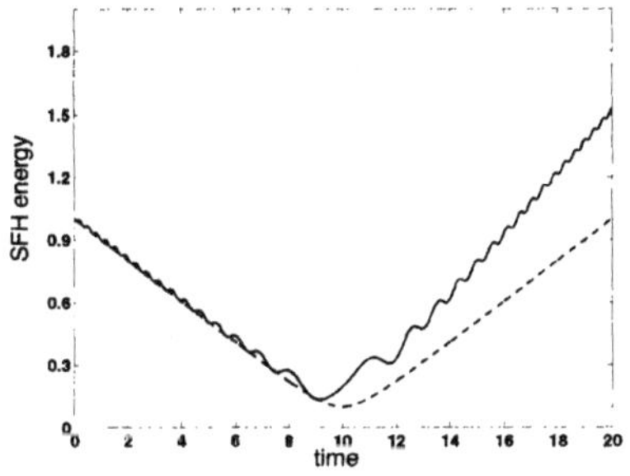

Fig. 6. Temporal evolution of normalized energy of SFH of acoustic waves $(E_{\mathbf{k}}(\tau)/E_{\mathbf{k}}(0))$ at $R = 1$; $\beta_0 = 10$; $\varphi_0 = 55^o$. Solid line corresponds to the real evolution of the energy with nonadiabatic behaviour in the vicinity of $\beta(\tau) = 0$. Dashed line corresponds to the equation $E_{\mathbf{k}}(\tau)/E_{\mathbf{k}}(0) = \omega(\tau)/\omega(0)$ (i.e. when the real/nonadiabatic behaviour in the vicinity of $\beta(\tau) = 0$ is ignored)

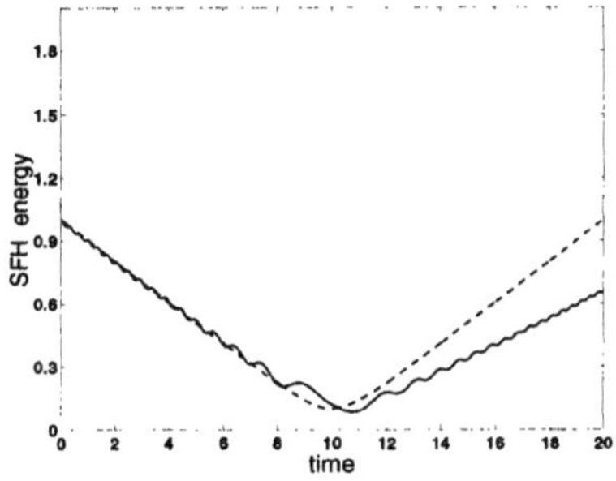

Fig. 7. Temporal evolution of normalized energy of SFH of acoustic waves $(E_{\mathbf{k}}(\tau)/E_{\mathbf{k}}(0))$ at $\beta_0 = 10$; $R = 1$; $\varphi_0 = 145^o$. Solid line corresponds to the real evolution of the energy with nonadiabatic behaviour in the vicinity of $\beta(\tau) = 0$. Dashed line corresponds to the equation $E_{\mathbf{k}}(\tau)/E_{\mathbf{k}}(0) = \omega(\tau)/\omega(0)$ (i.e. when the real/nonadiabatic behaviour in the vicinity of $\beta(\tau) = 0$ is ignored)

The evolution of the energy has an individual character when the SFH "enters" the nonadiabatic region. The actual course of the processes taking place in the nonadiabatic region depends crucially on the phase of the SFH at the "moment of entering" of this region. Hence after "leaving" the nonadiabatic region the energy $E_{\mathbf{k}}(\tau)$ in the first case (Fig. 6) is higher than before "entering" the nonadiabatic region. While in the second case (Fig. 7) just the opposite happens, i.e., the energy is lower when the SFH "leaves" the nonadiabatic region than before "entering" the nonadiabatic region. This difference in energies manifests itself during the subsequent behaviour of the SFH. In both cases, after leaving the nonadiabatic region one can apply (29a) and (29b), but the value of the coefficient C after the leaving is more in the case represented on Fig. 6 than in the case presented on Fig. 7.

It should be noted that the crucial dependence of SFH dynamics on the initial phase was noticed also for other kinds of waves [12]. It is the most probable that this is a characteristic feature of the nonadiabatic processes in shear flows and deserves separate consideration.

4 The Dynamics of the SFH of Vortices

In this Section we present the study of the dynamics of initially excited vortex mode disturbances in the shear flow. Particularly, evolution of individual SFH as well as localized package of vortex disturbances is described.

Linear dynamics of aperiodic vortex disturbances is studied in detail at low shear parameters ($R \ll 1$) in [3,5,6,17–20]. We concentrate the attention on the dynamics at the moderate shear parameters ($0.2 < R \leq 1$).

Numerical calculations of (13a)–(13c) are carried out for initially input vortex SFH (without admixes of acoustic wave $v_x(0) = v_x^{(V)}(0)$) that satisfy the condition $k_y(0)/k_x > 0$. The calculations show absolutely different dynamics of vortices at low and moderate shear rates. Results of calculations at $R = 0.05;\ 0.2;\ 0.4$ are presented on Figs. 8–10. As one can see from the Fig.8,

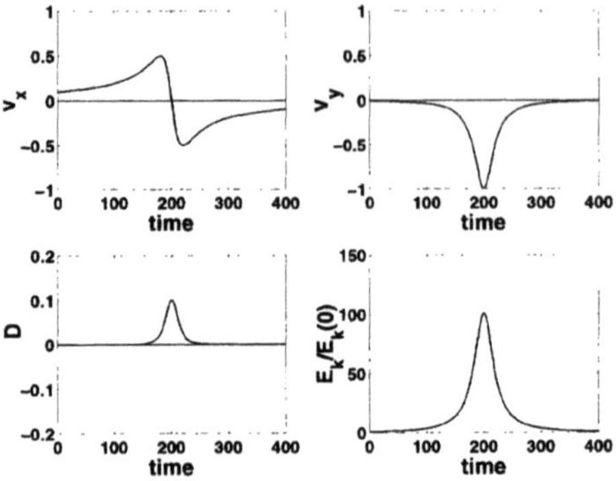

Fig. 8. Evolution of the SFH of disturbed quantities (v_x, v_y and D) and it's normalized energy ($E_{\mathbf{k}}/E_{\mathbf{k}}(0)$) at R=0.05. Initial values of disturbances correspond to the vortex SFH with wave-numbers $k_y(0)/k_x = 8$. The compressibility is negligible and only transient growth of SFH of aperiodic vortices takes place.

at low shear rates compressibility is negligible and only one phenomenon is occurred – transient growth of SFH of aperiodic vortices. It is obvious that the transient growth of vortex disturbances exists in the exponentially stable flow system. It is very important phenomenon that provides extraction of the shear energy and can promote the flow to the transition to turbulence [15,21,24]. But, it is well known phenomenon and was already studied in detail [3,5,6,17–20]. The physics of the transient growth is described in [29].

4.1 Conversion of Vortices to Waves

The calculations show the appearance of new phenomenon – conversion of vortices to waves – at moderate shear rates. The conversion becomes noticeable at $R = 0.2$ (see Fig.9) and even at $R > 0.3$ it is dominant in the shear flows dynamics (see Fig.10). Evolution of the SFH's of disturbed quantities (v_x, $v_x^{(V)}$, $v_x^{(W)}$, v_y, d) and its' normalized energy ($E_{\mathbf{k}}/E_{\mathbf{k}}(0)$) at $R = 0.4$ are demonstrated in Fig. 10. It turns out that linear dynamic of vortex is followed by the new phenomenon – it produces wave (see Fig. 10a–c). As

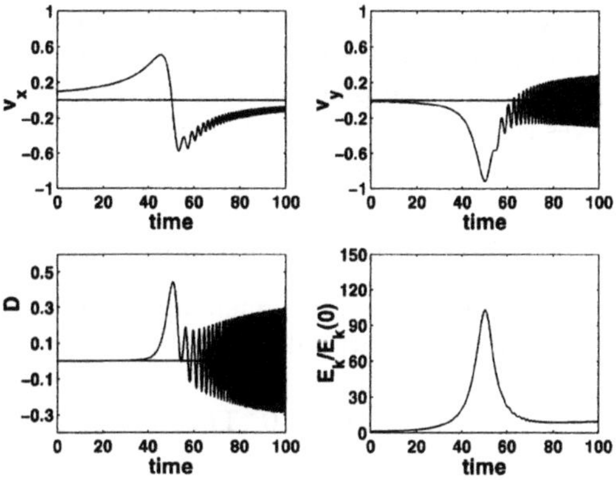

Fig. 9. Evolution of the SFH of disturbed quantities (v_x, v_y and D) and it's normalized energy ($E_{\boldsymbol{k}}/E_{\boldsymbol{k}}(0)$) at R=0.2. Initial values of disturbances correspond to the "pure" vortex SFH with wave-numbers $k_y(0)/k_x = 8$. The conversion of vortices to acoustic waves is noticeable.

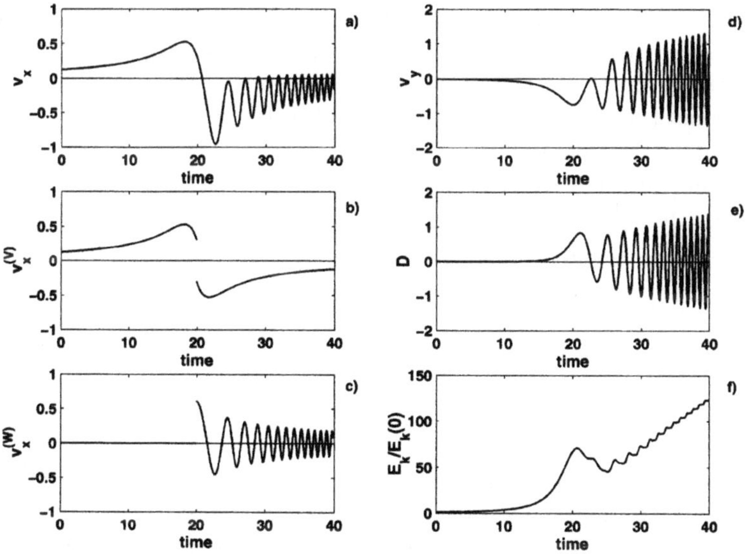

Fig. 10. Evolution of the SFH of disturbed quantities (v_x, $v_x^{(V)}$, $v_x^{(W)}$, v_y and D) and it's normalized energy ($E_{\boldsymbol{k}}/E_{\boldsymbol{k}}(0)$) at R=0.4. Initial values of disturbances correspond to the "pure" vortex SFH (without admixes of acoustic wave $v_x^{(W)} = 0$) with wave-numbers $k_y(0)/k_x = 8$. The SFH of vortex *abruptly* gives rise to the corresponding SFH of acoustic wave at $\tau = \tau^* = 20$.

it is seen from Fig. 10a,d,e,f, evolving in the shear flow, the SFH of vortex ($k_y(0)/k_x = 8$) gaining mean flow energy and amplifying remains aperiodic nature till $k_y(\tau)/k_x > 0$. At the point in time τ^*, at which $\beta(\tau^*) = 0$, the vortex abruptly gives rise to the corresponding SFH of acoustic wave. Abrupt character of the phenomenon is clearly seen from Figs. 10a–c.

4.2 The Nature of the Conversion Phenomenon

More insight into the nature of the phenomenon may be gained from mathematical treatment of (16a). The particular solution of the equation is directly associated with the parameter of inhomogeneity const and tends to the trivial solution ($v_x^{(V)} \to 0$) with const $\to 0$. This solution is odd function of argument $(\tau - \tau^*)$. Thus

$$v_x^{(V)}(\tau^* - 0) = -v_x^{(V)}(\tau^* + 0). \tag{30}$$

In addition, the first term in the left hand side of (16a), corresponding the flow compressibility, results the shift of $v_x^{(V)}$ from the zero value at $\tau \to \tau^* - 0$. This two facts lead to the discontinuity of the aperiodic vortex solution in the point $\tau = \tau^*$ ($v_x^{(V)}(\tau \to \tau^* - 0) \neq v_x^{(V)}(\tau \to \tau^* + 0)$). It is this discontinuity that results emergence of wave. Actually, v_x is a physical quantity and must be continues function of the argument τ. Therefore, initially input pure vortex disturbance ($v_x(0) = v_x^{(V)}(0)$) retains its' identity until the point τ^*. Passing the point $\tau = \tau^*$ from the left side with value $v_x^{(V)}(\tau \to \tau^* - 0)$, this solution cannot be turned in the particular solution of (16a) ($v_x^{(V)}(\tau \to \tau^* + 0)$) continuously. $v_x^{(V)}(\tau \to \tau^* - 0)$ is perceived from the right side as a sum of $v_x^{(V)}(\tau \to \tau^* + 0)$ and some solution of corresponding homogeneous equation:

$$v_x^{(V)}(\tau \to \tau^* - 0) = v_x^{(V)}(\tau \to \tau^* + 0) + v_x^{(W)}(\tau \to \tau^* + 0). \tag{31}$$

Amplitude of arisen acoustic wave SFH at τ^* may be defined from (18) and (30):

$$v_x^{(W)}(\tau \to \tau^* + 0) = 2v_x^{(V)}(\tau \to \tau^* - 0) \tag{32}$$

This is clearly seen from Figs. 10a–c, where evolution of v_x, $v_x^{(V)}$ and $v_x^{(W)}$ is given for initial disturbance satisfying the conditions $v_x(0) = v_x^{(V)}(0)$, $v_x^{(W)}(0) = 0$. Abrupt character of the acoustic wave emergence at $\tau = \tau^*$ is seen from Fig. 10c. Comparison of Figs. 10a–c supports (32).

As it is seen from Fig. 10, the energy of the SFH of vortices increases monotonously. At $\tau = \tau^*$ it abruptly gives rise to the acoustic wave. Thereafter, generated acoustic wave harmonic (satisfying condition $k_y(\tau)/k_x < 0$) gains the shear flow energy and is amplified (see Ref. [11] for details). While the vortex gives up energy back to the mean flow (see Ref. [3,5,6,17–20]). Therefore, after a relatively long time, the shear flow remains only SFH of acoustic wave.

4.3 Onto the Amplitude of the Generated Wave SFH at Conversion

The conversion strongly depends on the value of the shear parameter R. It does not appear at low shear rates ($R < 0.1$), becomes noticeable at $R = 0.2$ and dominates even at $R = 0.3$ in the shear flow dynamics. The cause of the actual absence of the conversion phenomenon at low shear rate is not the existence of a certain threshold, but the fact that the amplitude of the generated wave SFH is largely reduced at $R < 0.1$. According to the numerical calculations, the amplitude of the generated wave SFH at conversion is estimated by the following approximate equation:

$$v_x^{(W)}(\tau \to \tau^* + 0) = \mathrm{const}\frac{1}{1.2R}\exp\left(-\frac{1}{1.2R}\right), \tag{33}$$

where const is defined by (15).

The actual absence of the conversion phenomenon at low shear rates is clearly seen from the following figure plotted according to (33):

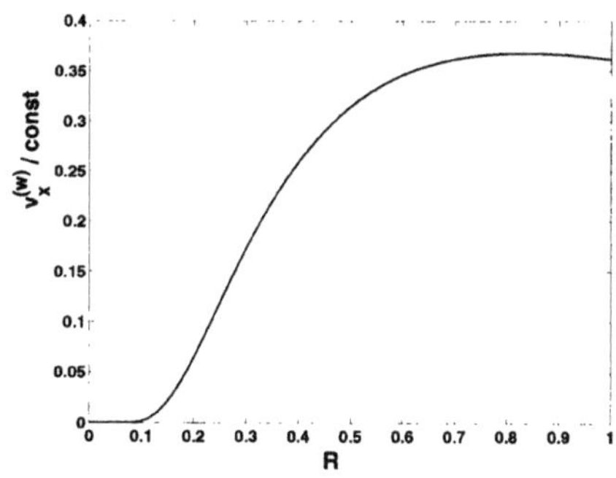

Fig. 11. Dependence of the amplitude of the generated acoustic wave SFH on R. The conversion does not appear at low shear rates ($R < 0.1$), but becomes noticeable at $R = 0.2$.

4.4 Demonstration of the Conversion Phenomenon by Direct Numerical Calculation

In Subsections 4.1-3 the vortex-wave conversion phenomenon is comprehended and described in a sophisticated case of separate SFH by means of the non-modal approach.

Now we demonstrate the conversion phenomena without any analytical approach – by direct numerical calculation of original (5a)-(5c). The conversion phenomenon is essentially linear and takes place at moderate shear rates. Therefore, in the numerical calculations we consider:

– small amplitude disturbances at which the nonlinear terms of (5a)-(5c) are negligible;
– flow with the moderate shear rate.

The initial flow has been composed from the superposition of mean flow with the shear rate $R \equiv A/k_{xc}c_s = 0.5$ and the package of 2D vortex mode disturbance (without acoustic admixes) localized in both, $\boldsymbol{k}$-space and $\boldsymbol{r}$-space. The centers of localization in the $\boldsymbol{k}$-space (k_{xc}, k_{yc}) are in points $(1,2)$ and $(-1,-2)$. Characteristic scales of the localization of the package along X axis (in the $\boldsymbol{r}$-space) and along K_x axis (in the $\boldsymbol{k}$- space) are $D_{\boldsymbol{r}} = 1/D_{\boldsymbol{k}} = 2/k_{xc}$.

Mean flow velocity shear leads to the fact that the vortex mode disturbances are not purely vortical, but involve aperiodic disturbances of density. Area of the localization in the $\boldsymbol{k}$-space of the package of the initial density disturbance is displayed on Fig. 12 by two dark spots. The symmetric location of the package is caused by the reality of the disturbance in the $\boldsymbol{r}$-space. The numerical calculation shows the following dynamics of the package in the $\boldsymbol{k}$-space. The dark spots of Fig. 12 (areas of the localization of the package) drifts parallel to the K_y axis in the course of time: the right spot – down, the left – symmetrically up. Clearly, this dynamics of the spots is caused by the drift of SFH described in the beginning of the lecture. The dynamics of the package in the $\boldsymbol{r}$-space is demonstrated on Fig. 13. Especially, the evolution of the density is shown at times $\tau = k_x c_s t = 0, 2, 4, 5, 6, 8, 10, 12$

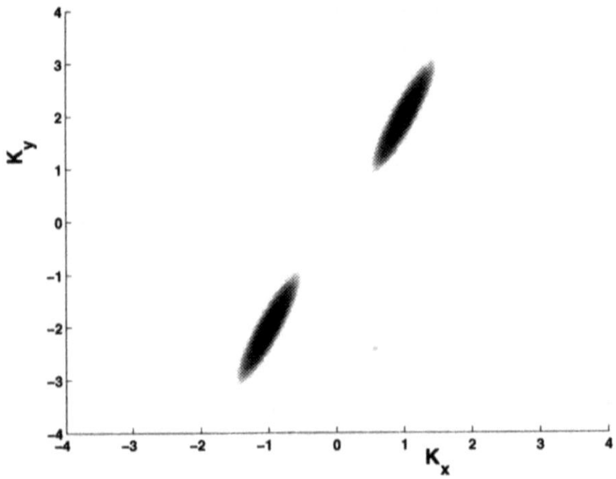

Fig. 12. Initial localized package of the vortex mode disturbance (without acoustic admixes). The centers of localization in the $\boldsymbol{k}$-space (k_{xc}, k_{yc}) are in points $(1,2)$ and $(-1,-2)$. Characteristic scales of the localization of the package along K_x axis is $D_k = k_{xc}/2$. The package initially is prolonged along the line $k_y = (k_{yc}/k_{xc})k_x$.

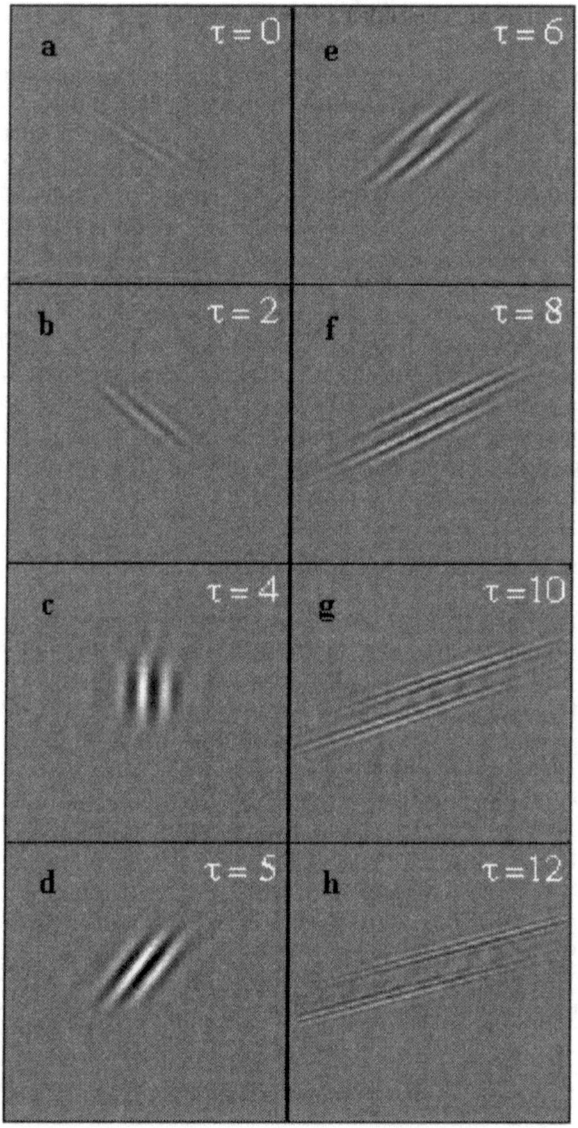

Fig. 13. The evolution of the density in the $\boldsymbol{r}$-space. X and Y axes are directed horizontally and vertically, respectively. The package initially is prolonged along the line $y = -(k_{xc}/k_{yc})x$. See text for detail.

on the a,b,c,d,e,f,g and h graphs, respectively. X and Y axes are directed horizontally and vertically, respectively.

In the beginning (up to $\tau < 4$), vortex disturbances gain the mean flow energy and remain aperiodic in time. Wave crests are turning around due to the effect of the shearing background (see Figs. 13a–c). Therewith, the package remains localized inasmuch as the group velocity of the vortex disturbances is zero. Disturbance dynamics is qualitatively different at $\tau > 4$, when the package (in the $\boldsymbol{k}$-space) crosses and passes the line $k_y = 0$ axis due to the drift. It gives rise to the acoustic wave mode disturbances. Disturbances acquire the acoustic wave features – have non-zero group velocity and are able to transport the energy. As a result the wave crests, in addition to the turning around a fixed point, undergo a translational motion. The widening of the wave-crest becomes apparent at longer times (see Figs. 13.e–h). This explicitly demonstrates the acoustic wave generation phenomenon.

This phenomenon may be also well seen from the package velocity field: the initial package of the mainly vortical (non-potential) disturbances acquire the potential (acoustic) wave one features.

4.5 Onto the Vortex-Wave Interaction

Vortex-wave interaction itself, including a vortex-acoustic interaction, is one of the basic dynamical process in compressible turbulent fluids. This phenomenon is also important for numerous engineering applications, such as acoustic destabilization of laminar flows [30,31], turbulence diagnostic and control [32], prediction of a form of sonic bangs produced by supersonic aircraft [33–35], underwater submarine signalling and detection [36] and so on.

Excitation of waves in the hydrodynamic medium is one of the principal problems in fluid mechanics. The theory of aerodynamic sound production has been introduced in the classical papers by Lighthill [37,38]. Qualitatively new approach to the problem is developed in [25,26] and described in this lecture. The described phenomenon (abrupt emergence of acoustic waves from vortices in flows with velocity shear) is essentially linear and takes place even when the Reynolds' stress is continuous and regular in time and space. Clearly it differs in principle from the Lighthill's stochastic excitation.

5 Mutual Transformation of Waves

Oscillatory systems are prevalent in the nature and play an important role in the wide variety of physical processes. It should be emphasized that these kinds of dynamical systems usually consist of coupled oscillators and have several degrees of freedom. At the same time, parameters of the oscillators in most cases are not constant but vary slowly in time [39]. Interplay of these two key aspects of dynamics has become one of the most important interdisciplinary problems in applied mathematics [27,39,40]. Hydrodynamic and plasma flows constitute such oscillatory systems, where coupling, as a rule, is associated with nonlinear processes [41], which ensures the mutual transformations of different wave modes. In plasma physics, a linear coupling phenomenon is also known: mutual transformation of different kind of plasma waves arising due to a *spatial inhomogeneity of thermodynamic variables* of the medium. For example, existence of the density inhomogeneity induces coupling between MHD waves [42] or the transformation of compressional-type waves into electromagnetic-type waves (and vice-versa) propagating across a density discontinuity [43], etc. There is the other mechanism of the linear mutual transformation of different wave modes arising in flows due to the *kinematic inhomogeneity – velocity shear* [13,14,22]. In this section we describe the physics of this velocity shear induced transformation.

5.1 Mechanical Analogy

The mutual transformation of waves is a resonant phenomenon and its physics can be easily understood by mechanical analogy of the oscillating system. Let us consider two weakly coupled pendulums, lengths of which are slowly (adiabatically) varying in time. The latter causes the temporal variation of the characteristic frequencies of the pendulums $\omega_1 = \omega_1(t)$ and $\omega_2 = \omega_2(t)$. Equations describing the oscillations of such pendulums are the following:

$$\frac{\partial^2 X_1}{\partial t^2} + \omega_1^2(t) X_1 = \chi(t) X_2, \qquad \frac{\partial^2 X_2}{\partial t^2} + \omega_2^2(t) X_2 = \chi(t) X_1. \tag{34}$$

where $\chi(t)$ is the inter-pendulum coupling coefficient, while X_1 and X_2 are oscillating variables of the pendulums. If the frequencies of these pendulums strongly differ from one another, then despite of the coupling, there is virtually no exchange of energy between them. Effective energy exchange starts when the frequencies of the oscillators close to one another. Necessary conditions for the effectiveness of the energy exchange are [27]:

1. existence of the "degeneration region," where $|\omega_1^2(t) - \omega_2^2(t)| \leq |\chi(t)|$,
2. slow traversal of the degeneration region: $|\partial\omega_1(t)/\partial t|, |\partial\omega_2(t)/\partial t| \ll |\chi(t)|$.

If only the first pendulum oscillates initially, then because of the change in its length, the characteristic frequencies $\omega_1(t)$ and $\omega_2(t)$ can come close to one another and the conditions 1. and 2. will be satisfied during a certain limited interval of time. Remarkable (and possible the main) part of the vibrational energy of the first pendulum is transferred to the second pendulum. As a result, strong oscillations of the second pendulum start. In the process, the first pendulum may stop completely.

5.2 Qualitative Description of the Resonance Transformation of Waves

The above scenario may straightforwardly be applied in the analysis of linear coupling and resonant transformation of waves. It is clear that resonant

transformation of waves in the flow is expected if:
- there are two or more wave branches;
- wave frequency varies in time.

There is only one wave branch – acoustic wave – in the hydrodynamic flow considered in the lecture. Second branch of disturbances in the flow is aperiodic vortices with zero frequency. The resonance coupling of disturbances is impossible in this case. Consequently, the presented hydrodynamic equations does not contain/describe the mutual transformation of waves. And, only the qualitative picture of the phenomenon is described below.

Besides the acoustic wave branch, the second wave branch – internal gravity wave – exists in stratified hydrodynamic flow. Therefore, the existence of the resonant mutual transformation of acoustic and internal gravity waves is expected in the stratified hydrodynamic flow (the phenomenon is studied in paper [14]).

As it was shown in paper [13] mutual transformation of waves may be realized also in MHD shear flows, where there exist at least three wave branches: slow magnetosonic, fast magnetosonic and Alfven waves. Let us analyze the process of the transformation on an abstract example when two wave branches does exist in the shear flow system. The relative disposition of dispersion curves of the wave branches should be different depending on the parameters of the system. The dispersion curves are far from each other (as it is shown on Fig.14) or have the pieces situated nearby one another (as it is shown on Fig.15). In the second case, the degeneration region exists and the transformation is expected.

What is the reason of the time variation of wave frequencies in the shear flows, and what is the result of the time variation?

Wave frequency is a certain function (specified by its dispersion) of the wave number. The variation of $k_y(t)$ in time leads to the variation of the fre-

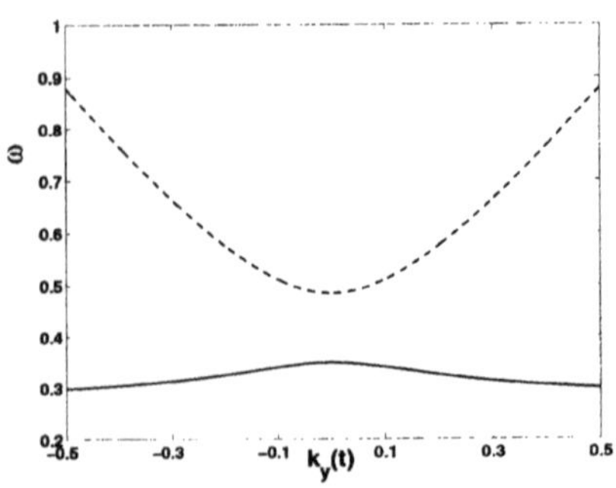

Fig. 14. Abstract example when: two wave branches exist in the flow system; dispersion curves of the wave branches are far from each other.

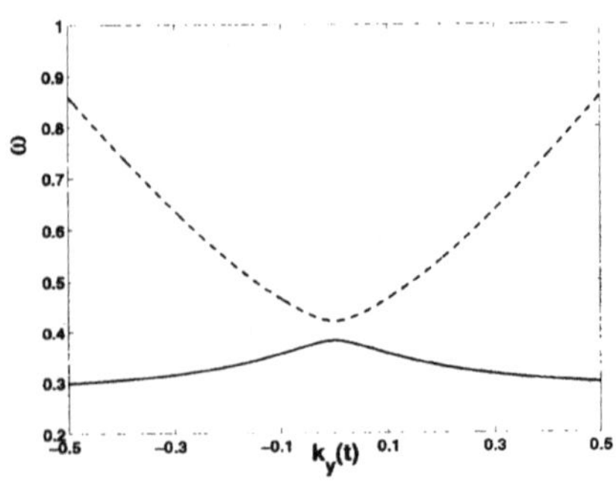

Fig. 15. Abstract example when: two wave branches exist in the flow system; dispersion curves of the wave branches have the pieces nearby one another.

quency of SFH – the wave "slides along" the dispersion curve of the branch. As a result, for certain parameters of the system, the dispersion curves have pieces passing nearby one another and wave frequencies are closely related in a limited time interval. It is this fact which causes the mutual transformation of the waves in the case of a quite slow variation of wave frequencies. Let us illustrate this process by Fig. 16. Suppose the wave harmonic from

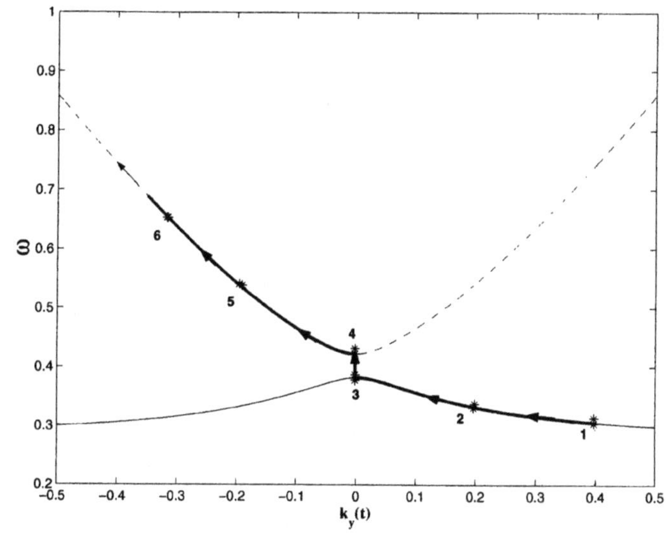

Fig. 16. Schematic representation of the wave transformation. See text for detail.

the lower branch with wave number $k_y(0) = 0.4$ is excited initially – the wave is "situated" in point **1** initially. In due course, with the variation of $k_y(t)$, the wave "slides along" the dispersion curve ($\mathbf{1} \to \mathbf{2} \to \mathbf{3}$) and its frequency is changed. In the vicinity of the point **3** the piece of the dispersion curve of another/upper wave branch is situated, i.e. it exists the degeneration region of the dispersion curves. Consequently, the frequency of the upper branch in point **4** is closely related to the frequency in point **3**. The resonance transformation of wave takes place ($\mathbf{3} \to \mathbf{4}$). Then the transformed wave (i.e. wave from the upper branch) "slides along" the upper dispersion curve ($\mathbf{4} \to \mathbf{5} \to \mathbf{6} \to \ldots$).

Thus, existence of the degeneration regions of dispersion curves initiates the energy exchange between the SFH in the shear flow, while the noticeability of the process is provided by the "slow slide" of the SFH over the dispersion curves. Obviously, the latter condition means a low shear of the flow ($R \ll 1$).

Evidently, this phenomenon should be realized in MHD and fusion plasma devices, where the existence of shear and numerous wave modes ensures the action of the mechanism for a wide range of system parameters. For the same reasons, such processes may also be important in terrestrial and astrophysical compressible hydrodynamic shear flows.

6 Conclusion

In the present lecture we have concentrated on the study of the 2D shear flow inasmuch as the third dimension does not carry in a qualitative novelty.

Below we itemize concluding remarks of this lecture:
– notwithstanding a century history of investigation not all types of shear flows successfully passed analysis of basic linear processes;
– the difficulties are caused by the non self-adjoint character of operators of the modal approach;
– impressive progress in the understanding of shear flow phenomena was enforced by the extensive usage of the nonmodal approach since the 1990s. Especially, it was found the *novel linear* channels of energy exchange in shear flows. According to them:
(i) energy exchange between acoustic waves and mean flow occurs in shear flows;
(ii) conversion of vortices to acoustic waves takes place;
(iii) phenomenon of mutual transformation of waves takes place.

• Phenomenon **(i)**
– Acoustic waves are able to extract the shear flow energy as intensively as vortex disturbances do. However, in the latter case the energy exchange is *effective* only for those SFH that satisfy the condition $|k_y(\tau)/k_x| \lesssim 1$ – for larger values of $|k_y(\tau)/k_x|$ the energy exchange efficiency between mean flow and vortices heavily decreases (see Fig. 8). While in the case of acoustic waves, the intensity of energy exchange with the mean flow is substantial *even* at $|k_y(\tau)/k_x| \gg 1$. Note that (when the shear parameter is positive) the SFH of acoustic waves with $k_y(\tau)/k_x > 0$ give energy to the background flow and weaken, while with $k_y(\tau)/k_x < 0$ gain energy from the flow (see Fig. 5);
– Phase displacements of different physical variables (v_y and d relative to v_x) appear at $R \neq 0$ (see Figs. 3,4);
– different disturbed physical variables of acoustic wave SFH (v_x, v_y and d) propagates with different group velocities (no valid unique group velocity may be defined);
– The basis of the theory of geometrical acoustic necessitates re-examination inasmuch as this theory is built upon the identity of phases of disturbed physical quantities;
– In the flows with low shear rates ($R \ll 1$) the evolution of the SFH of acoustic waves is highly adiabatic, while at moderate shear rates ($R \simeq 1$), the adiabatic character of the SFH evolution is preserved only for $|\beta(\tau)| \gg 1$ (see Figs. 5-7);

– At the nonadiabatic stage of evolution the energy exchange process ($|\beta(\tau)|\leq 1$, $R\simeq 1$) has rather complicated character and strongly depends on the phase of the SFH at the moment of "entering" of the nonadiabatic region (see Figs. 6,7).

• Phenomenon **(ii)**

– the phenomenon of the vortex conversion to waves is non-resonant by nature;
– it takes place even when the Reynolds' stress is continuous and regular in time and space;
– clearly it differs in principle from the Lighthill's stochastic excitation [37,38];
– the emergence of wave SFH from vortex SFH has abrupt character in time (see Fig. 10);
– the amplitude of the generated wave SFH is largely reduced at low shear rates. The conversion phenomena becomes noticeable at $R = 0.2$ and even at $R = 0.3$ dominates in the shear flow dynamics (see (33) and Fig. 11).

• Phenomenon **(iii)**

– the phenomenon of mutual transformation of waves is resonant by nature;
– it takes place if at least two wave modes/branches exists in the shear flow system;
– it takes place at low shear rates ($R \ll 1$) if dispersion curve of wave branches have pieces situated nearby.

Acknowledgments

The author would like to thank the organizing committee of Cargese summer school (where this lecture was given) for the warm hospitality. The author is immensely grateful to Drs. G. R. Khujadze and A. G. Tevzadze for valuable help in the preparing of the text of this lecture. The author is indebted to co-authors of papers that are forming the basis of the lecture. This research was supported by the joint grant INTAS-Georgia97 #0504.

References

1. S. C. Reddy, P. J. Schmid and D. S. Hennigson, SIAM J. Appl. Math. **53**, 15 (1993).
2. T. Kato, *Perturbation Theory for Linear Operators* (Springer-Verlag, New York 1976).
3. S. C. Reddy and D. S. Henningson, J. Fluid Mech. **252**, 209 (1993).
4. Lord Kelvin (W. Thomson), Phil. Mag. **24**, Ser. **5**, 188 (1887).
5. L. H. Gustavsson, J.Fluid Mech. **224**, 241 (1991).
6. K. M. Butler and B. F. Farrell, Phys. Fluids A **4**, 1637 (1992).
7. S. A. Balbus and J. H. Hawley, Astrophys. J. **400**, 610 (1992).
8. W.O. Criminale and P.G. Drazin, Studies in Applied Mathematics **83**, 123 (1990).
9. W.O.Criminale, T.L.Jackson, D.G.Lasseigne and R.D.Joslin, J.Fluid Mech., **339**, 55 (1997).

10. G. D. Chagelishvili, A. D. Rogava and I. N. Segal, Phys. Rev. E **50**, R4283 (1994).
11. G. D. Chagelishvili, G. R. Khujadze, J. G. Lominadze and A. D. Rogava, Phys. Fluids. **9**, 1955 (1997).
12. G. D. Chagelishvili, R. G. Chanishvili, T. S. Khristov and J. G. Lominadze, Phys. Rev. E **47**, 366 (1993).
13. G. D. Chagelishvili, R. G. Chanishvili, J. G. Lominadze and A. G. Tevzadze, Phys. Plasmas **4**, 259 (1997).
14. A. D. Rogava, S. Mahajan, Phys. Rev. E, **55**, 1185 (1997).
15. J. S. Baggett, T. A. Driscoll and L. N. Trefethen, Phys. Fluids **7**, 833 (1995).
16. A. G. Tevzadze, Phys.Plasmas, **5**,1557, (1998).
17. K. Moffatt,'The interaction of turbulence with strong wind shear', In: *Atmospheric Turbulence and Radio wave Propagation*, A. M. Yaglom and V. I. Tatarskii (editors), (Nauka Press, Moscow, 1967).
18. A. D. D. Craik and W. O. Criminale, Pros. R. Soc. Lond. A**406**, 13 (1986).
19. P. Marcus and W. H. Press, J. Fluid Mech. **79**, 525 (1977).
20. J. G. Lominadze, G. D. Chagelishvili and R. G. Chanishvili, Pis'ma Astron. Zh. **14**, 856 (1988) [Sov. Astron. Lett. **14**, 364 (1988)].
21. G. D. Chagelishvili, R. G. Chanishvili, J. G. Lominadze and I. N. Segal,'Turbulization of Free Shear Flows - 3D Perturbations', In: *Proceedings of the fourth International Conference on Plasma Physics and Controlled Nuclear Fusion, Toki, Japan 17-20 November 1992* (ESA SP–351, 1993), p.23.
22. G. D. Chagelishvili, A. D. Rogava and D. G. Tsiklauri, Phys. Rev. E **53**, 6028 (1996).
23. G. D. Chagelishvili and O. G. Chkhetiani, JETP Lett. **62**, 301 (1995).
24. T. Gebhardt and S. Grossmann, Phys. Rev. E **50**, 3705 (1994).
25. G.D. Chagelishvili, A.G. Tevzadze, G. Bodo, S.S. Moiseev, Phys. Rev. Lett., **79**,3178, (1997).
26. G.D. Chagelishvili, A. G. Tevzadze, G. Bodo, P. Rossi, G.T. Gogoberidze, 'Vortex-wave conversion in high shear flows', In: *Advances in Turbulence. VIII, Barcelona, Spain, June 27-30,2000*, ed. by C. Dopazo, (CIMNE, 2000), pp. 737–740.
27. K. Magnus, *Schwingungen* (B. G. Teubner, Stuttgart)(1976).
28. Ya. B. Zel'dovich and A. D. Mishkis *Elementi Prikladnoi Matematiki* Nauka, Moscow, (1972).
29. G. D. Chagelishvili, R. G. Chanishvili and J. G. Lominadze, JETP Lett. **63**, N6 (1996).
30. E. G. Broadbent and D. Moore, Philos. Trans. R. Soc. London, Ser. A **290**, 353 (1979).
31. C. M. Ho and P. Huerre, Annu. Rev. Fluid Mech. **16**, 365 (1984).
32. F. Lound and C. Rojas, Physica D **37**, 508 (1989).
33. S. C. Crow, Fluid Mech. **37**, 529 (1969).
34. J. E. Williams and M. S. Howe, J. Fluid Mech. **58**, 461 (1973).
35. A. D. Pierce, J. Acoust. Soc. Am. **49** 906 (1971).
36. L. E. Kinsler and A. R. Frey, *Fundamentals of Acoustics*, 2nd ed. (Wiley, New York, 1962).
37. M. J. Lighthill, Proc. R. Soc. London Ser. A **211**, 564 (1952).
38. M. J. Lighthill, Proc. R. Soc. London Ser. A **222**, 1 (1954).
39. A.J. Lichtenberg, A.M. Lieberman, *Regular and Chaotic Dynamics*, (Springer-Verlag, New York) (1982).

40. F.C. Hoppensteadt, *Analysis and Simulations of Chaotic Systems*, (Springer-Verlag, New York) (1993).
41. R.C. Davidson, *Methods in Nonlinear Plasma Physics*, (Academic Press, New York) (1972).
42. S.S. Moiseev, V.R.Smilianski, Magn.Gidrodin. **2**, 23 (1965).
43. A.H. Kritz, D. Mintzer, Phys. Rev. **117**, 382 (1960).

Singular Eigenfunctions and an Integral Transform for Shear Flow

Philip J. Morrison

Department of Physics and Institute for Fusion Studies
The University of Texas, Austin, TX 78712, USA

Abstract. Euler's equation linearized about a shear flow equilibrium is solved by means of a novel invertible integral transform that is a generalization of the Hilbert transform. The integral transform provides a means for describing the dynamics of the continuous spectrum that is well-known to occur in this and other systems. The results are interpreted briefly in the context of infinite dimensional Hamiltonian systems theory, which serves as a unifying principle.

1 Introduction

This paper is about the classic problem of inviscid incompressible shear flow in a channel. The main contribution is an integral transform that is of general utility for describing physical systems with a continuous spectrum. The transform was previously developed with collaborators in a series of papers pertaining to the Vlasov–Poisson [14,12,15] and shear flow [1–3] systems. The reader is referred to these works for greater detail.

In Sect. 2 we describe Euler's equations in the channel geometry and discuss the equilibria of interest. The integral transform is applicable to the linear dynamics about these equilibria. We describe the normal mode approach and its associated difficulties because of the critical level singularity. Singular eigenfunctions are constructed. In Sect. 3 we introduce the transform, its inverse, and some identities, which are then used to obtain a solution of our problem. In Sect. 4 we conclude by offering some comments about the underlying Hamiltonian structure of our problem and why it serves as a unifying principle.

2 Shear Flow

The two-dimensional scalar vorticity equation can be obtained from the incompressible Navier–Stokes equations by neglecting viscosity and restricting to two spatial dimensions, say (x, y). This allows the introduction of a streamfunction, ψ, in terms of which the velocity field and the scalar vorticity can be written as $\mathbf{v} := (-\partial\psi/\partial y, \partial\psi/\partial x)$ and $\omega = \nabla^2\psi$, respectively. The resulting equation of motion for the single dynamical variable, ω, can be written as

$$\frac{\partial \omega}{\partial t} + \mathbf{v} \cdot \nabla \omega = 0 . \tag{1}$$

There are several notations in use for the term $\mathbf{v} \cdot \nabla\omega$:

$$\mathbf{v} \cdot \nabla\omega = [\psi, \omega] = J(\psi, \omega) = \frac{\partial\psi}{\partial x}\frac{\partial\omega}{\partial y} - \frac{\partial\psi}{\partial y}\frac{\partial\omega}{\partial x}\,. \tag{2}$$

We will use the notation $[\psi, \omega]$.

For shear flow in a channel, the domain D is an infinitely long channel of finite width, $(x, y) \in \mathbb{R} \times [-1, 1]$. The usual boundary condition of no normal flow at the boundaries located at $y = \pm 1$ will be assumed. This amounts to $\psi(x, y = \pm 1) =$ constant. We state no restrictions on the streamwise flow at $x = \pm\infty$. The only remaining requirement for well-posedness is a suitable initial condition $\omega(x, y, t = 0) = \mathring{\omega}(x, y)$. Global existence of solutions for this system was proven long ago. (See, for example, [6], a highly cited reference in this regard.)

By straightforward formal manipulation it can be shown that (1) conserves the energy functional

$$H[\omega] = \frac{1}{2}\int_D |\nabla\psi|^2\, \mathrm{d}y\mathrm{d}x\,, \tag{3}$$

a quantity that will be seen to be the Hamiltonian functional for the infinite degree-of-freedom Hamiltonian description given in Sect. 4.

2.1 Equilibrium and Linearization

Equilibrium solutions of (1) satisfy

$$[\omega_e, \psi_e] = 0\,, \qquad \nabla^2\psi_e = \omega_e\,. \tag{4}$$

The solutions of (4) corresponding to shear flow in a channel depend only on y, and thus $\omega_e(y) = -U'(y) = \psi_e''(y)$, where prime denotes $\mathrm{d}/\mathrm{d}y$. However, the equilibrium velocity profile $U(y)$ is in general an arbitrary function of y. For convenience we assume it is monotonic for $y \in [-1, 1]$. Most importantly, we assume that the linear dynamics about U supports no instability. This can be assured by Rayleigh's criterion or by more sophisticated means (see [5,2]). We rule out instability and stipulate that the only spectrum of our problem is the continuous spectrum.

Setting $\psi = \psi_e + \delta\psi$ and $\omega = \omega_e + \delta\omega$ and expanding (1) to first order, gives

$$\frac{\partial\delta\omega}{\partial t} + U\frac{\partial\delta\omega}{\partial x} - U''\frac{\partial\delta\psi}{\partial x} = 0\,, \tag{5}$$

with $\nabla^2\delta\psi = \delta\omega$. This linear integro-differential equation is our main object of study. It requires the boundary condition $\delta\psi(x, y = \pm 1) = 0$ and a suitable initial condition. (For mathematical results pertaining to this system see [17,20].)

The linear dynamics of (5) conserves the following functional:

$$H_L[\delta\omega] = \frac{1}{2}\int_D \left[\frac{U}{U''}(\delta\omega)^2 - \delta\omega\delta\psi\right] \mathrm{d}y\mathrm{d}x\,, \tag{6}$$

which is the total amount of energy contained in a perturbation away from the equilibrium state. In Sec. 4 we will see that this functional is the Hamiltonian for the linear dynamics.

For convenience we introduce the Fourier modal representation,

$$\delta\psi = \psi_k(y,t)\exp(ikx)\,, \qquad \delta\omega = \omega_k(y,t)\exp(ikx) \tag{7}$$

where k is the streamwise wavenumber, which could be summed over if desired, but in the sequel we will fix k and consider the problem with only a single mode. Equation (5)becomes

$$\frac{\partial\omega_k}{\partial t} + ikU\omega_k - ikU''\psi_k = 0 \tag{8}$$

with

$$\psi_k(y,t) = \int_{-1}^{1} \mathcal{K}_k(y,y')\,\omega_k(y',t)\,\mathrm{d}y'\,, \tag{9}$$

where

$$\mathcal{K}_k(y,y') = \begin{cases} \sinh[k(y-1)]\sinh[k(y'+1)]/k\sinh[2k] & \text{for } y > y', \\ \sinh[k(y'-1)]\sinh[k(y+1)]/k\sinh[2k] & \text{for } y \le y'. \end{cases} \tag{10}$$

The quantity ψ_k inherits its boundary conditions from $\delta\psi$, namely, $\psi_k(y = \pm 1, t) = 0$.

Because the channel is infinite in x, $k \in \mathbb{R}$. In acoustic as well as in quantum mechanics there is a continuous spectrum associated with the domain being infinite in extent. We emphasize that this is not the continuous spectrum that causes the major difficulty in shear flow and other problems of continuous media. This other continuous spectrum arises from a singularity in the equation, a singularity with a location dependent upon the eigenvalue. We discuss this difficulty below in Sec. 2.2, where k will be assumed fixed.

2.2 Singular Eigenmodes and the Continuous Spectrum

Now we search for eigenvalues of (8) in the usual way. More precisely, we are interested in the spectrum of a linear operator $\mathcal{L}_k$, which is defined by rewriting (8) as follows:

$$\frac{\partial\omega_k}{\partial t} = \mathcal{L}_k\,\omega_k\,. \tag{11}$$

Assuming

$$\omega_k(y,t) = \hat{\omega}_k(y,c)\exp(-ikct)\,, \qquad \psi_k(y,t) = \hat{\psi}_k(y,c)\exp(-ikct)\,, \tag{12}$$

we seek the eigenfunction $\hat{\omega}_k(y, c)$ associated with an eigenvalue that we choose to write in terms of the phase velocity c. With (12), (8) and (9) become

$$(U - c)\,\hat{\omega}_k - U''\hat{\psi}_k = 0\,, \tag{13}$$

and

$$\hat{\psi}_k(y, c) = \int_{-1}^{1} \mathcal{K}_k(y, y')\,\hat{\omega}_k(y', c)\,\mathrm{d}y'\,, \tag{14}$$

respectively. Equation (13) is essentially Rayleigh's equation, which is usually written in terms of $\hat{\psi}_k$ as follows:

$$(U - c)\,(\hat{\psi}_k'' - k^2\hat{\psi}_k) = U''\hat{\psi}_k\,. \tag{15}$$

The eigenvalue problem associated with (15) has a feature that makes it different from eigenvalue problems encountered in quantum mechanics or other fluid mechanics problems that give rise to dispersion relations. The feature is the presence of the singularity that occurs where the phase velocity matches the equilibrium velocity, i.e. at values of y such that $U = c$. This singularity differs from that of Legendre's equation or the equations for other special functions because the singularity depends upon the eigenvalue c. In fact there is a continuous correspondence between values of c and points in the channel y_c (where $U(y_c) = c$), and this is the origin of the continuous spectrum.

This singularity, which in fluid mechanics is called a critical layer or critical level, has a long history dating to Rayleigh [16], and has since been considered by many researchers. One approach is to give up the search for eigenmodes and to use the Laplace transform to solve the initial value problem [4,7,17]. Another is to change the physics by adding some form of viscosity or nonlinearity. We will stick to the eigenvalue problem and solve it by following a procedure introduced by Van Kampen in the context of plasma physics [22].

The difficulty arises in part because the eigenfunctions associated with singular problems like this one do not exist in a usual Hilbert space, i.e. are not square integrable. In fact one must resort to the theory of distributions or generalized functions (see e.g. [8]) in order to obtain the eigenfunctions. In the calculus of generalized functions, an equation of the form $xf(x) = 0$ has the solution $f(x) = C\delta(x)$, where C is a constant and $\delta(x)$ is Dirac's delta function. Van Kampen's procedure is to employ this device to obtain a homogeneous solution to (13) that is added to a particular solution that incorporates the Cauchy principal value. Applying his procedure to the present problem gives the solution

$$\hat{\omega}_k = \mathcal{P}\left(\frac{U''\hat{\psi}_k}{U - c}\right) + \mathcal{C}_k\delta(U - c)\,, \tag{16}$$

where $\mathcal{P}$ signifies the Cauchy principal value and $\mathcal{C}_k$ is (as yet) an arbitrary constant.

By writing the vorticity eigenfunction in the form of (16), we explicitly display the singular nature of the mode. Moreover, we see that the eigenvalue c does not depend upon the wave number k, i.e. there is no dispersion relation, and that any value of c in the range of U is acceptable. This suggests the continuous spectrum. (A proof of the continuous spectrum requires analyzing the complement of the point spectrum of $\mathcal{L}_k$ in a Banach space setting.)

Although (16) gives a general form for the eigenfunction, it remains indeterminant for two reasons: because we do not have a value for the constant $\mathcal{C}_k$ and even more so because the function $\hat{\psi}_k$ is unknown. The latter problem is addressed by inserting $\hat{\omega}_k$ as given by (16) into (14). This yields the following inhomogeneous integral equation:

$$\hat{\psi}_k(y, y_c) = \mathcal{P}\int_{-1}^{1} \mathcal{K}_k(y, y')\frac{U''(y')\hat{\psi}_k(y', y_c)}{U(y') - c}\, dy' + \frac{\mathcal{C}_k}{|U_c'|}\mathcal{K}_k(y, y_c)\,, \tag{17}$$

where for convenience we have chosen to write $\hat{\psi}_k$ with the argument y_c instead of c (recall we have assumed U is monotonic so $c = U(y_c)$ establishes a unique correspondence) and $U_c' := U'(y_c)$. The kernel of this integral equation is singular at the critical layer, which adds to the difficulty, but we have yet to our second point of indeterminancy, the specification of $\mathcal{C}_k$

Because the problem is linear we are free to scale the amplitude of the eigenfunction, and the value of the constant $\mathcal{C}_k$ can be set by a normalization procedure. A convenient one is (cf. [9,10,18])

$$\Xi_k = \int_{-1}^{1} \hat{\omega}_k \, dy\,, \tag{18}$$

where the modal amplitude Ξ_k can be taken to be any desired function of the eigenvalue c (or y_c). Inserting the (16) into (18) gives

$$\Xi_k = \frac{\mathcal{C}_k}{|U_c'|} + \mathcal{P}\int_{-1}^{1} \frac{U''(y')\hat{\psi}_k(y', y_c)}{U(y') - U(y_c)}\, dy'\,. \tag{19}$$

Solving (19) for $\mathcal{C}_k$ and inserting the result into (17) gives finally

$$\hat{\psi}_k(y, y_c) = f_k(y) + \int_{-1}^{1} \mathcal{F}_k(y, y'; y_c)\, \hat{\psi}_k(y', y_c)\, dy'\,, \tag{20}$$

where the inhomogeneous term $f_k(y) := \Xi_k \mathcal{K}_k(y, y_c)$ and the kernel is defined by

$$\mathcal{F}_k(y, y'; y_c) = \left[\frac{\mathcal{K}_k(y, y') - \mathcal{K}_k(y, y_c)}{U(y') - U(y_c)}\right] U''(y')\,. \tag{21}$$

This kernel is regular at the critical level, and thus the choice of normalization (18) has regularized the singular integral equation (17). Note that the quantity y_c appears only as a parameter.

The procedure above resulted in an eigenfunction, but unfortunately one that cannot be written down explicitly because it is given in terms of the solution to the (albeit regular) integral equation (20). The good news is this only needs to be done once for a given profile $U(y)$ in order to obtain the eigenfunctions. The solutions $\hat{\psi}_k(y, y_c)$ should be viewed as a special function associated to a given profile. Once $\hat{\psi}_k(y, y_c)$ is known a general solution to the initial value problem can be obtained by summation. In the next section we view this in a more general context by using the singular eigenfunctions to construct an integral transform that can in turn be used to effect a coordinate change that dramatically simplifies the problem.

3 The Integral Transform–Coordinate Change

Now we turn to the description of our integral transform solution. Rather than constructing solutions by multiplying the singular eigenfunction $\hat{\omega}_k$ by an amplitude that is selected according to the initial condition, multiplying the result by e^{-ikct}, and finally integrating over c (or y_c), we construct an integral transform that involves only the y-dependence and makes no assumptions about time. This approach is Hamiltonian in spirit in that one solves the problem by transforming to coordinates that make the problem simple. This procedure is, of course, what is usually done with conventional integral transforms and we will view this problem in that light.

3.1 Integral Transform Pair

A superposition of the singular eigenfuntions (16), each of which is labeled by y_c, with amplitudes given by $\Lambda_k(y_c, t)$ suggests the following transform:

$$G_k[\Lambda_k](y,t) := \int_{-1}^{1} \mathcal{G}_k(y, y_c)\Lambda_k(y_c, t)\, \mathrm{d}y_c \tag{22}$$

where the transform kernal is given by

$$\mathcal{G}_k(y, y_c) := A_k(y)\delta(y - y_c) + \mathcal{P}\frac{U''(y)\psi_k(y, y_c)}{U(y) - U(y_c)}\,, \tag{23}$$

with

$$A_k(y) := 1 - \mathcal{P}\int_{-1}^{1} \frac{U''(y')\psi_k(y', y)}{U(y') - U(y)}\, \mathrm{d}y'\,, \tag{24}$$

and $\psi_k(y, y_c)$ a solution to (20). The transform of (22) takes the function $\Lambda_k(y_c, t)$ into a function $\omega_k(y, t) := G_k[\Lambda_k](y, t)$.

Alternatively, we can rewrite (22) by inserting (23) into (22) to obtain the following equivalent form:

$$G_k[\Lambda_k](y,t) := A_k(y)\Lambda_k(y_c, t) + \mathcal{P}\int_{-1}^{1} \frac{U''(y)\psi_k(y, y_c)}{U(y) - U(y_c)}\Lambda_k(y_c, t)\, \mathrm{d}y_c\,, \tag{25}$$

whence we see that G_k is the sum of a piece that merely multiplies the function to be transformed by $A_k(y)$ with a piece that is a generalization of the Hilbert transform (see e.g. [21,19]).

The inverse of (22), subject to the conditions we have placed on U (monotonicity and no discrete spectrum), is given by the following:

$$\widehat{G}_k[\omega_k](y_c,t) := \int_{-1}^{1} \widehat{\mathcal{G}}_k(y_c,y)\omega_k(y,t)\,\mathrm{d}y\,, \tag{26}$$

where

$$\widehat{\mathcal{G}}_k(y_c,y) = \frac{1}{A_k^2(y_c)+B_k^2(y_c)}\left[A_k(y_c)\delta(y-y_c) + \mathcal{P}\frac{U''(y_c)\psi_k(y,y_c)}{U(y)-U(y_c)}\right], \tag{27}$$

and where

$$B_k(y) := -\pi\psi_k(y_c,y_c)\frac{U''(y)}{U'(y)}\,. \tag{28}$$

Note that on the right hand sides of (22) and (26), the functions $\Lambda_k(y_c,t)$ and $\omega_k(y,t)$ are functions of y_c and y, respectively, and the explicit time dependence appears only as a parameter. We include this time dependence to emphasize that the integral transforms are coordinate changes; the actual time dependence will be determined by the equation of motion.

If $\widehat{G}_k$ is to be the inverse of the transform G_k, then $\widehat{G}_k[G_k[\Lambda_k]] \equiv \Lambda_k$, which follows if

$$\int_{-1}^{1} \widehat{\mathcal{G}}_k(y_c',y)\mathcal{G}_k(y,y_c)\,\mathrm{d}y = \delta(y_c - y_c')\,. \tag{29}$$

This inverse relation can also be viewed as a relation that verifies the completeness of the continuous spectrum. Similarly, there exists a reciprocal completeness relation,

$$\int_{-1}^{1} \widehat{\mathcal{G}}_k(y_c,y')\mathcal{G}_k(y,y_c)\,\mathrm{d}y_c = \delta(y-y')\,. \tag{30}$$

Just as the Fourier and other integral transforms posses useful identities, there are a variety of identities possessed by G_k and $\widehat{G}_k$. We record two such identities that will be used below:

$$\widehat{G}_k[U\omega_k](y_c,t) = U(y_c)\widehat{G}_k[\omega_k](y_c,t) + \frac{U''(y_c)}{A_k^2(y_c)+B_k^2(y_c)}\mathcal{P}\int_{-1}^{1}\omega_k(y,t)\psi_k(y,y_c)\,\mathrm{d}y\,, \tag{31}$$

and

$$\widehat{G}_k[U''\psi_k](y_c,t) = \frac{U''(y_c)}{A_k^2(y_c)+B_k^2(y_c)}\mathcal{P}\int_{-1}^{1}\omega_k(y,t)\psi_k(y,y_c)\,\mathrm{d}y\,, \tag{32}$$

where (32) is valid if ω_k is related to ψ_k according to (14).

The above and other related transform formulas are verified in [1,3] and their counterparts in the Vlasov-Poisson context are proved rigorously in [12].

3.2 Solution

Now we are in a position to use the transform pair to solve (8). Let $\Lambda_k(y,t) := \widehat{G}_k[\omega_k]$ and then $\omega_k(y,t) = G_k[\Lambda_k]$. Transforming (8) with $\widehat{G}_k$ gives

$$\frac{\partial \Lambda_k}{\partial t} + ikc\Lambda_k = 0\,, \tag{33}$$

where use has been made of both (31) and (32), with the obvious cancellation, and we have set $c = U(y_c)$. This simple form is obtained because the transform has been designed especially for (8); it is a coordinate change in which the equation becomes trivial to solve. The solution to (33) is clearly $\Lambda_k = \mathring{\Lambda}_k e^{-ikct}$, which upon transforming back gives

$$\omega_k(y,t) = G\left[\widehat{G}[\mathring{\omega}_k]e^{-ikct}\right]\,. \tag{34}$$

Here we have written $\mathring{\Lambda}_k$ in terms of $\mathring{\omega}_k(y) := \omega_k(y, t = 0)$ by means of $\mathring{\Lambda}_k = \widehat{G}[\mathring{\omega}_k]$.

The solution (34) is exact, but not entirely explicit because it requires $\psi_k(y, y_c)$, which in turn requires the solution of the integral equation (20). This is a fairly easy numerical exercise because (20) is a regular Fredholm integral equation. We reiterate that the fact that this calculation only needs to be done once for each flow profile, $U(y)$. After $\psi_k(y, y_c)$ is obtained, (34) gives the solution for all initial conditions.

4 Hamiltonian Interpretation

It has been know for a long time that the equations for continuous media without dissipation effects are an infinite dimensional Hamiltonian system (see e.g. [11] for review). An advantage of this Hamiltonian structure is that systems that possess it have phenomena in common. Thus when approaching a new problem one can appeal to a large body of lore. For example, it is known that all stable finite degree-of-freedom Hamiltonian systems can be reduced, by a coordinate change, to a system of uncoupled oscillators. That is, the Hamiltonian can be transformed into the following:

$$H(p,q) = \sum_k \frac{\nu_k}{2}\left(q_k^2 + p_k^2\right) = \sum_k \nu_k\, J_k\,, \tag{35}$$

where the second equality corresponds to action-angle variables. The integral transform of Sec. 3 is in fact the essence of a canonical transformation that does this for the infinite dimensional fluid system, which because of the continuous spectrum takes the form

$$H_L = \int\!\!\int \nu_k(y_c) J_k(y_c, t)\, \mathrm{d}y_c \mathrm{d}k\,. \tag{36}$$

In this concluding section, we briefly describe how this transpires, and refer the reader to [3] for details.

There are various ways of describing the Hamiltonian structure of fluids; the one we adopt is that of [13] in term of the noncanonical Poisson bracket. In the present context, the upshot of this is that (1) can be written as follows:

$$\frac{\partial \omega}{\partial t} = \{\omega, H\} = [\omega, \psi] \,, \tag{37}$$

where the noncanonical Poisson bracket is given by

$$\{A, B\} = \int_D \omega \left[\frac{\delta A}{\delta \omega}, \frac{\delta B}{\delta \omega} \right] \mathrm{d}y\mathrm{d}x \,, \tag{38}$$

and the Hamiltonian functional H is given by (3). (Note $\delta A/\delta\omega$ is the functional or variational derivative.)

The linear dynamics obtains its Hamiltonian structure from that of the full dynamics given in (37) by linearizing. This procedure involves linearizing the bracket as well as the Hamiltonian. Upon inserting $\omega = \omega_e + \delta\omega$ into the bracket and retaining the lowest order piece, we obtain

$$\{A, B\}_L = \int_D \omega_e \left[\frac{\delta A}{\delta\delta\omega}, \frac{\delta B}{\delta\delta\omega} \right] \mathrm{d}y\mathrm{d}x \,, \tag{39}$$

where $\delta A/\delta\delta\omega$ is the functional derivative with respect to the linear dynamical variable, $\delta\omega$. The linear dynamics of (5) can be written as

$$\frac{\partial \delta\omega}{\partial t} = \{\delta\omega, H_L\}_L \,. \tag{40}$$

where H_L, the Hamiltonian for the linear dynamics, is given by (6).

Our desire is to transform (6) into (36), but to do this requires first writing the Hamiltonian description (40) in terms of canonically conjugate variables. Once this is achieved the transformation to action-angle variables can be effected. Performing these calculations is beyond the scope of this paper, but we mention in closing that this can be done by constructing a mixed variable generating functional that incorporates the integral transform of Sec. 3. The details of this along with other Hamiltonian insights are given in [3].

References

1. N.J. Balmforth and P.J. Morrison: 'Singular eigenfunctions for shearing fluids I'. *Institute for Fusion Studies, University of Texas, Austin,* Report No. **692** (1995). Available as NTIS document DE95008952INZ. See http://www.ntis.gov.
2. N.J. Balmforth and P.J. Morrison: Stud. Appl. Math. **102**, 309 (1999)

3. N.J. Balmforth and P.J. Morrison: 'Hamiltonian Description of Shear Flow'. In *Large-Scale Atmosphere-Ocean Dynamics I + II.* ed. by J. Norbury and I. Roulstone (Cambridge, Cambridge 2002) pp. 117–142
4. K.M. Case: Phys. Fluids **3**, 143 (1960)
5. P.G. Drazin and L.N. Howard: Adv. in Appl. Mech. **9**, 1 (1966)
6. D.G. Ebin and J.E. Marsden: Ann. Math. **92**, 102 (1970)
7. L. Engevik: 'On the stability problem in hydrodynamics'. Dept. Applied Math. Bergen, Norway, Pub. No. **11** (1966).
8. I.M. Gel'fand: *Generalized Functions* (Academic, New York 1964)
9. L.P.J. Kamp: J. Phys. A: Math. Gen. **24**, 2029 (1991)
10. K.L. Kowalski and D. Feldman: J. Math. Phys. **2**, 499 (1961)
11. P.J. Morrison: Rev. Mod. Phys. **70**, 467 (1998)
12. P.J. Morrison: Trans. Theory Stat. Phys. **29**, 397 (2000)
13. P.J. Morrison and J.M. Greene: Phys. Rev. Lett. **45**, 790 (1980)
14. P.J. Morrison and D. Pfirsch: Phys. Fluids B **4**, 303 (1992)
15. P.J. Morrison and B. Shadwick: Acta Phys. Pol. **85**, 759 (1994)
16. J.W.S. Rayleigh: *Theory of Sound* (Macmillan, London 1896) Art. 369
17. S.I. Rosencrans and D.H. Sattinger: J. Math. Phys. **45**, 289 (1966)
18. Z. Sedláček: J. Plasma Phys. **6**, 187 (1971)
19. E.M. Stein and G. Weiss: *Introduction to Fourier Analysis on Euclidean Spaces* (Princeton, New Jersey 1971)
20. S.A. Stepin: Izvestiya: Math. **60**, 1293 (1996)
21. E.C. Titchmarsh: *Introduction to the Theory of Fourier Integrals* (Clarendon, Oxford 1937)
22. N.G. Van Kampen: Physica **21**, 949 (1955)

Structural Acoustics with Mean Flow

Nigel Peake

DAMTP, University of Cambridge,
Silver St., Cambridge CB3 9EW U.K.

Abstract. The basic problem to be considered concerns the response of a fluid-loaded elastic plate which is driven by a harmonic point forcing in the presence of mean flow in the fluid. In order to enforce causality a rather sophisticated technique needs to be applied, and some surprising results are found, including absolute instability and the existence of modes which contradict the usual radiation condition of outgoing group velocity. Also, it has been known for some time that this sort of system supports waves with negative activation energy. All of these effects are absent in the problem of a fluid-loaded plate in still fluid. In these notes the details of these intriguing phenomena will be described, with particular emphasis on the technique for determining causal solutions of initial value problems, and on the energetics of the problem. Extensions to include plate curvature, steady shear in the flow, and nonlinear effects will be presented.

1 Basic Problem

The problem which will we consider in these notes is an apparently very simple one. This analysis has been carried out by Brazier-Smith & Scott [3], and perhaps in more depth by Crighton & Oswell [8], and full details can be found there. Consider a thin elastic plate of thickness h (see figure 1), aligned in its undisturbed state with the x axis. The y axis is perpendicular to the plate and the z axis is in the spanwise direction (the spanwise dependence will be suppressed in what follows, reducing the system to two dimensions). In $y < 0$ below the plate we assume there is a vacuum, but in $y > 0$ we have a fluid which has a uniform flow speed U parallel to the plate. The plate is driven into unsteady motion by a line force at $x = 0$, which exerts a force $F(t)\delta(x)$ on the plate. Assuming that all motions are small amplitude, the plate deflection can then be described purely by a transverse deflection $\eta(x,t)$, which satisfies that standard equation of motion of a thin elastic plate, i.e.

$$m\frac{\partial^2\eta}{\partial t^2} + B\frac{\partial^4\eta}{\partial x^4} = F\delta(x) - p(x,0,t). \tag{1}$$

Here m is the mass per unit area of the plate, $B = Eh^3/12(1-\nu^2)$ is the bending stiffness, with E the Young's modulus and ν Poisson's ratio. The term $-p(x,0,t)$ on the right of (1) is the so-called 'fluid loading', and of course corresponds to the hydrodynamics pressure in the fluid. Note that in this linearised approximation we evaluate this pressure on the undisturbed plate location $y = 0$. The motion in the fluid can be described with varying degrees of sophistication, but here we suppose that it is inviscid and irrotational, so that any perturbation to the uniform stream is described by a velocity potential $\phi(x, y, t)$, satisfying Laplace's equation

$$\nabla^2 \phi = 0. \tag{2}$$

The hydrodynamic pressure can then be obtained via the linearised Bernoulli equation

$$p = -\rho_0 \left(\frac{\partial \phi}{\partial t} + \frac{\partial \phi}{\partial x} \right), \tag{3}$$

where ρ_0 is the density of the fluid. Finally, we need to provide an interface condition between the fluid and the plate, and this boundary condition simply corresponds to continuity of normal velocity, i.e.

$$\frac{\partial \phi}{\partial y} = \frac{\partial \eta}{\partial t} + U \frac{\partial \eta}{\partial x}. \tag{4}$$

The aim here is to study causal solutions of the above system. Specifically, we will suppose that the system is at first undisturbed and that the forcing is turned on at $t = 0$ (so that $F(t) = 0$ for $t < 0$), and then aim to analyse the time evolution and in particular the long-time behaviour. When this is done we will find some exceedingly surprising behaviour, including: absolute instability at sufficiently high flow speeds; the existence of negative energy waves, which are destabilised by dissipation; and neutral modes which do not satisfy the usual radiation condition of out-going group velocity at infinity.

Before analysing our equations in detail, some comment on the magnitude of the fluid loading is in order. Consider first the plate in a vacuum (so the right hand side of equation 1 becomes zero), then clearly the plate supports simple 'bending waves'. If these bending waves have frequency ω then their corresponding wavenumber is

$$\left(\frac{m\omega^2}{B} \right)^{\frac{1}{4}}. \tag{5}$$

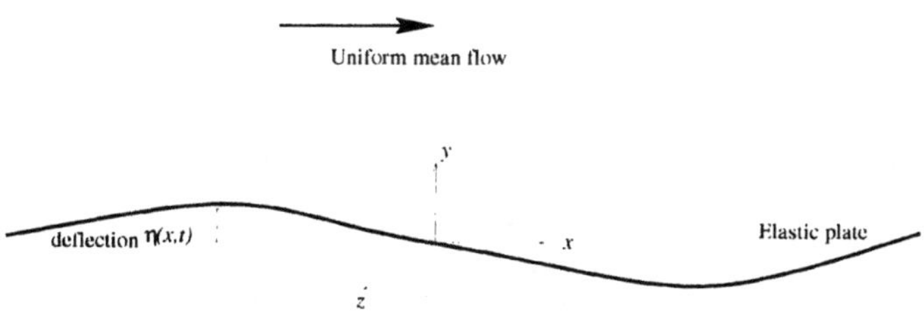

Fig. 1. Diagram of fluid-loaded plate with mean flow in $y > 0$.

The strength of the fluid loading can then be estimated by comparing the mass of fluid within one bending wavelength of the plate with the mass of the plate, all per unit length and per unit span. The ratio of the mass of the fluid to the mass of the plate is then proportional to

$$\frac{\rho_0 B^{\frac{1}{4}}}{m^{\frac{5}{4}} \omega^{\frac{1}{2}}}. \tag{6}$$

If we now choose units in which $m = B = 1$ (by scaling lengths by m/ρ_0 and time by $m^{\frac{3}{2}}/(\rho_0 B^{\frac{1}{2}})$), it then follows that this ratio is just $1/\omega^{\frac{1}{2}}$, so that for fluid loading to be important we will need ω to be small. Of course, the mean flow we are trying to study will inevitably only have an important effect if the fluid loading is heavy, and therefore we will be considering small values of nondimensional frequency ω in everything that follows. Note also that the dimensional steady flow speed is $U\sqrt{B/m}$, where U is now the nondimensional flow speed.

2 Solution Technique

The procedure for determining the causal solution of initial-value problems was set out authoritatively by Briggs [4] and Bers [2]. By taking Fourier

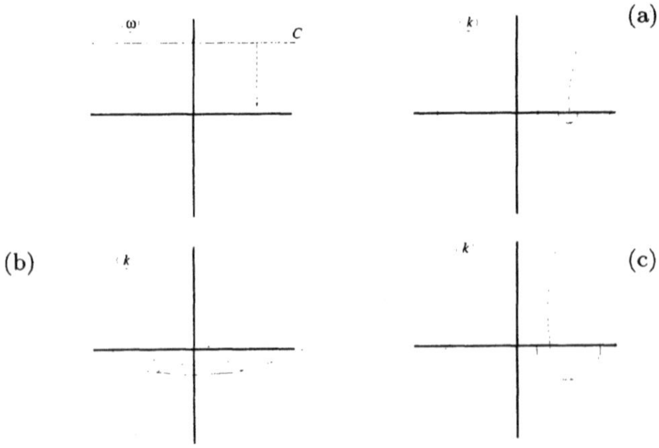

Fig. 2. Motion of k branches of $D(k, \omega) = 0$ as the temporal contour C is lowered onto real axis: (a) a single neutral mode k for a single ω on C; (b) a range of convectively unstable modes for ω on C; (c) a single mode with anomalous propagation. In (a) and (c) the dotted line denotes the path of the single mode as $Im(\omega)$ is reduced to zero, while in (c) the dotted line represents the k branch for real ω. In each case the solid line is the deformed k inversion contour.

Transforms in t and x, it follows that the linear response is given by

$$\eta(x,t) = \frac{1}{4\pi^2}\int_C \Psi(x,\omega)\bar{F}(\omega)\exp(-\mathrm{i}\omega t)\mathrm{d}\omega$$
$$\Psi(x,\omega) = \int_{-\infty}^{\infty} \frac{\exp(\mathrm{i}kx)}{D(k,\omega)}\mathrm{d}k. \tag{7}$$

Here $\bar{F}(\omega)$ is the time transform of the forcing, and $D(k,\omega)$ is the so-called dispersion function. The result (7) is a very general one, and all that changes between different problems is the form of $D(k,\omega)$. For our problem of a plate with mean flow it is easy to show that

$$D(k,\omega) = \omega^2 - k^4 + \frac{(\omega - kU)^2}{|k|}, \tag{8}$$

where $|k| = k$ if $Re(k) > 0$ and $|k| = -k$ if $Re(k) < 0$. The spatial inversion contour is chosen to be the real k axis. The crucial aspect of (7) is the choice of the temporal inversion contour C. Since for causality we require $\eta(x,t) = 0$ for $t < 0$, we must chose C to lie above all singularities in the ω plane. Since we know that $\bar{F}(\omega)$ is analytic in the upper half of the ω plane, it follows that we must choose C to lie above all ω-zeros of the dispersion function for real k (and indeed above all branch points as well). In certain problems this is simply not possible (such as the Kelvin-Helmholtz instability of a vortex sheet), when there is no maximum temporal growth rate, but this is certainly not the case for our problem.

The idea now is to attempt to deform C downwards in the ω plane towards the real axis, but all the time that one is doing this there are poles in the k plane corresponding to the k-zeros of $D(k,\omega)$ for ω on C. These poles will move as C is deformed, and the spatial inversion contour must be deformed to avoid any of the poles (or branch points) crossing it, thereby retaining an analytic solution. A number of possible deformations are required, and these are set out as follows.

1. One possibility (figure 2a) is that a k root, or roots, originate in the upper half of the k plane, and more monotonically onto the real k axis from above as C is lowered onto the real ω axis. The k inversion contour must then be deformed just below the corresponding pole, or poles. For $x > 0, \Psi(x,\omega)$ is evaluated by closing the contour in the upper half plane, leading to a residue contribution of the form (from $k = k_p$, say)

$$\frac{e^{ik_p x}}{\partial D/\partial k(k_p,\omega)} \, . \tag{9}$$

This corresponds to a neutral wave downstream of the driver. Moreover, since

$$k_p(\omega + i\epsilon) \approx k_p(\omega) + i\epsilon\left(\frac{\partial\omega}{\partial k}\right)^{-1} + \dots\ , \tag{10}$$

it follows that this neutral mode has positive group velocity (i.e. $\partial\omega/\partial k > 0$). The group velocity, and hence the direction of energy propagation, is therefore directed away from the driver at $x = 0$. Of course, for $x < 0$ the spatial inversion contour is closed in the lower half plane, and no contribution is found from this pole.

2. Another possibility (figure 2b) is for k roots originating from the upper half plane to move downwards and cross the real k axis as C is lowered, so that, once C lies on the real ω axis, there exists a range of real ω such that the dispersion relation $D(k, \omega) = 0$ has k roots with negative imaginary part. In this case the k inversion contour must be deformed below the real axis to avoid a pole crossing. In $x > 0$, residue contributions are again found, but now these modes grow spatially as $x \to +\infty$ and therefore correspond to convective instability.
3. Yet another possibility is shown in figure 3. Here k roots of $D(k, \omega) = 0$ originate in opposite halves of the k plane when C is high in the ω plane (i.e. for a given range of frequencies, $D(k, \omega)$ has two spatial branches). As C is lowered, the two branches move together in the k plane, and coalesce at the point $k = k_p$ when $\omega = \omega_p$. Providing that ω_p has a positive imaginary part, this means that the spatial contour can no longer be deformed to avoid pole crossing, and is said to be "pinched". The pinch point is a double root of $D(k, \omega) = 0$, and it therefore follows that in the neighbourhood of the pinch

$$k - k_p \propto (\omega - \omega_p)^{\frac{1}{2}} , \tag{11}$$

so that $\omega = \omega_p$ is a branch point of the dispersion function in the ω plane. Introducing a branch cut to join $\omega = \omega_p$ to infinity through the lower half plane, and now deforming C down onto the real ω axis, but wrapped around the branch cut, it is straightforward to show that in the long-time limit $t \to +\infty$ the plate response is dominated by the contribution from the pinch, and in particular that

$$\eta(x, t) \propto \frac{\overline{F}(\omega_p)}{t^{\frac{1}{2}}} e^{ik_p x - i\omega_p t} . \tag{12}$$

Since by construction we have that ω_p possesses a positive imaginary part, it now follows that $\eta(x, t)$ grows exponentially in time for all x. This is absolute instability, and corresponds to the idea that disturbances introduced into a system may grow *in situ*, rather than growing as they are convected away downstream, as is the case in the more conventional case of convective instability. It should be noted that $\partial\omega/\partial k = 0$ at the pinch point, corresponding to the fact that energy is not convected away. However, the zero group velocity criterion is only necessary, and not sufficient, for the occurrence of absolute instability; what is needed as well, as the foregoing discussion makes clear, is for the corresponding modal coalescence to be formed from modes which originate in opposite halves of the k plane. Modes coalescing from the same half of the k plane do not pin the k inversion contour, and therefore do not stop C being deformed downwards. Finally, we note here that in absolute instability the flow chooses its own response frequency, ω_p. It is now well-established that absolute instability leads to a number of self-excited oscillations, including Karman vortex shedding behind a bluff body and the break-up of hot jets. Strong evidence has also been put forward for absolute instability playing a rôle in boundary-layer transition to turbulence on rotating disks (see Lingwood [11]).

4. As a final possibility we mention here (and will discuss in greater detail presently) the case in which a k mode originates in the upper half plane, then moves downwards as C is deformed, crosses the real k axis when the imaginary part of ω is still positive, and then turns around in the k plane and comes to rest on the real k axis once ω is real. This motion is shown in figure 2c. Because the mode originated in the upper half of the k plane, the spatial inversion contour is deformed below the mode, and the residue contribution from the mode is therefore picked up in $x > 0$ only. However, as equation (10) makes clear, this mode has $\partial\omega/\partial k < 0$, and therefore has group velocity directed towards the driver. The well-known radiation condition, due to Rayleigh and developed by Lighthill [10], is that the group velocity is always directed away from the driver, contradicting the prediction made using our causal analysis. We will return to this point later.

3 Solution for Fluid-Loaded Plate with Mean Flow

All four of the mathematical phenomena described above can be found for the dispersion function given in equation (8), and full details are given by Crighton & Oswell [8]. Absolute instability is found to occur for flow speeds U in excess of about 0.0742 (in non-dimensional units) – for steel in water this critical speed translates to about 15m/s. For $U < 0.0742$ the ω contour C can be deformed all the way down onto the real axis, resulting in the dispersion diagram shown in figure 4. For $0 < \omega < \omega_s$ the system possesses a convective instability (as in point 2 above), in addition to neutral modes. For $\omega_s < \omega < \omega_b$ the system possesses four neutral modes; three of these modes are conventional, in that their group velocity points away from the driver, as in point 1 above; but the neutral mode with $k_b < k < k_s$ has anomalous behaviour as in point 4. For $\omega > \omega_b$ the system behaves conventionally, and indeed for larger ω the effects of mean flow (i.e. $U \neq 0$) become unimportant. All the interesting features mentioned here arise directly as a result of the presence of the mean flow, and we note from figure 4 that they occur at low values of the non-dimensional frequency. This indicates, as would be expected, that the mean flow only has a significant effect under conditions of heavy fluid loading.

The prediction of a mode with anomalous group velocity, in contradiction of the Rayleigh-Lighthill condition, might seem controversial, but in fact Lighthill's [10] proof relies on the twin assumptions that the driver is the only source of energy in the problem and that the sign of the group velocity determines the sign of the energy flux. For the fluid-loaded plate this is, of course, not the case, because the plate both extracts energy from the mean flow and, as will be seen later, possesses waves of negative energy. The idea of energy moving into the driver is then completely reasonable (just as it would be in the case, say, of two point forces of size F_0 and $2F_0$ in zero mean flow, with the direction of net energy transfer being towards the smaller force). This does indicate, however, the importance of adopting a robust and systematic approach to determining the long-time limit of the solution of initial-value problems. In particular, consideration of what might be called 'local' properties of the dispersion function, $D(k, \omega)$, such as group velocity, will potentially miss the features described here. What is needed instead is a careful analysis of $D(k, \omega)$ throughout both temporal and spatial planes.

3.1 Wave Energy

In order to understand more clearly the way in which the mean flow can act as a source of energy additional to the driver, we now consider the wave energy associated with a neutral mode in our problem. The concept of wave energy in this sort of situation was first introduced by Benjamin [1] and Landahl [9], and

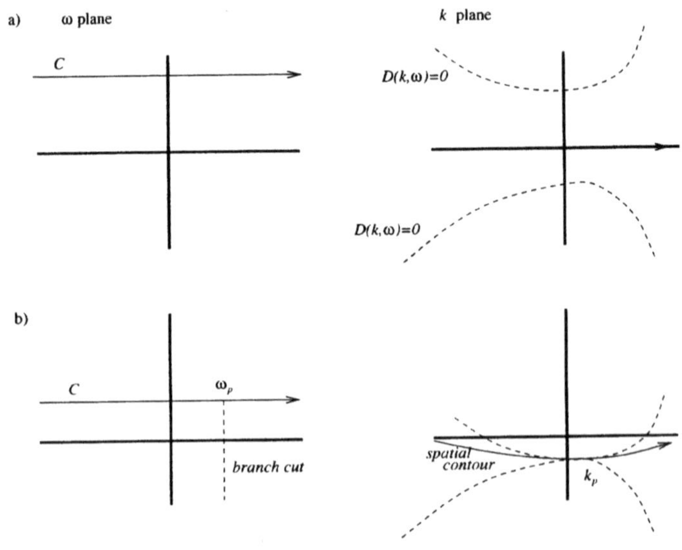

Fig. 3. Absolute instability - pinch point at $k = k_p$ for $\omega = \omega_p$ (from [16]).

corresponds to the work done in creating a neutral wave of a given amplitude at time t, starting from rest at $t = -\infty$ and building the amplitude up slowly. The wave energy per unit amplitude, E, can be shown to be related to the frequency derivative of our dispersion function – specifically, when averaged over a period, $2\pi/\omega$, of the carrier wave Cairns [5] shows that

$$E = \frac{1}{4}\omega\frac{\partial D}{\partial \omega} . \tag{13}$$

Now the definition of wave energy suggests that E can be either positive or negative, depending on whether the excitation of the neutral mode increases or decreases the system energy. $E > 0$ are positive energy waves, the conventional case, and $E < 0$ are termed negative energy waves (NEW) which have the unusual property of being destabilised by the introduction of dissipation. This latter point follows because dissipation will act to reduce the system energy, which is equivalent to increasing the amplitude of a NEW.

For the fluid-loaded plate with mean flow, it is easy to show that $\partial D/\partial\omega > 0$ for all $\omega > \omega_b$, so that these neutral modes have $E > 0$. However, $\partial D/\partial\omega < 0$ for $\omega < \omega_b$, and it therefore follows that we have NEWs on the <u>lower</u> branch of the neutral curve for $k > k_s$ and $\omega > 0$ (i.e. in figure 4 for $k_s < k < k_0$). The NEWs, as already stated, will be destabilised by introducing dissipation, and if, for instance, we add dashpots to the plate, providing a drag force $c_d\partial\eta/\partial t$,

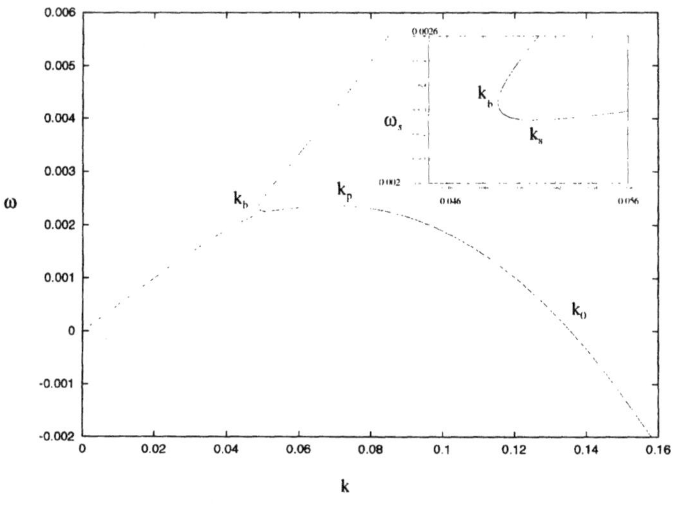

Fig. 4. Solutions of $D(k,\omega) = 0$ for real ω. The dotted line between $k = k_s$ and $k = 0$ corresponds to the real part of the convectively unstable mode. The inset corresponds to the region in the neighbourhood of $k = k_s$ (from [15]).

it can be shown that the amplitude of the NEW is proportional to

$$\exp\left(-\frac{\omega^2 c_d}{E}t\right) . \tag{14}$$

Since $E < 0$ this corresponds to exponential growth. We will see in due course how the nonlinear evolution of destabilised NEW can be considered.

Another useful concept when describing the energetics of our system is the energy flux, J, of a neutral mode. We define

$$J = \pm E\frac{\partial\omega}{\partial k} ,$$

with $\pm$ corresponding to $x > 0$ and $x < 0$ respectively, so that $J > 0$ means that energy is carried away from the driver. On the upper branch of our neutral curve (i.e. above $\omega = \omega_b$) we have the conventional behaviour $J > 0$. However, both $J > 0$ and $J < 0$ are possible on the lower branch. For instance, the mode with $k_b < k < k_s, \omega_s < \omega < \omega_b$ is a NEW but has $\partial\omega/\partial k < 0$, so that $J > 0$, and therefore has a positive energy flux. In contrast, the mode with $k_s < k < k_p$ with $\partial\omega/\partial k > 0$ is also a NEW, and therefore has $J < 0$, so that it carries energy towards the driver. In fact, one can measure the net energy flux into or out of the driver by calculating the line admittance, A_0, of the driver. Crighton & Oswell [8] show that the real part of A_0 is negative for $\omega < \omega_p$, demonstrating that in this low frequency range the driver absorbs energy from the flow.

A further point is that an energy equation for the fluid-loaded plate in the control volume $\eta < y < \infty$, $x_1 < x < x_2$ can be derived in the form

$$\frac{d}{dt}\left(T_f + T_p + V_p\right) = F(t)\frac{\partial\eta}{\partial t} + [\text{FLUX}] . \tag{15}$$

Here, T_f and T_p correspond to the kinetic energy in the fluid and in the plate and V_p is the potential energy in the plate. The first term on the right hand side of equation (15) corresponds to the rate of working of the driver, while the second term is the net energy flux entering a control volume, containing contributions from both fluid and plate energy fluxes (and a coupled component). The key point is that both T_p and V_p are positive definite, but T_f is given by

$$T_f = \frac{1}{2}\int_{x_1}^{x_2}\int_{\eta}^{\infty}\left\{(U + \phi_x)^2 + \phi_y^2 - U^2\right\} \mathrm{d}y \ \mathrm{d}x . \tag{16}$$

The interpretation of T_f is that it is the difference between the kinetic energy of the total flow and the kinetic energy of the steady mean flow, both of which are individually divergent. It is now clear that T_f is not necessarily positive, since there is a cross term $U\phi_x$ between the mean flow and the unsteady perturbation. This further confirms the idea that the mean flow can act as a source of energy, since it is quite possible to have $T_f < 0$, depending on the phase of ϕ_x.

4 Modifications to the Basic Model

We will now proceed to describe a series of modifications to our simple model of uniform mean flow past a flat plate, with the aim of making the model more physically realistic. The aim at each stage is to determine whether or note the unusual features described so far are robust and therefore likely to be observed in practice.

4.1 Plate Curvature

In order to consider the effects of plate curvature, consider a hollow cylindrical shell with axial, uniform mean flow outside it. In the limit in which the cylinder radius goes to infinity the simple two-dimensional problem of Crighton & Oswell [8] is regained, but at finite radius the shell possesses an additional hoop stiffness, not present in two dimensions. Full details of the governing equations and analysis are given in Peake [14], and the analytical (but cumbersome) dispersion relation resulting from these equations is also presented. However, very considerable simplification of this dispersion relation is possible when the normalised flow speed, U, is assumed small (a good approximation for steel in water), in which case it turns out that the dispersion relation reduces down to

$$\frac{(\omega - Uk)^2}{|k|} + \omega^2 - k^4 - \frac{12(1-\nu^2)}{\rho^2 a^2} = 0 \ , \qquad (17)$$

where ν is the Poisson ratio in the shell, $\rho = \rho_0/\rho_s$ is the ratio of the fluid and solid densities and $a = r\rho/h$, where r is the cylinder radius and h is the shell thickness. Here we have adopted a preferred asymptotic limit $a = O(U^{-5/3})$, necessitating $k = O(U^{2/3}), \omega = O(U^{5/3})$. By comparing this result to the flat-plate dispersion relation (8), it is easy to see that the plate curvature has introduced just one extra term, which is formally equivalent to introducing a spring foundation to a simple plate in two dimensions.

Either the exact dispersion relation, or the approximation given above, can now be investigated for absolute instability. The small U approximation yields an analytic boundary for absolute instability in $a-U$ space (essentially by searching for a triple root of (17)), in the form of absolute instability for flow speed U in excess of the critical value given implicitly by

$$\rho^2 a^2 = \frac{(1-\nu^2)5^{5/3}2^{8/3}}{U^{8/3}[U^{2/3} - U_0^{2/3}]} \ . \qquad (18)$$

Here, $U_0 \approx 0.0742$ is the critical flow speed for absolute instability, so that as the cylinder radius $a \to \infty$ the two-dimensional result is regained. It is instructive to consider the critical values of U for finite a: if we set $a = 100$, corresponding to a cylinder radius of 780 times the plate thickness for steel in water, then the critical value of U is 0.87, which is exceedingly high (about 170 m/s for steel in water). Conversely, to achieve absolute instability at $U = 0.1$ (about 20 m/s), we would require a cylinder of radius 6×10^4 plate thicknesses, a degree of flatness which seems unlikely to be realised in practice. Hence, we might safely conclude that in most practical situations the extra stiffness induced by the plate curvature will act to eliminate the occurrence of absolute instability.

Below the absolute instability boundary we can consider real ω, and the sequence of dispersion diagrams which results from varying a is shown in figures 5–7. For $a > a_1(U)$ (figure 5), the roots of $D(k, \omega) = 0$ look very much like the two-dimensional case, but with an extra branch of neutral modes extending from zero frequency. The convective instability is still present, and lies between the two branch points $k = k_a$ and $k = k_b$. As a is reduced, k_a and k_b move together, and coalesce when $a = a_1(U)$. Using approximation (17), it is easy to show that

$$a_1(U) = \left[\frac{48(1-\nu^2)}{\rho^2 U^4}\right]^{\frac{1}{2}} . \tag{19}$$

Once the two branches have coalesced (figure 6), it turns out that the anomalous modes with group velocity directed towards the driver cease to exist (since their existence is intimately linked to the presence of the branch point $k = k_b$). For realistic parameter values it turns out that a_1 is very large indeed, indicating that these anomalous modes may well be eliminated by plate curvature in many practical situations.

As a is reduced still further the turning point $k = k_p$ moves downwards, and at a critical value $a = a_2(U)$ crosses the axis $\omega = 0$ (and eventually

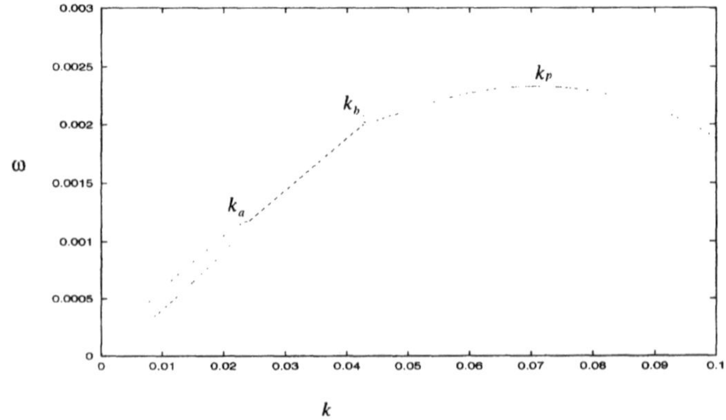

Fig. 5. Solutions of $D(k, \omega) = 0$ for the circular cylinder with flow (taken from [14]). Here we have $U = 0.05$ and $a = 25000$.

collides with a second turning point and disappears for a yet smaller value of a). Again using approximation (17) for small U, it can be shown that

$$a_2(U) = \frac{2^{7/3}(1-\nu^2)^{\frac{1}{2}}}{\rho U^{4/3}} . \tag{20}$$

This is significant, because $a = a_2(U)$ marks the lower boundary for the existence of Negative Energy Waves; for $a < a_2$ (figure 7), the lower branch of modes has $\omega < 0$ and therefore positive wave energy. In practice, the value of $a_2(U)$ is likely to be considerably smaller than $a_1(U)$, suggesting that NEWs could well be important for curved plates.

4.2 Nonlinear Motion

Our next aim is to consider how the results of Crighton & Oswell might be modified by consideration of nonlinear effects (especially with regard to the nonlinear evolution and amplitude saturation of a destabilised NEW). Full details are given in Peake [15]. Because we are interested in situations of heavy fluid loading, it will prove necessary to include nonlinear terms in both the plate and the fluid.

The plate equation is taken to be

$$\frac{\partial^2}{\partial x^2}\left[\frac{\eta_{xx}}{(1+\eta_x^2)^{\frac{3}{2}}}\right] + \eta_{tt} - \frac{\tau}{s}\left(\int_0^s \eta_{x'}^2 \mathrm{d}x'\right)\eta_{xx} + \epsilon c_d \eta_t = -p(x, y=\eta, t) . \tag{21}$$

Here the first term on the left hand side corresponds to an exact prescription of the curvature, while the third term is the nonlinear tension induced by bending. The quantity s corresponds to some multiple of the wavelength (we

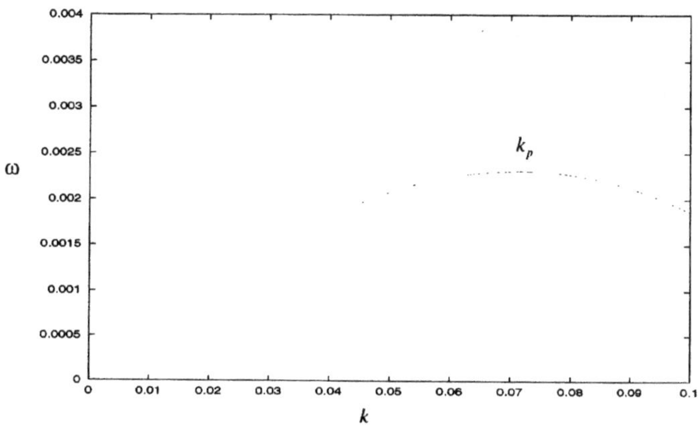

Fig. 6. As in figure 5, but with $a = 20000$ (taken from [14]).

could, for instance, assume that s contains many waves, and that the interval of length s is bounded by fixing points which prevent longitudinal motion). The coefficient τ is proportional to the square of the solid-fluid density ratio, and is therefore large. The fourth term on the left hand side of equation (21) corresponds to the drag force induced by dashpots attached to the plate, with drag coefficient ϵc_d, $\epsilon \ll 1$. These dashpots are included in order to destabilise the NEWs. On the right hand side of equation (21) note that the hydrodynamic pressure is evaluated on the genuine plate surface, rather than on the mean location $y = 0$.

We still consider irrotational, incompressible fluid, so that we still have Laplace's equation in the nonlinear régime. However, the pressure must now be determined from the full Bernoulli equation

$$p = -\phi_t - U\phi_x - \frac{1}{2}\left(\phi_x^2 + \phi_y^2\right) \quad , \tag{22}$$

while at the fluid-place boundary the continuity of vertical velocity condition yields

$$\phi_y = \eta_t + (U + \phi_x)\,\eta_x \tag{23}$$

In order to determine the nonlinear evolution of a NEW, it proves convenient to introduce the slow time $T = \epsilon t$, and to suppose that $\tau = \epsilon^{-2}\overline{\tau}$, with $\overline{\tau} = O(1)$. Then writing

$$\eta(x,t) = \epsilon[AE + \text{c.c.}] + \dots \quad , \tag{24}$$

with $A = A(T)$ and

$$E = \exp\left(\mathrm{i}kx - \frac{\mathrm{i}}{\epsilon}\int^T \omega(T')\mathrm{d}T'\right), \tag{25}$$

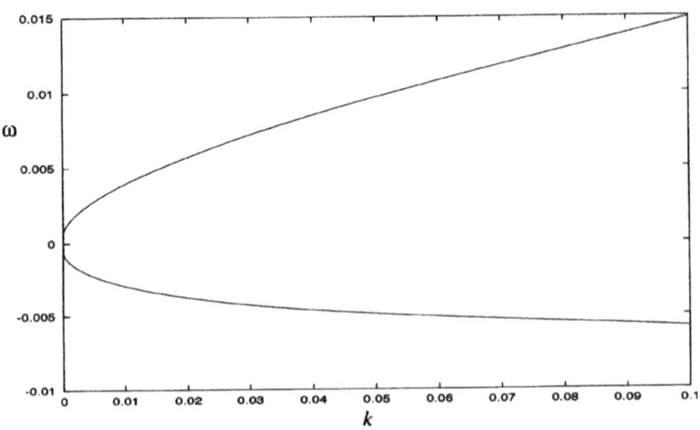

Fig. 7. As in figure 5, but with $a = 4000$ (taken from [14]).

(note that we must allow for slow frequency modulation), and equating powers of ϵ in the above equations, we find at leading order the amplitude-dependent dispersion relation

$$D^{\mathcal{NL}}(k,\omega,|A|) \equiv \omega^2 - k^4 + \frac{(\omega - Uk)^2}{|k|} - 2k^4\overline{\tau}|A|^2 = 0 \ . \qquad (26)$$

Note that because the coefficient of the nonlinear-tension is large, the tension appears at leading order, and takes the form of an augmentation of the elastic restoring fore in the plate. At the next order we find the evolution equation

$$\frac{\mathrm{d}|A|}{\mathrm{d}T} + \frac{c_d\omega^2}{4E_{NL}}|A| = 0 \ , \qquad (27)$$

where

$$E_{NL} = E + \frac{(k+1)\omega^2\overline{\tau}k^3|A|^2}{4E} \ . \qquad (28)$$

Note now that if we eliminate all terms involving the square of the amplitude in the above then we are left simply with ω constant, and that A evolves simply as in (14), which is the linear instability of the NEW (recall $E < 0$). However, when the nonlinear terms are included it becomes clear that $|A|$ saturates as $T \to \infty$ (see figure 8), and indeed that $\omega \to 0$ as $T \to \infty$. It therefore follows that the rôle of dissipation has been to shift our system from a state of equilibrium in which the plate is undisturbed to another state of equilibrium in which the plate is buckled, but static. As the NEW initially grows the tension in the plate increases until, together with the elastic restoring force in the plate it balances the destabilising fluid forces. As the forces come closer to balance the dashpot do less work and hence, since the rate of working of the dashpots is proportional to ω^2, the frequency of the motion decreases until the plate is static, when all the forces balance.

Nonlinear analysis of the neutral modes on the plate can also be considered. Suppose that the dashpots are removed and that the nonlinear tension is also neglected (for instance, by imagining that it is relaxed to zero over an infinite plate length). Then by writing a neutral mode on the lower branch of the dispersion diagram given in figure 4 as

$$A(X,T)e^{-i\omega t + ikx}, \qquad (29)$$

where X and T are slow space and time scales (possibly including a Galilean transform into the frame moving with the group velocity of the mode), an evolution equation can be determined. This takes the form of a general Nonlinear Schrodinger (NLS) equation

$$\frac{D_{kk}}{2}\frac{\partial^2 A}{\partial X^2} - iD_\omega\frac{\partial A}{\partial T} = lA|A|^2 \ ,$$

where l is a positive constant which can easily be determined. The ω derivative of $D(k,\omega)$, i.e. D_ω, is negative all along the lower branch of the dispersion curve, while D_{kk} changes sign from positive near $k = k_s$ to negative near $k = k_p$. The signs of the coefficients in the NLS equation are crucial; for $D_{kk} > 0$ this is the so-called 'minus' form of NLS, and it is generally accepted that for initial data of compact support the solution diffuses to zero at $T \to \infty$. In contrast, for $D_{kk} < 0$ we have the 'plus' form of NLS, and now general initial data will form into solitary-wave solutions (Zakharov & Shabat [18]) of the general type

$$A(X,T) = a\left|\frac{D_{kk}}{l}\right|^{\frac{1}{2}} \operatorname{sech}\left[a\left(X + \frac{bD_{kk}}{D_\omega}T\right)\right] \exp\left[\mathrm{i}bX - \frac{\mathrm{i}D_{kk}}{2D_\omega}(a^2 - b^2)T\right] , \tag{30}$$

where a and b are arbitrary. These are envelope solitons, in that there is a small correction to the carrier wave phase as well as the characteristics amplitude modulation. It turns out that, although formally of weakly nonlinear amplitude, the coefficient in front of these solitons is potentially large (for small U), indicating that they might be observable in practice. Of course, it is only the waves towards the right hand side of the lower branch of the dispersion curve which can form these solitary waves – the other neutral modes decay away under nonlinear action.

4.3 Effects of a Boundary Layer

One other modification, which we mention only very briefly here, is to return to the two-dimensional, linear problem but replace the uniform mean flow

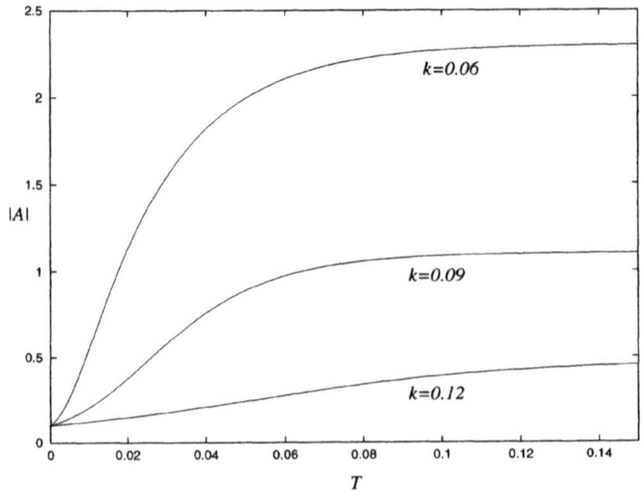

Fig. 8. Evolution of $|A(T)|$ for various choices of wavenumber k (from [15]).

with a boundary layer profile. This problem has been studied before, by Carpenter & Garrad [6], [7], but in the context of the present analysis full details are given by Lingwood & Peake in [12]. Specifically, we suppose that the steady velocity parallel to the plate is of the form

$$Uf(y/\epsilon) \tag{31}$$

where $f(0) = 0$ to satisfy the no-slip condition on the plate, $f \to 1$ as $y \to \infty$ to match onto a uniform outer flow of speed U, and ϵ here is the boundary layer thickness. The plate equation is unchanged but, of course, the unsteady fluid motion must account for the steady shear (assuming inviscid perturbations, this leads to Rayleigh's equation), and also the boundary condition on $y = 0$ is simply that the fluid normal velocity must equal $\partial\eta/\partial t$, since the convective term now vanishes on the plate.

Full details of the solution of this system are given in Lingwood & Peake [12], and we only make a few brief points here:

1. For the simplest case of a piecewise-linear profile (e.g. $f = y/\epsilon$ for $y < \epsilon, f = 1$ for $y > \epsilon$), the results as $\epsilon \to 0$ seem to agree with what is seen with uniform flow.
2. For a profile with curvature, and hence a critical layer, the limit $\epsilon \to 0$ seems to be much less clear. In particular, it appears that there are more saddle points than observed for uniform flow, and that these saddle points interact with each other to change the topology of $D(k, \omega)$. The absolute instability boundary, for instance, moves to exceedingly high values of U as $\epsilon \to 0$;
3. Modes with anomalous group velocity are still present for a Blasius boundary layer. For instance, it is possible to find a mode with $Im(k) < 0$ when $Im(\omega) = 0$, which moves downwards and then upwards in the k plane as $Im(\omega)$ is increased. This is therefore a convective instability located downstream of the driver, whereas naive considerations based on group velocity would imply incorrectly that the mode is located upstream and is damped!

5 Conclusion

In these notes we have attempted to show how careful analysis of the initial-value problem can lead to novel and unexpected results. A number of other extensions, including to the dynamics of fluid-loaded composite structures [17] and to plates with multiple supports [13] have also been completed. Full details of all these calculations can be found in the references cited.

References

1. Benjamin, T.B. The threefold classification of unstable disturbances in flexible surfaces bounding inviscid flows. J. Fluid Mech. **16**, 436–450 (1963).
2. Bers, A. Space-time evolution of plasma instabilities - absolute and convective. In 'Handbook of Plasma Physics', ed M.N. Rosenbluth, R. Z. Sagdeev. 1:451–517. North-Holland (1983).
3. Brazier-Smith, P.R. & Scott, J.F. Stability of fluid flow in the presence of a compliant surface. Wave Motion **6**, 547–560 (1984).
4. Briggs, R.J. *Electron-stream Interaction with Plasmas* . Monograph No. 29. MIT Press, (1964).
5. Cairns, R.A. The role of negative energy waves in some instabilities of parallel flows. J. Fluid Mech. **92**, 1–14 (1979).
6. Carpenter, P.W. & Garrad, A.D. The hydrodynamic stability of flow over Kramer-type compliant surfaces. Part 1. Tollmien-Schlichting instabilities. J. Fluid Mech. **155**, 465–510 (1985).
7. Carpenter, P.W. & Garrad, A.D. The hydrodynamic stability of flow over Kramer-type compliant surfaces. Part 2. Flow-induced surface instabilities. *J. Fluid Mech.* **170**, 199–232 (1986)
8. Crighton, D.G. & Oswell, J.E. Fluid loading with mean flow. I. Response of an elastic plate to localized excitation. Phil. Trans. R. Soc. Lond. A **335**, 557–592 (1991).
9. Landahl, M.T. On the stability of a laminar incompressible boundary layer over a flexible surface. *J. Fluid Mech.* **13**, 609–632 (1962).
10. Lighthill, M.J. Studies on magnetohydrodynamic waves and other anisotropic wave motions. Phil. Trans. R. Soc. Lond., **252**, 397-430 (1960).
11. Lingwood, R.J. Absolute instability of the boundary layer on a rotating disk. J. Fluid Mech., **299**, 17–33 (1995).
12. Lingwood, R.J. & Peake, N. On the causal behaviour of flow over a compliant wall. J. Fluid. Mech., **396**, 319–344 (1999).
13. Metcalfe, P. Smith's Prize essay, University of Cambridge (2001).
14. Peake, N. On the behaviour of a fluid-loaded cylindrical shell with mean flow. J. Fluid. Mech., **338**, 387-410. (1998).
15. Peake, N. Nonlinear stability of a fluid-loaded elastic plate with mean flow. J. Fluid Mech., to appear (2001).
16. Peake, N. Analytical methods for fluid-structure interaction problems. In 'Fluid-structure interactions in acoustics', Ed. D. Habault. CISM 396. Springer.
17. Peake, N. & Sorokin S.V. On the behaviour of fluid-loaded sandwich panels with mean flow. J. Sound Vib., to appear (2001).
18. Zakharov, V.E. & Shabat, A.B. Exact theory of two-dimensional self-focusing and one-dimensional self-modulating waves in nonlinear media. *Sov. Phys. J.E.T.P*, **34**, 62–69 (1972).

Nonlinear Acoustics and Acoustic Chaos

Werner Lauterborn

Georg-August-Universität Göttingen, Drittes Physikalisches Institut, Göttingen, Deutschland

Abstract. The very basic standard elements of nonlinear acoustics are presented: The origin of nonlinearity, equations of state, simple nonlinear waves and shock wave formation. Sonoluminescence is chosen as a modern example of nonlinear acoustics. Some notations from nonlinear dynamics are introduced needed for understanding chaotic dynamics and acoustic chaotic systems.

1 Introduction

Acoustics belongs to those areas of physics, where nonlinear phenomena were observed first. All musical instruments including the human voice belong to nonlinear acoustics as they produce sound waves and these, moreover, are not simple sinusoidal waves. Every generation process is necessarily nonlinear as new things appear that are not proportional to already existing ones. Propagation, on the other hand, normally is linear and only starts to become nonlinear at high intensities. As mankind has started to speak early in its history and also is supposed to have made music soon, nonlinear acoustics belongs to the oldest parts of science. Thus Sorge in 1745 and Tartini in 1754 report that when two tones of frequency f_1 and $f_2 > f_1$ are presented to the ear at high intensities, additionally the difference frequency $f_2 - f_1$ is heard. For a long time there was a debate, whether these tones are generated in the ear or are already present before in the propagating medium. Today we know that in this case the ear generates the difference frequency. But we also know today that a difference frequency can be generated in a nonlinear medium during propagation of a sound wave. Under suitable circumstances generally a series of combination frequencies $f_{mn} = mf_1 + nf_2$ with $m, n = \pm 1, \pm 2, \ldots$ can be generated. A special case is a single harmonic wave of frequency f propagating in a nonlinear medium. It generates the higher harmonics $f_m = mf$. This leeds to a steepening of the wave and not seldom to the formation of shock waves. Nonlinear acoustics and shock waves are the subject of many a book, congress proceedings and survey articles [1-11].

Lately, the general theory of nonlinear dynamics came across the interesting finding that nonlinear dynamical systems may not just show the above simple scenario of combination tones, but a complicated dynamics resembling stochastic behaviour. It is then called deterministic chaos or, in the context of acoustics, *acoustic chaos* [12].

2 Origin of Nonlinearity

All acoustic phenomena necessarily become nonlinear at high intensities. That can be demonstrated when looking at the propagation of a harmonic sound wave in air. In Fig. 1 a (linear) sound wave is represented graphically. In the upper diagram the sound pressure p is plotted versus location x. The static pressure p_{stat} serves as reference line around which the sound pressure oscillates. This pressure normally is about 1 bar. In the lower diagram the density distribution is given schematically in black and white for the pressure range of the sound wave. Supposed the sound pressure amplitude is increased

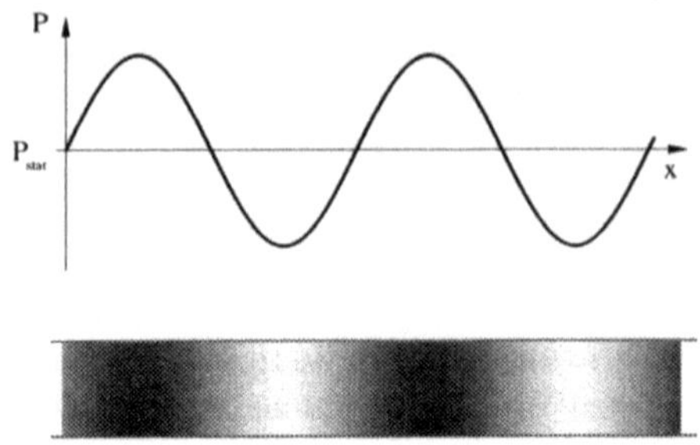

Fig. 1. Graphic representation of a sound wave

steadily, then a point is reached where the pressure amplitude attains 1 bar giving a pressure of zero in the sound pressure minimum. Below that it is impossible to go as there can be no less than no air molecules in the minimum. However, air can be compressed to more than 1 bar pressure by just increasing the force. Therefore, beyond 1 bar sound pressure amplitude no harmonic wave can exist at 1 bar static pressure. It must become nonlinear and contain harmonics. The basic reason of this nonlinearity is the unsymmetry between compression and expansion in a medium.

3 Equation of State

The compression and expansion of a medium is described by the equation of state. For that an interrelation between three quantities is needed, usually pressure p, density ϱ and entropy s. Often the variation of entropy can be neglected in acoustic phenomena and the calculations can be done with constant entropy. The equations are then called isentropic.

Already an ideal gas is nonlinear and obeys the isentropic equation of state

$$p = p_0 \left(\frac{\varrho}{\varrho_0} \right)^{\gamma} \tag{1}$$

with

$$\gamma = \frac{c_p}{c_v} \, . \tag{2}$$

Here γ is the quotient of the specific heat c_p at constant pressure and c_v at constant volume.

For liquids often an empirical formula is used, the Tait equation, that obviously has beeen constructed similarly to the equation of state of the ideal gas:

$$p = P \left(\frac{\varrho}{\varrho_0} \right)^{\gamma_L} - Q \, . \tag{3}$$

or also

$$\frac{p + Q}{p_\infty + Q} = \left(\frac{\varrho}{\varrho_0} \right)^{\gamma_L} \, . \tag{4}$$

The two quantities Q and γ_L have to be fitted to the experimental pressure-density curve. For water $\gamma_L = 7$, $P = p_\infty + Q = 3001$ bar and $Q = 3000$ bar is used. (Attention: γ_L in the Tait equation is not the quotient of the specific heats).

There is another way of writing the equation of state where the pressure is developed into a Taylor series as a function of density and entropy. In the isentropic case it reads

$$p - p_0 = \left(\frac{\partial p}{\partial \varrho} \right)_{\varrho=\varrho_0} (\varrho - \varrho_0) + \frac{1}{2} \left(\frac{\partial^2 p}{\partial \varrho^2} \right)_{\varrho=\varrho_0} (\varrho - \varrho_0)^2 + \ldots \tag{5}$$

or

$$p - p_0 = A \, \frac{\varrho - \varrho_0}{\varrho_0} + \frac{B}{2} \left(\frac{\varrho - \varrho_0}{\varrho_0} \right)^2 + \ldots \tag{6}$$

with

$$A = \varrho_0 \left(\frac{\partial p}{\partial \varrho} \right)_{\varrho=\varrho_0} = \varrho_0 c_0^2 \, , \tag{7}$$

$$B = \varrho_0^2 \left(\frac{\partial^2 p}{\partial \varrho^2} \right)_{\varrho=\varrho_0} \, . \tag{8}$$

Here c_0 is the sound velocity in the unperturbed medium. The higher order terms of the Taylor series often can be neglected. Then B/A is introduced as *nonlinearity parameter*, a pure number. For liquids, this number varies astonishingly little, mostly between 3 and 10. For water under normal conditions it is about 5.

It is easy to show that for gases we have

$$\frac{B}{A} = \gamma - 1 \, , \tag{9}$$

so that B/A $= 0.67$ for a monoatomic gas and B/A $= 0.40$ for a diatomic gas (air). Gases are thus much less nonlinear than liquids, by about a factor of ten.

4 Simple Nonlinear Waves

In the linear case the wave equation is obtained for the quantities pressure $p - p_0$, density $\varrho - \varrho_0$ and each component of the particle velocity $\mathbf{u} - \mathbf{u}_0$ (p_0 ambient pressure, ρ_0 density at rest, $\mathbf{u}_0$ constant velocity, often $\mathbf{u}_0 = 0$) from the equations for a compressible liquid. In the spatially onedimensional case, when φ denotes one of the perturbational quantities, one gets

$$\varphi_{tt} - c_0^2 \varphi_{xx} = 0 \; , \tag{10}$$

where c_0 is the propagation velocity of the perturbation and given by $c_0^2 = (\partial p / \partial \varrho)_{\varrho = \varrho_0}$. A simple way to incorporate the influence of nonlinearities of the medium without much mathematical effort consists in considering the propagation velocity no longer as constant. One proceeds as follows. As nonlinear waves do not superpose interactionfree and waves running in opposite direction thus would cause problems, the wave equation above is written as

$$\left(\frac{\partial}{\partial t} - c_0 \frac{\partial}{\partial x}\right) \left(\frac{\partial}{\partial t} + c_0 \frac{\partial}{\partial x}\right) \varphi = 0 \tag{11}$$

and only one part, i.e. a wave running in one direction only, is taken, for instance:

$$\varphi_t + c_0 \varphi_x = 0 \; . \tag{12}$$

The general solution of this equation is

$$\varphi(x,t) = f(x - c_0 t) \; . \tag{13}$$

The function f can be a quite general function of the argument $x - c_0 t$.

A nonlinear extension that can also be justified mathematically can then be written as

$$\varphi_t + c(\varphi) \varphi_x = 0 \; , \tag{14}$$

whereby now the propagation velocity $c(\varphi)$ is a function of the perturbation φ. This way the simplest propagation equation for nonlinear waves is obtained. It already leads to the problem of breaking waves and the occurrence of shock waves.

It should be noted that an extension of the linear propagation equation (12) by introducing a dependance of c_0 on a parameter, for instance on ω, does not lead to a nonlinear equation, but a still linear one:

$$\varphi_t + c(\omega) \varphi_x = 0 \; . \tag{15}$$

This equation describes *dispersive* waves, that is waves, whose propagation velocity may be different for different frequencies contained in the wave.

A solution to the nonlinear propagation equation (14) can be given in an implicit way:

$$\varphi(x,t) = f(x - c(\varphi) t) \; , \tag{16}$$

as can be proved immediately by insertion. This seems of little use to really calculate the propagation of the perturbation profile. But the equation allows for a simple physical interpretation, that is, that the quantity φ propagates with the velocity $c(\varphi)$. This leads to a simple geometric construction for the propagation of a perturbation (Fig. 2). To this end we consider the initial

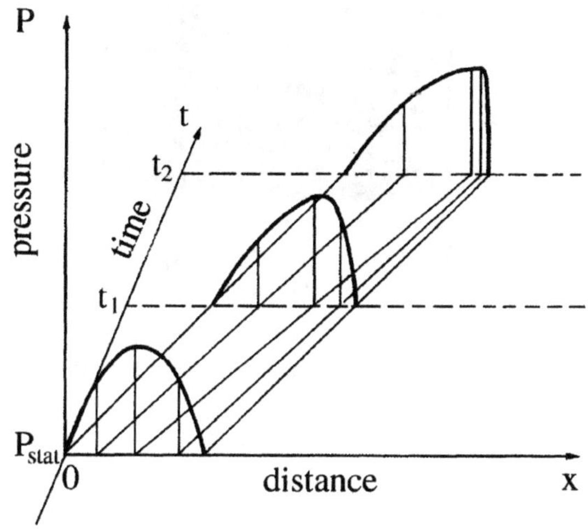

Fig. 2. Characteristics and the construction of the wave form for nonlinear wave propagation (after Beyer [3])

value problem

$$\varphi(x, 0) = f(\xi) \ . \tag{17}$$

To each ξ-value belongs a value $f(\xi)$ that propagates with the velocity $c(f(\xi))$:

$$\frac{\mathrm{d}x}{\mathrm{d}t} = c(f(\xi)) \tag{18}$$

or

$$x = \xi + c(f(\xi))t \ . \tag{19}$$

This is the equation of a straight line in the (x, t)-plane that crosses $x = \xi$ with the derivative $c(f(\xi))$. Along this straight line φ stays constant. These lines are called *characteristics*. They were introduced by Riemann. That way the solution of (hyperbolic) partial differential equations can be transformed to the solution of ordinary differential equations. Also numerically that is of great advantage.

The steepening of a wave front can be measured by sending a strong sound wave through the then nonlinear medium and monitoring the wave with a microphone. Figure 3 shows the steepening of a sound wave in air after three meters of propagation. The wave of frequency 2 kHz was guided in a plastic tube. As the horizontal axis denotes time, the steepening is on the left flank, the flank of the propagation direction. It passes the microphone first.

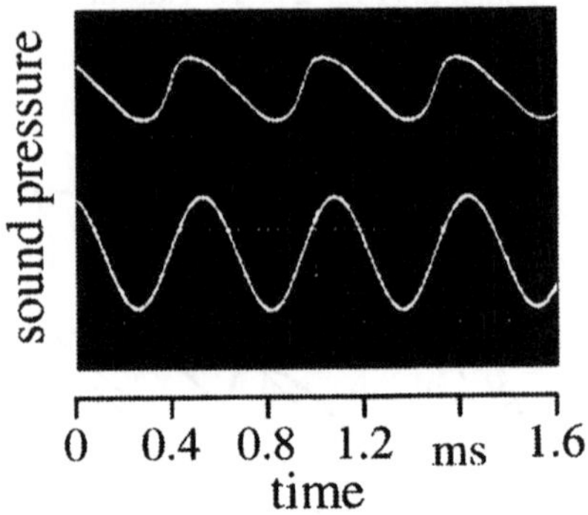

Fig. 3. Form of the sound wave at the launch site (below) and after three meters of propagation (above). Measurement done at the Drittes Physikalisches Institut, Göttingen

5 Shock Waves

The characteristics of Fig. 2 must cross. According to the geometrical construction the profile of the wave becomes multivalued. This is easily envisaged with water waves. With pressure or density waves, however, there is only one pressure or one density at one place. The theory thus is too simplified and must be enlarged with the help of physical arguments and the corresponding mathematical formulation. Shortly before overturning the gradients of pressure and density become very large, and it is known that then the damping effects, brought about by viscosity and heat conduction, can no longer be neglected. When these effects are taken into account, these waves do no longer turn over. Instead, a thin zone is formed, a *shock wave*, in which the values of φ (pressure, density, temperature, etc.) vary rapidly.

It has been found that damping effects have not neccessarily to be included into the theory, but that an extended damping free theory can be developed by introducing certain shock conditions that connect the variables across a discontinuity. A description of this theory can be found in the book of Whitham [1]. The result can be described in a simple construction. To this end we look at Fig. 4, where the profile of the wave has been determined according to the methods of characteristics for a time where the characteristics already have crossed. The shock wave then has to be inserted in such a way that the areas between the shock wave and the wave profile to the left and the right

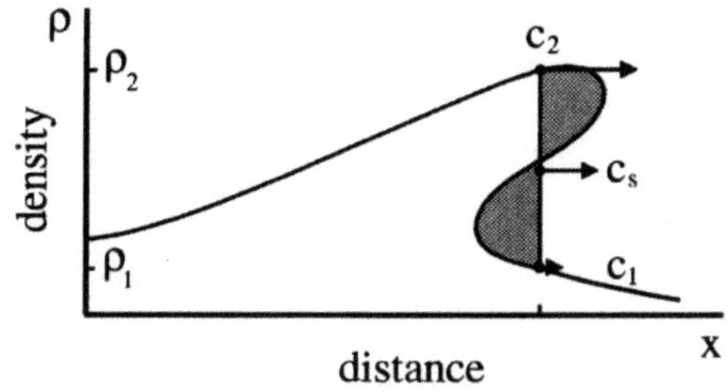

Fig. 4. Condition for the location of the shock wave for an overturning wave

are equal. Moreover, it can be shown that the velocity of the shock wave is approximately given by

$$c_s = \frac{1}{2}(c_1 + c_2) \ , \tag{20}$$

where c_1 and c_2 are the velocities of the sound wave before and behind the shock, respectively.

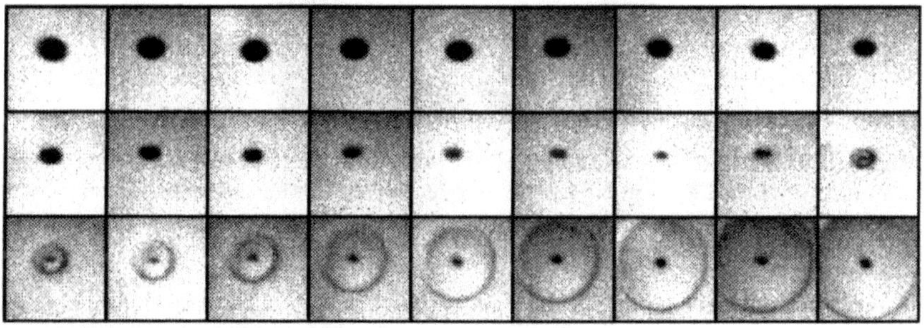

Fig. 5. Example of a shock wave. Photographic sequence of the collapse of a spherical bubble in water generated by a laser pulse. 20.8 million frames per second. Size of the pictures is 1.5 × 1.8 mm (after Ohl et al. [13])

An example of a shock wave is given in Fig. 5. A bubble has been formed in water by focusing the light of a laser pulse into the liquid [13]. The sequence has been taken at 20.8 million frames per second in the vicinity of the first collapse of the bubble. Despite the high velocity of sound in water of about 1500 m/s the shock wave only moves slowly from frame to frame because of the extremely high framing rate. Bubble and shock wave have been illuminated from the back in parallel light. The light is deflected at the bubble and at the density variation of the shock wave, so that the bubble is visible as a dark spot and the shock wave as a dark ring that expands.

6 Sonoluminescence

Sound waves not only give rise to shock waves but may become the driving agent to the emission of light, a phenomenon called sonoluminescence. It is mediated by bubbles in a liquid that are driven to strong oscillations and collapse [14–22]. The detection of the conditions where a single bubble can be trapped in the pressure antinode of a standing sound wave (see [23]) has furthered this field considerably. In Fig. 6 a basic arrangement for single

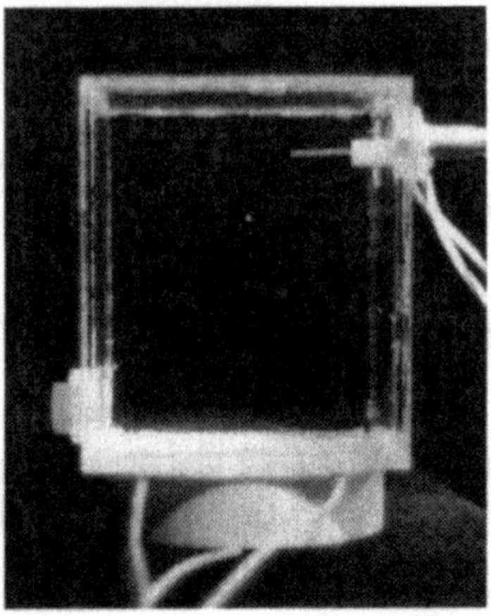

Fig. 6. Photograph of a rectangular container with light-emitting bubble. There is only one transducer glued to the bottom plate of the container (Courtesy of R. Geisler)

bubble sonoluminescence (SBSL) can be seen where a bubble is trapped in a rectangular container. The sound field is produced by a circular disc of piezoelectric material glued to the bottom of the container. The bright spot in the upper part of the container is the light from the bubble. It is of blueish white color. The needle sticking out into the container in the upper left part of the figure is a platinum wire for producing a bubble via a current pulse. The bubble is then driven via acoustic radiation forces into its position where it is trapped almost forever. The bubble exerts large oscillations that can be photographed [24]. Figure 7 shows a sequence of the bubble oscillation of a light emitting bubble. A full cycle is shown after which the oscillation repeats again and again. Indeed, this repetition has been used to sample the series of photographs by shifting the instant of exposure by 500 ns from frame to frame with respect to the phase of the sound wave. The camera is looking directly into the illuminating light source (back illumination). The bubble then shows dark on a bright background because the light is deflected off its

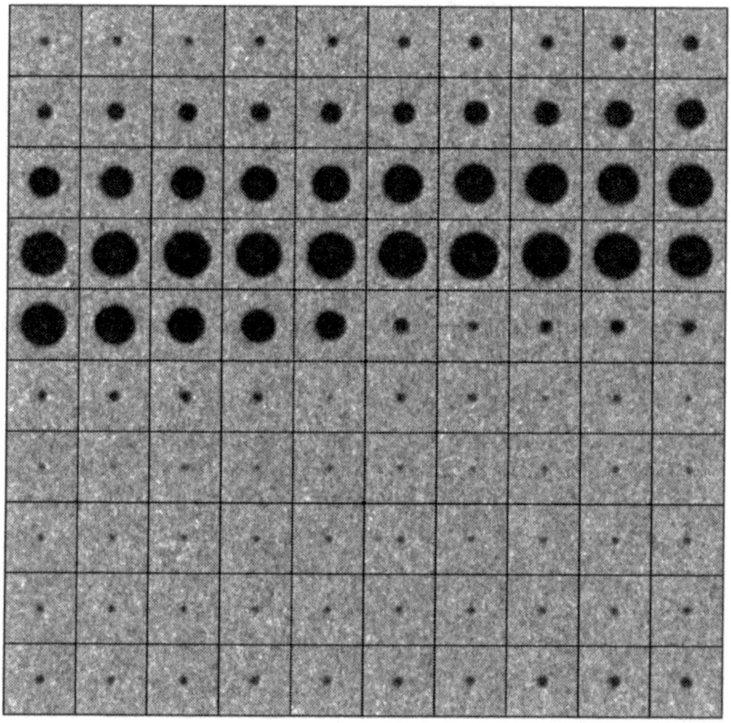

Fig. 7. Photographic series of a trapped, sonoluminescing bubble driven at 21.4 kHz and a pressure amplitude of 132 kPa. The interframe time is 500 ns. Picture size is 160 μm × 160 μm (Courtesy of R. Geisler)

surface. The white spot in the middle of the bubble results from the light passing the spherical bubble undeflected.

Upon the first large collapse a shock wave is radiated (Fig. 8). At 30 ns interframe time the twelve frames given cover 330 ns, less than one interframe time of Fig. 7. The shock wave becomes visible via deflection of the illuminating light. The bubble expands at a much slower time scale than the speed of the shock wave and is seen as the tiny black spot left in the middle of the ring of the shock wave.

The bubble oscillation as measured photographically can be compared with theoretical models for the motion of a bubble in a liquid. A comparison is shown in Fig. 9. The overall curve is reproduced well albeit the steep collapse is not resolved experimentally. For details the reader is referred to [25].

There is the question whether shock waves are also emitted into the interior of the bubble [26–29]. Theoretically, shock waves occur in certain models, when the collapse is of sufficient intensity. Recently, the problem has been

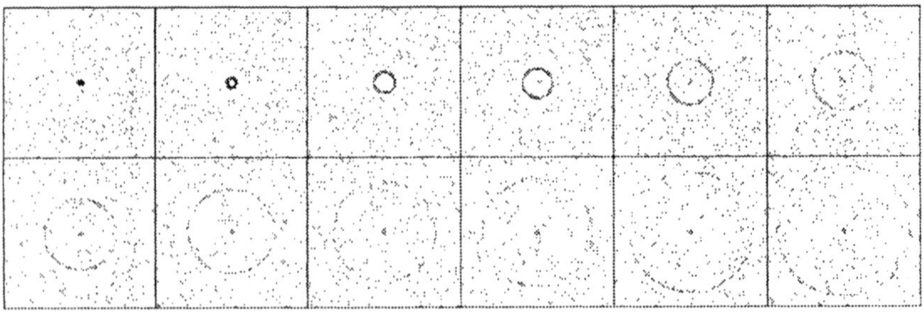

Fig. 8. Shockwave from a trapped, sonoluminescing bubble driven at 21.4 kHz and a pressure amplitude of 132 kPa. The interframe time is 30 ns. Picture size is 160 μm × 160 μm (Courtesy of R. Geisler)

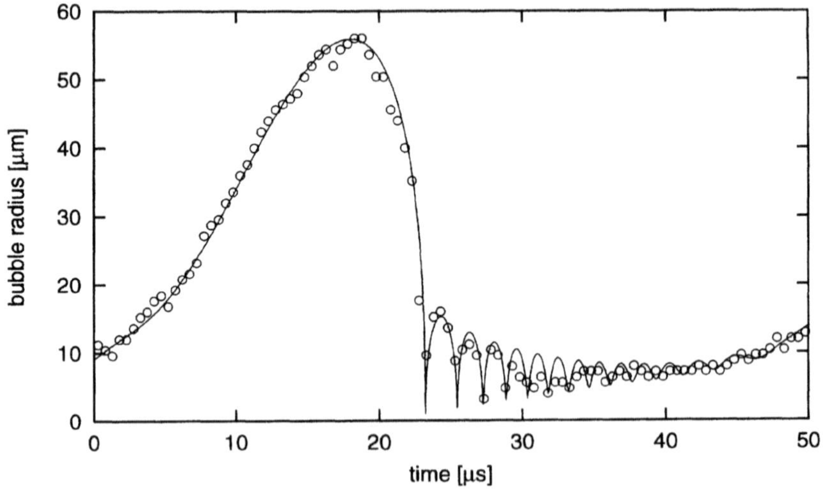

Fig. 9. Radius-time curve of a trapped bubble in a water-glycerine mixture derived from photographic observations. A numerically calculated curve (Gilmore model) is superimposed on the experimental data points (open circles). The calculation is based on the following parameters: driving frequency $f_0 = 21.4$ kHz, ambient pressure $p_0 = 100$ kPa, driving pressure $p_0 = 132$ kPa, vapor pressure $p_v = 0$, equilibrium radius $R_n = 8.1$ μm, density of the liquid $\rho = 1000$ kg/m^3, viscosity $\mu = 0.0018$ Ns/m^2 (measured) and surface tension $\sigma = 0.0725$ N/m. The gas within the bubble is assumed to obey the adiabatic equation of state for an ideal gas with $\kappa = 1.2$ (Measured points courtesy of R. Geisler, calculation courtesy of T. Kurz)

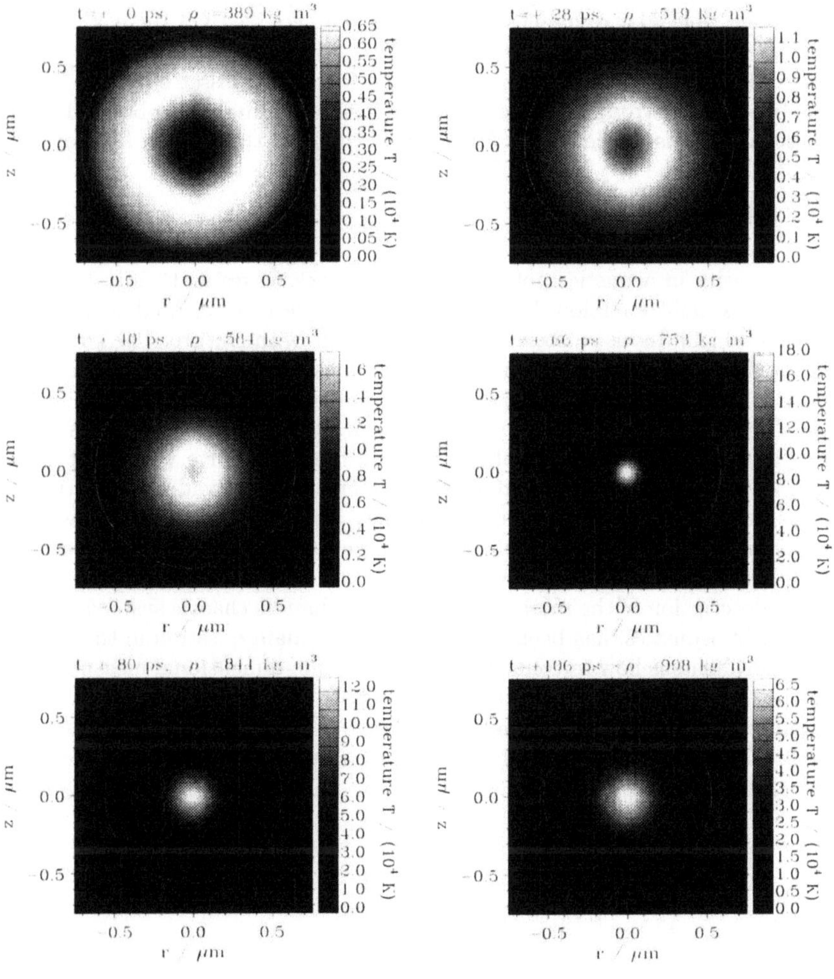

Fig. 10. Temperature distribution inside a collapsing sonoluminescence bubble filled with argon, 1 million particles, radius of the bubble at rest 4.5 μm, sound pressure amplitude 130 kPa, sound field frequency 26.5 kHz, total time covered 106 ps (Courtesy of B. Metten [32])

approached by molecular dynamics calculations [30–32]. As the number of molecules in a small bubble is relatively small, around 10^9 to 10^{10} molecules, this approach promises progress for the question of internal shock waves. Figure 10 gives a graphics view on the internal temperature distribution inside a collapsing sonoluminescence bubble with near adiabatic conditions (reflection of the molecules at the inner boundary of the bubble). The liquid motion and the inner molecular dynamics calculations were coupled via the gas pressure at the interface that was determined from the molecular dynamics of the one million particles and inserted into the Rayleigh-Plesset equation for the bubble wall. A strong focusing of energy inside the bubble is observed under these conditions. More details are given in the PhD thesis of B. Metten [32].

7 Chaos Physics

Nonlinearity in acoustics not only leads to shock waves and, as just demonstrated, to light emission, but to even deeper questions of nonlinear physics in general [33]. In his curiosity, man wants to know the future. The knowledge of the future may also be of help for better mastering one's life by taking the proper provisions. The question therefore arises: How far can we look into the future and where are the difficulties in doing so? It is the question of predictability or unpredictability. In the attempt to solve this question we find deterministic systems, stochastic systems and, nowadays, chaotic systems. The last ones are special in that they combine deterministic laws that nevertheless give unpredictable output. They thus form a link between deterministic and stochastic systems.

A description of the methods developed to handle chaotic systems, in particular in acoustics, has been given in [12]. The main question in the context of experiments is how to connect the typically one-dimensional measurements (a time series) with the high-dimensional state space of theory (given by the number of variables describing the system). This is done by what is called *embedding* (Fig. 11) [34,35] and has led to a new field called *nonlinear time series analysis* [36–38].

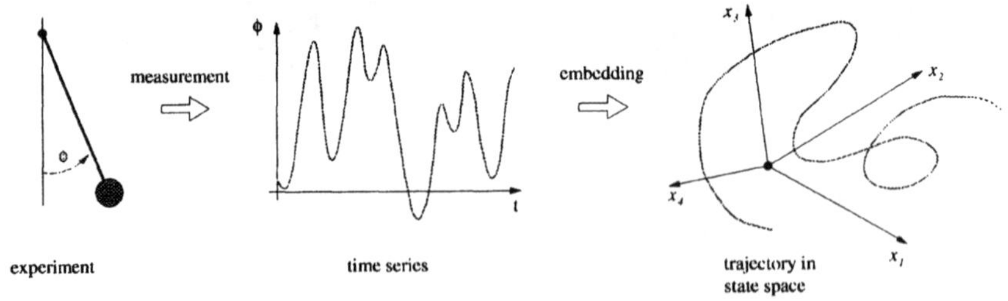

Fig. 11. Visualization of the embedding procedure

The embedding procedure runs as follows. Let $\{p(kt_s), k = 1, 2, \ldots, N\}$ be the measured time series, then an embedding into a n-dimensional state space is achieved by grouping n elements each of the time series to vectors $\boldsymbol{p_k}^{(n)} = (p(kt_s), p(kt_s + t_l), \ldots, p(kt_s + (n-1)t_l), k = 1, 2, \ldots, N - n$, in the n-dimensional state space. There t_s is the sampling interval at which samples of the variable p are taken. The delay time t_l is usually taken as a multiple of t_s, $t_l = lt_s, l = 1, 2, \ldots$, to avoid interpolation. The sampling time t_s, the

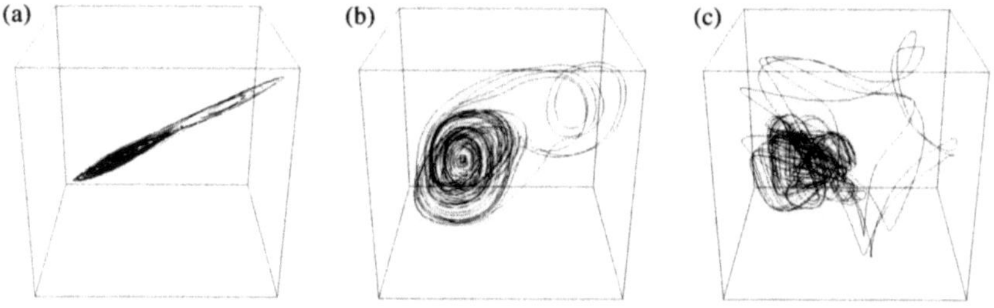

Fig. 12. Three embeddings with different delay times t_l for experimental data obtained from a chaotically oscillating, periodically driven pendulum

delay time t_l and the embedding dimension n all must be chosen properly according to the problem under investigation. Figure 12 demonstrates the effect of different delay times with experimental data from a chaotically oscillating periodically driven pendulum. When the delay time t_l is too small then from one embedding vector to the next there is little difference and the embedded points all lie along the diagonal in the embedding space (Fig. 12a). On the other hand, when the delay time t_l is too large, the reconstructed trajectory obtainded by connecting successive vectors in the embedding space becomes folded and the reconstruction becomes fuzzy (Fig. 12c). The embedding dimension n usually is not known beforehand when an unknown system is investigated for its properties. Therefore the embedding is done for increasing n until some criterion is fulfilled. The basic idea for a criterion is that a structure in the point set obtained by the embedding should appear. Then some law must be at work that generates it. The task after embedding thus is to find and characterize the structure and to find the laws behind it. The characterization of the point set obtained can be done by determining its static structure (e.g. its dimension) and its dynamic structure (e.g. its expansion properties via the dynamics of nearby points).

The dimension easiest to understand is the pointwise dimension. It is defined at a point P of a point set M by $N(r) \sim r^D$, $r \to 0$, where $N(r)$ is the number of points of the set M inside the ball of radius r around P (Fig. 13). Examples for a line and a surface give the right integer dimensions one and two (Fig. 14), because for a line we have $N(r) \sim r^1$ for $r \to 0$ and thus $D = 1$, and for a surface we have $N(r) \sim r^2$ for $r \to 0$ and thus $D = 2$.

Chaotic or strange attractors usually give a fractal (noninteger) dimension. The pointwise dimension may not be the same in this case for each point P. The distribution of dimensions then gives a measure of the inhomogeneity of the point set. Figure 15 gives an example of a fractal point set. It has been

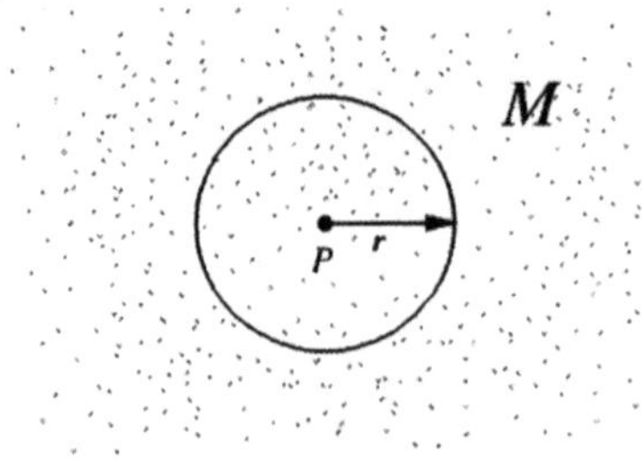

Fig. 13. Notations for the definition of the pointwise dimension

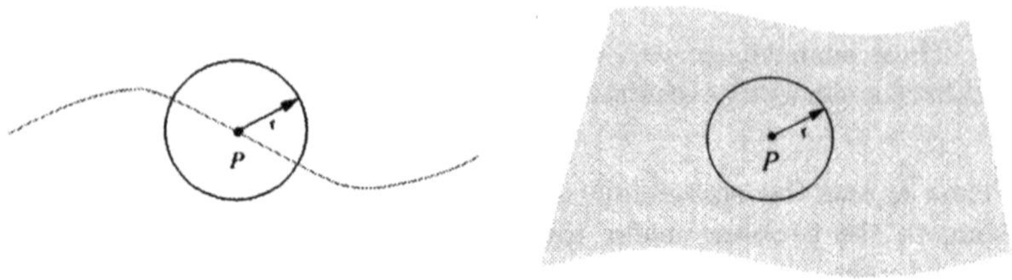

Fig. 14. Example of pointwise dimensions for one-dimensional (a curve) and two-dimensional (a surface) point sets

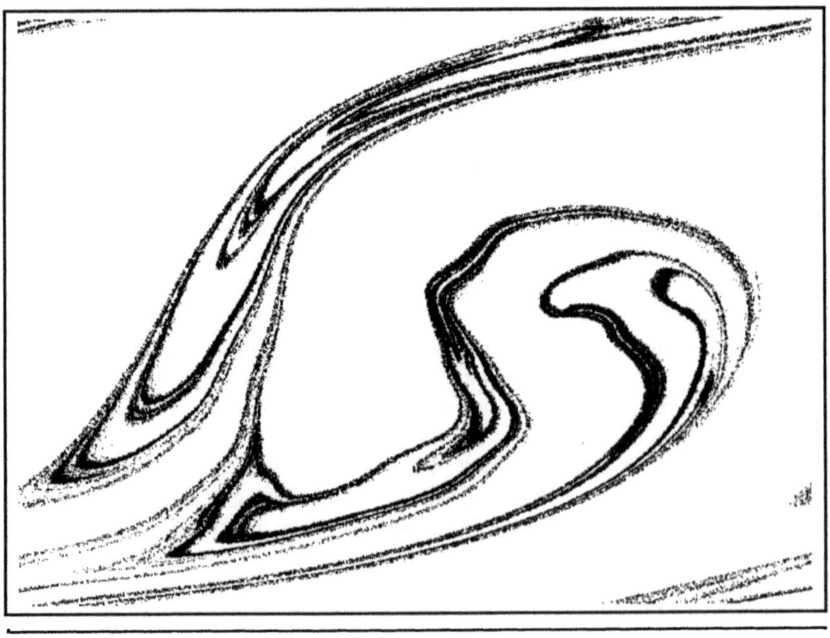

Fig. 15. The point set of a strange attractor of an experimental, driven pendulum

obtainded by sampling the angle and the angular velocity of a pendulum at a given phase of the periodic driving for parameters in a chaotic regime.

To investigate the dynamic properties of a (strange or chaotic) set the succession of the points must be retained. As chaotic dynamics is brought about by a stretching and folding mechanism and shows sensitive dependence on initial conditions (see e.g. [12]) the behaviour of two nearby points is of interest: Do they separate or approach under the dynamics? This behaviour is quantified by the notion of the Lyapunov exponent. In Fig. 16 the calculation scheme [39] is depicted. It gives the maximum Lyapunov exponent via the

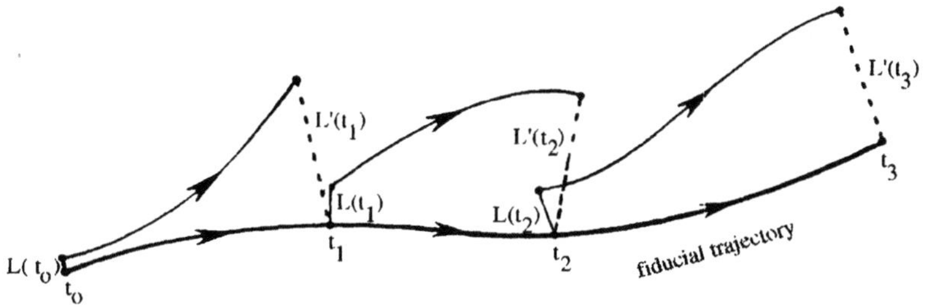

Fig. 16. Notations for the definition of the largest Lyapunov exponent

formula

$$\lambda_{max} = \frac{1}{t_m - t_0} \sum_{k=1}^{m} \log_2 \frac{L'(t_k)}{L(t_{k-1})} \quad \left[\frac{\text{bits}}{\text{s}}\right]. \tag{21}$$

As space do not permits to dig deeper into the operations possible with embedded data, just a graph depicting the possible operations in nonlinear time series analysis is given in Fig. 17. Starting from the time series obtained by some measurement the usual way is to do a linear time series analysis, for instance by calculating the Fourier transform or some correlation. The new way for chaotic systems proceeds via embedding to a reconstructed state space. There may be some filtering involved in between, but this is a dangerous operation as the results are often difficult to predict. For the surrogate data operation and nonlinear noise reduction the reader is referred to [36–38]. From the characterization operations we have mentioned here the dimension estimation and the largest Lyapunov exponent. Various statistical analyses can be done and the data can also be used for modelling, prediction and controlling the system. Overall, nonlinear time series analysis is a new diagnostic tool for describing (nonlinear) systems.

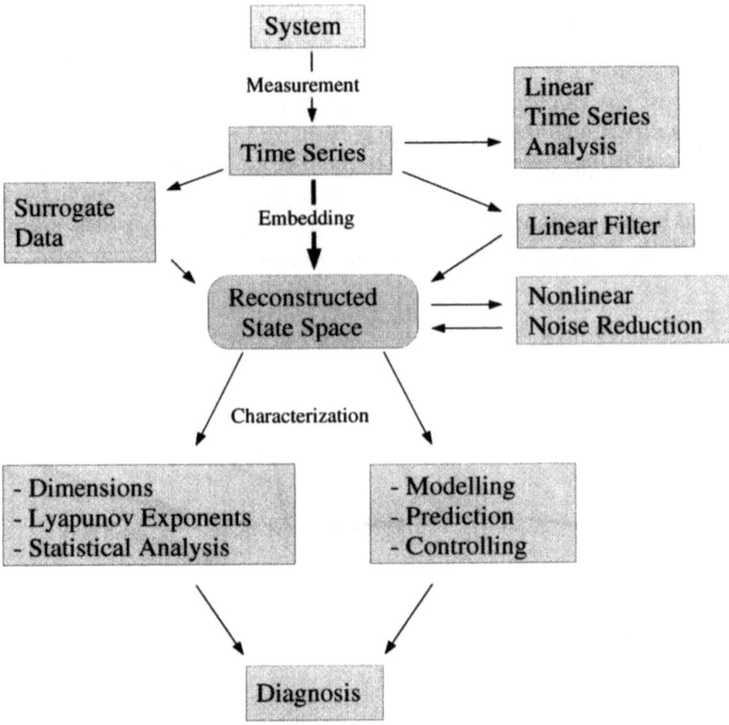

Fig. 17. Operations in nonlinear time series analysis

8 Acoustic Chaos

One of the first examples of chaotic dynamics and where the above methods were first applied was acoustic cavitation [40,41,33]. To investigate this phenomenon, the rupture of liquids by sound waves and the phenomena associated with it, an experimental arrangement as depicted in Fig. 18 was used. A cylindrical transducer of piezoelectric material is used to generate a strong sound field in water. A typical transducer was of 76 mm length, 76 mm inner diameter, and 5 mm wall thickness submerged in a water filled glass container. Cylinders of different size were available driven at different frequencies between 7 kHz and 23 kHz. A standing acoustic wave is set up inside the cylinder having its maximum pressure amplitude along the axis of the transducer. A hydrophone monitors the sound output of the liquid. Some electronics and a computer surround the experiment to drive the cylinder and to store the sampled pressure data from the hydrophone. When cavitation is fully developed, a dentritic filamentary bubble pattern is set up in the cylinder (Fig. 19).

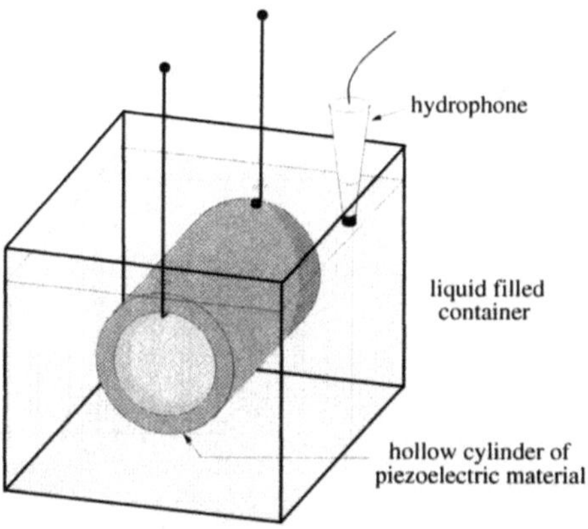

Fig. 18. Cylindrical transducer submerged in water for bubble oscillation, cavitation noise, and sonoluminescence studies

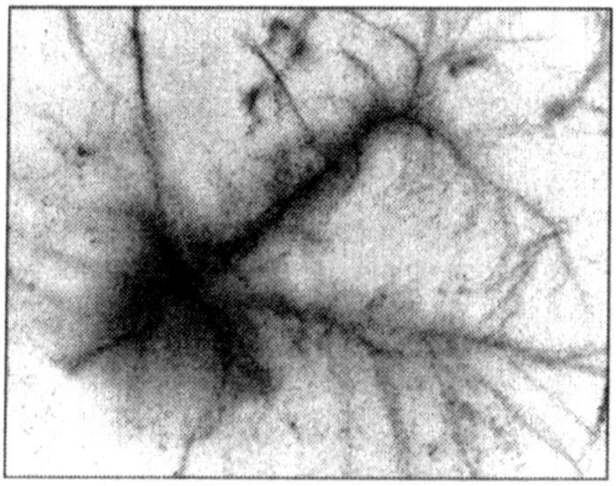

Fig. 19. Example of the filamentary bubble pattern (acoustic Lichtenberg figure) inside the cylinder in the projection along the axis of the transducer (Courtesy of A. Billo)

When the sound field amplitude is raised the whole bubble field undergoes a period doubling cascade to chaos [42,25], a strong indication that chaotic dynamics really has been reached and that the broadband spectrum then encountered (Fig. 20) is not of stochastic origin. The cascade of period-doubled sound in the liquid has its origin in the cascade of period-doubled oscillations of the nonlinearly oscillating bubbles [25], and thus has the chaotic sound.

Embedding of the sound data into spaces of increasing dimension yields a structure with low fractal dimensions between two and three (Fig. 21). This is a quite surprising result in view of the thousands of oscillating bubbles of different sizes. Indeed, new results on the superposition of the sound output of a population of driven bubbles have indicated that the low dimensionality may be a result of the (weak) synchronization of their motion from the periodic driving and the scale of observation [43].

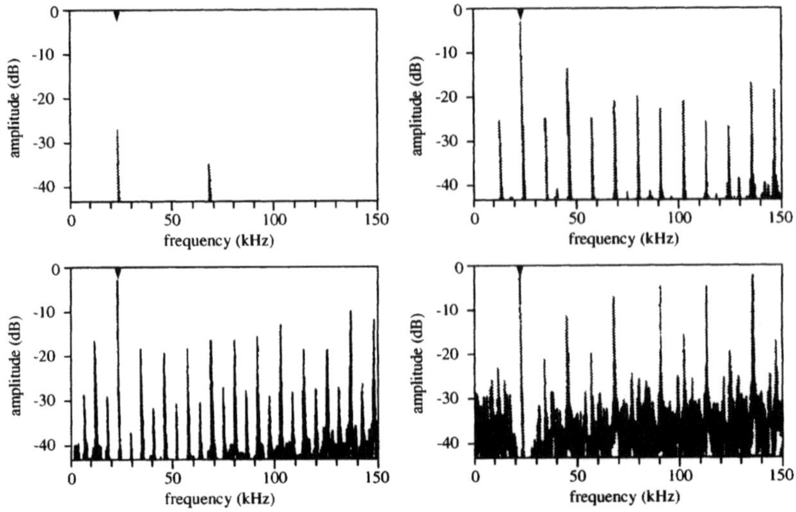

Fig. 20. Power spectra of cavitation noise taken at a driving frequency of $f_0 = 23$ kHz, marked in the figures by a small triangle. The driving pressure was increased from one graph to the next, giving first a periodic, anharmonic signal (*top left*), then subharmonic motion with lines at multiples of $f_0/2$ (period-2 motion, *top right*), then period-4 motion with lines at multiples of $f_0/4$ (*bottom left*) and finally a broadband spectrum indicative of chaotic behaviour (*bottom right*)

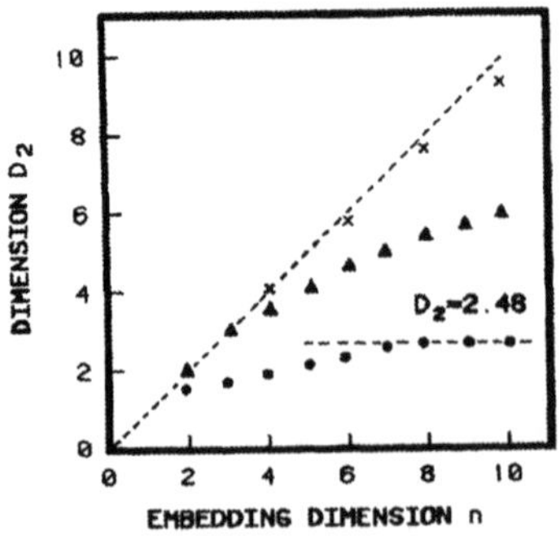

Fig. 21. Dimension estimation by embedding the sound pressure data into spaces of increasing dimension

Acknowledgment

I would like to thank the organizers for having made an exceptionally productive summer school possible and my numerous coworkers that collected the data reported in the course of time.

References

1. G.B. Whitham: *Linear and Nonlinear Waves* (Wiley, London 1974)
2. M.F. Hamilton, D.T. Blackstock, eds.: *Nonlinear Acoustics* (Academic Press, San Diego 1998)
3. R.T. Beyer: *Nonlinear Acoustics* (Acoustical Society of America, Woodbury 1997)
4. L.D. Rozenberg, ed.: *High-Intensity Ultrasonic Fields* (Plenum Press, New York 1971)
5. O.V. Rudenko, S.I. Soluyan : *Theoretical Foundations of Nonlinear Acoustics* (Consultants Bureau, London 1977)
6. N.S. Bakhvalov, Y.M. Zhileikin, E.A. Zabolotskaya: *Nonlinear Theory of Sound Beams* (Am. Inst. Physics, Melville 1987)
7. M.F. Hamilton, D.T. Blackstock, eds.: *Frontiers of Nonlinear Acoustics* (Elsevier, London 1990)
8. H. Hobaek, ed.: *Advances in Nonlinear Acoustics* (World Scientific, Singapore 1993)
9. R.J. Wei: *Nonlinear Acoustics in Perspective* (Nanjing University Press, Nanjing 1996)
10. W. Lauterborn, T. Kurz, eds.: *Nonlinear Acoustics at the Turn of the Millennium* (Am. Inst. Physics, Melville 2000)
11. S. Makarov, M. Ochmann: Acustica **82**, 579 (1996) **83**, 197 (1997) **83**, 827 (1997)
12. W. Lauterborn, U. Parlitz: J. Acoust. Soc. Am. **84**, 1975 (1988)
13. C.D. Ohl, A. Philipp, W. Lauterborn: Ann. Physik **4**, 26 (1995)
14. C. E. Brennen: *Cavitation and Bubble Dynamics* (Oxford University Press, Oxford 1995)
15. T.G. Leighton: *The Acoustic Bubble* (Academic Press, London 1994)
16. J.R. Blake, J. M. Boulton-Stone, and N. H. Thomas, eds.: *Bubble Dynamics and Interface Phenomena* (Kluwer, Dordrecht 1994)
17. F. R. Young: *Cavitation* (McGraw-Hill, London 1989)
18. K.S. Suslick, ed.: *Ultrasound: Its Chemical, Physical and Biological Effects* (VCH Publishers, New York 1988)
19. L. van Wijngaarden, ed.: *Mechanics and Physics of Bubbles in Liquids* (Martinus Nijhoff Publishers, The Hague 1982)
20. W. Lauterborn, ed.: *Cavitation and Inhomogeneities in Underwater Acoustics* (Springer, Berlin 1980)
21. E.A. Neppiras: Physics Reports **61**, 159 (1980)
22. H.G. Flynn: 'Physics of acoustic cavitation'. In: *Physical Acoustics*, ed. by W.P. Mason (Academic Press, New York 1964) pp. 57-172
23. D.F. Gaitan, L.A. Crum, C.C. Church, R.A. Roy: J. Acoust. Soc. Am. **91**, 3166-3183 (1992)

24. Y. Tian, J.A. Ketterling, and R.E. Apfel: J. Acoust. Soc. Am. **100**, 3976-3978 (1996)
25. W. Lauterborn, T. Kurz, R. Mettin, C.D. Ohl: Advances in Chemical Physics **110**, 295 (1999)
26. E. Heim: 'Über das Zustandekommen der Sonolumineszenz'. In: *Proceedings of the Third International Congress on Acoustics, Stuttgart, 1959*, ed. by L. Cremer (Elsevier, Amsterdam 1961) pp. 343-346
27. C.C. Wu, P.H. Roberts: Proc. Roy. Soc. Lond. A **445**, 323-349 (1994)
28. W.C. Moss, D.B. Clarke, and D.A. Young: Science **276**, 1398-1401 (1997)
29. V.Q. Vuong, A.J. Szeri, D.A. Young: Phys. Fluids **11**, 10 (1999)
30. B. Metten, T. Kurz, W. Lauterborn: 'Molecular dynamics simulation versus continuum mechanics for a sonoluminescing bubble'. In: *Modelling Collective Phenomena in Complex Systems*, ed. by P.L. Garrido, J. Marro, R. Toral, Europhysics Conference Abstracts **22**, 268-269 (1998)
31. B. Metten, W. Lauterborn: 'Molecular dynamics approach to single-bubble sonoluminescence'. In: *Nonlinear Acoustics at the Turn of the Millennium*, ed. by W. Lauterborn, T. Kurz (Am. Inst. Physics, Melville 2000) pp. 429-432
32. B. Metten: Molekulardynamik-Simulationen zur Sonolumineszenz. PhD Thesis, Georg-August Universität, Göttingen (2000)
33. W. Lauterborn, T. Kurz, U. Parlitz: Int. J. Bifurcation and Chaos **7**, 2003 (1997)
34. N.H. Packard, J.P. Crutchfield, J.D. Farmer, R.S. Shaw: Phys. Rev. Lett. **45**, 712 (1980)
35. F. Takens: 'Detecting strange attractors in turbulence'. In: *Dynamical Systems and Turbulence*, ed. by D.A. Rand, L.S. Young (Springer, Berlin 1981) pp. 366-381
36. H.D.I. Abarbanel: *Analysis of Observed Chaotic Data* (Springer, New York 1996)
37. H. Kantz, T. Schreiber: *Nonlinear Time Series Ananlysis* (Cambridge University Press, Cambridge 1997)
38. U. Parlitz: 'Nonlinear Time Serie Analysis'. In: *Nonlinear Modeling – Advanced Black Box Techniques*, ed. by J.A.K. Suykens, J. Vandewalle (Kluwer Academic Publishers, Dordrecht 1998) pp. 209-239
39. A. Wolf, J.B. Swift, H.L. Swinney, J.A. Vastano: Physica D **16**, 285 (1985)
40. W. Lauterborn, E. Cramer: Phys. Rev. Lett. **47**, 1445 (1981)
41. W. Lauterborn, J. Holzfuss: Int. J. Bifurcation and Chaos **1**, 13 (1991)
42. W. Lauterborn, A. Koch: Phys. Rev. A **35**, 1774 (1987)
43. S. Luther, M. Sushchik, U. Parlitz, I. Akhatov, W. Lauterborn: 'Is cavitation noise governed by a low-dimensional chaotic attractor?'. In: *Nonlinear Acoustics at the Turn of the Millennium*, ed. by W. Lauterborn, T. Kurz (Am. Inst. Physics, Melville 2000) pp. 355-358

Subject Index

absolute instability 252
acoustic
- cavitation 280
- chaos 280–282
- energy 10–11, 75, 85, 163–164
adiabatic invariant 220
analogies table 208
aperiodic vortices 216

Biot-Savart law 14, 76
blade-vortex interaction 66–74
Born approximation 126, 129, 177

causality 251
channel 239
characteristics 269
completeness (of the continuous spectrum) 244
continuous spectrum 240–243
convective instability 252
conversion phenomenon 224
critical layer 241
Curle's equation 58
curvature effect 257–259
cyclotron damping 207

Debye series 162
distribution function 197
Dyson equation 147

eigenfunction 211
eigenvalue 210
elastic plate 248
energy exchange 221
entropy equation 4
equation
- Fredholm integral 245
- kinetic 194
- Nonlinear Schrodinger 261
- of state 266–267
- Orr-Sommerfeld 196
- Rayleigh 196, 241
- dispersion 139, 163
- eikonal 163
- Helmholtz 43, 153, 170
- Lighthill *see* Lighthill
- parabolic 171–172

fluid loading 248

geometrical acoustics 160–168, 218
Green's function 8, 17–20, 43
group velocity 252

hamiltonian
- functional 239
- structure 245
high-speed train 94
Hilbert transform 244
Howe's acoustic analogy 40

instability (acoustic) 201

Landau damping 194
leading edge 80
Lighthill
- acoustic analogy 4–7, 31–33
- eighth power laws 35
- equation 33
- stress tensor 32

mass operator 147
multiple scattering 143–159

negative energy waves 199, 249, 254–256
noncanonical Poisson bracket 246

nonlinear effect 259–262
nonlinear waves 268–270
normal operators 211
numerical calculations 227

optical theorem 129
oscillator 245
oscillators 230
over-reflection 200

phase speed 194

random media 143, 144
ray path 165, 168
Rayleigh criterion 239
resonance 194, 207, 231
Rossby waves 193

scattering of sound 122–131
self-adjoint 211
shallow water 135–139
shear flow 239
shear rate 217
shock wave 270–271
singularity 194, 241
solitary wave 262
sonoluminescence 272–276
spatial Fourier harmonics 211
spinning vortex 41
surface wave 132–135

total enthalpy 20, 39
trailing edge 64, 78
turbulence spectrum 173–175

velocity
– group velocity 133, 134, 164
– phase velocity 133, 164
vortex sheet 199
vortex sound 14–17, 61, 119–121

wave-vortex interactions 213

Zel'dovich effect 207

MIX
Papier aus verantwortungsvollen Quellen
Paper from responsible sources
FSC® C105338

If you have any concerns about our products,
you can contact us on
ProductSafety@springernature.com

In case Publisher is established outside the EU,
the EU authorized representative is:
Springer Nature Customer Service Center GmbH
Europaplatz 3, 69115 Heidelberg, Germany

Printed by Libri Plureos GmbH
in Hamburg, Germany